GAUGE FIELDS

[illegible]

GAUGE FIELDS

N. P. Konopleva
and
V. N. Popov

Translated from the Second Russian Edition
and Edited by
N. M. Queen
Department of Mathematical Physics
University of Birmingham, England

harwood academic publishers
chur • london • new york

Published under license by:

Harwood Academic Publishers GmbH
Poststrasse 22
CH-7000 Chur, Switzerland

Editorial Office for the United Kingdom:
61 Grays Inn Road
London, WC1X 8TL

Editorial Office for the United States of America:
Post Office Box 786
Cooper Station
New York, NY 10276

Originally published in Russian in 1980 as КАЛИБРОВОЧНЫЕ ПОЛЯ by Atomizdat

Library of Congress Cataloging in Publication Data
Konopleva, N.P. (Nelli Pavlovna)
Gauge fields

Translation of Kalibrovochnye polia.
Bibliography: p.
Includes index.
1. Gauge fields (Physics) I. Popov, V. N. (Viktor Nikolaevich) II. Queen, N.M. III. Title.
QC793.3F5K6613 530.1′43 81-20209
ISBN 3-7186-0045-5 AACR2

Printed in the United States of America

N. P. KONOPLEVA AND V. N. POPOV

PREFACE TO THE SECOND RUSSIAN EDITION

Since the appearance of the first edition of this book, which was devoted largely to the construction of the classical and quantum theory of gauge fields, interest in unified theories of the various interactions has grown appreciably. This is due primarily to the solution of two major theoretical problems which stood in the way of the construction of realistic gauge models of elementary particles: 1) renormalizability of gauge theories; 2) the origin of the masses of the vector particles. The mechanism of spontaneous breaking of local gauge symmetry which had already been proposed by Higgs in 1964 not only made it possible to assign mass to the quanta of gauge fields, but also ensured the renormalizability of the resulting theory of massive fields. The latter was demonstrated by 't Hooft in 1971 for the example of the Weinberg—Salam model (1967), which provides a unified description of the weak and electromagnetic interactions. The correctness of the Weinberg—Salam model was confirmed experimentally by the discovery in 1973 of neutral currents, which were predicted by this model in first-order perturbation theory. Subsequently, a large number of unified gauge models of the strong, weak, and electromagnetic interactions required the existence of new quarks possessing a new quantum number ("charm"), as well as new types of elementary particles. In 1974 bound systems of two charmed quarks $c\bar{c}$ — the mysterious ψ particles — were actually observed experimentally. At the present time, these particles, which manifest themselves as extremely narrow long-lived resonances, have already been well studied. There has even appeared a spectroscopy of the family of ψ particles, which are excited states of the $c\bar{c}$ system.

In 1977 an analogous system of two b quarks ($b\bar{b}$) was discovered — the Υ particle. Charmed mesons and baryons in which the new quantum number is not compensated were also discovered. Many unified gauge models of the weak and electromagnetic interactions predicted the existence of heavy leptons, and for a long time this was regarded as an argument against such models. However, in 1976 the heavy τ lepton with mass ~1.8 GeV was discovered experimentally. Thus, unified gauge models of the interactions are leading to a new physics of elementary particles, which is rich in discoveries. Therefore a solution to the problem of finding a unified description of all forms of interaction (strong, weak, electromagnetic, and gravitational) is not only of mathematical interest, but is becoming practically essential. For the first time since the creation of quantum electrodynamics, unified gauge models of the weak and electromagnetic interactions provide a theory in which calculations can be carried through to completion to arbitrary order of perturbation theory.

Asymptotically free gauge models of the strong interactions are free from ultraviolet divergences and ensure "confinement" for quarks in the infrared region. The next problem is to include quantum gravity in the unified scheme of interactions. An idea which is very promising in this respect is to make use of dual models ("strings") in conjunction with gauge invariance, and possibly also supergravity.

The classical theory of gauge fields is being developed with equal success. The nonlinearity of the classical equations of non-Abelian gauge fields has given birth to a new industry among theoreticians. We have in mind the study of particle-like solutions of these equations (solitons, kinks, monopoles, and vortices). Particle-like solutions possess a new type of charge — topological charge, which one can attempt to associate with the quantum numbers that characterize the elementary particles. Therefore the theory of gauge fields raises the question of the relationship between classical and quantum physics in a new way. Unfortunately, the volume of this book does not enable us to give a sufficiently complete treatment of all the problems. However, we present here the basic mathematical apparatus (with the exception of renormalization theory): the Lagrangian and geometrical formulations of the classical theory of gauge fields, and the quantum theory using the method of functional integration. In addition, we analyze the role of the principles of relativity and symmetry in the construction of a physical theory.

We shall make use of contemporary mathematical methods in the book: the variational formalism and Noether's theorems — in the Lagrangian formulation of field theory invariant with respect to an infinite group (Chapter II); the coordinate-free method of exterior forms on a manifold and the concept of a fiber space — in the analysis of the geometrical picture of interaction (Chapter III); the path-integral method — in the construction of a quantum theory of gauge fields (Chapter IV). In particular, we shall show that the classical theory of a gauge field can be regarded as an aspect of geometry, and in this sense we have a realization of the profound physical and philosophical idea of Einstein that the geometry of space-time does not in itself exist, since it is determined by the interaction of physical bodies. In other words, each form of interaction creates its own geometry.

This book is based on original work of the authors, and it also contains a survey of the most important results on gauge fields by Soviet and foreign authors.

All the chapters of the book are relatively self-contained and may be read independently. The first chapter is introductory in character. To make the exposition of the other chapters more accessible, it introduces, in particular, geometrical and physical terminology in parallel. For an understanding of the remaining chapters, it is desirable to be acquainted with group theory, Riemannian geometry, and field theory at the level of courses given in physics and mathematics departments at Universities. Chapters I—III and the Preface were written by N. P. Konopleva, and Chapter IV by V. N. Popov.

The authors are grateful to Academicians M. A. Markov, L. D. Faddeev, and A. G. Iosif'yan for supporting the second edition of this book and for valuable remarks.

TRANSLATOR'S PREFACE

This is a translation of the second Russian edition of the book _Kalibrovochnye polya_, which was published in Moscow in 1980 as an updated version of the first edition of 1972. While no attempt has been made to revise the text of the translation, a number of minor misprints and erroneous references have been corrected, and many references to Russian translations of works published in the West have been replaced by the references to the original sources.

N. M. Queen

CONTENTS

CHAPTER I. INTERACTION OR GEOMETRY?

§1. PRINCIPLES OF RELATIVITY, GEOMETRY, AND INTERACTION

Introduction. In the 1960s a peculiar situation arose in elementary-particle theory: on the one hand, there was not a single experimental fact for which a theoretical basis could not be found, and on the other hand, there was no consistent theory which provided a unified description of the entire diversity of properties and species of elementary particles. The gulf between the "internal" symmetries (hypercharge, isospin, etc.) and the "external" (space-time) symmetries of elementary particles was felt particularly acutely. It became increasingly clear that the construction of a unified theory of interactions would require modification of the fundamental principles on which physical theories are based and would lead to the use of new ideas about the structure of space-time and the nature of elementary-particle interactions.

The symmetry properties of elementary particles are usually formulated in terms of the invariants of the symmetry groups* of space-time, which specify the principle of relativity of the theory (for example, Lorentz invariance), and the internal symmetry groups (for example, isospin invariance of the strong interactions, which is a consequence of the fact that the nuclear forces are independent of the electric charges of the particles). Thus, the problem of finding a natural unification of the internal and external symmetries is intimately related to the use of new principles of relativity and symmetry in elementary-particle theory. Such a fundamental principle is the requirement

---

*A group is a class of transformations (or operations) over the elements of the given set satisfying the following conditions (axioms): 1) the product of two transformations A and B (two transformations performed in succession) gives some transformation C from the same class, i.e., $A \cdot B = C$; 2) this law of multiplication is associative, i.e., $A \cdot (B \cdot C) = (A \cdot B) \cdot C$; 3) an identity transformation E is defined; 4) each of the transformations A has an inverse transformation A^{-1}, i.e., $A \cdot A^{-1} = E$. A group is said to be finite if its transformations depend on a finite number of numerical parameters, and infinite if the transformations of the group depend on a finite number of functions or on an infinite number of parameters.

of local invariance of the theory, and it is this that is related to the ideas of universal interactions and gauge fields.

The work of Yang, Mills, Utiyama, and Sakurai,[1-3] who first discussed gauge fields, was based on the assertion that all the internal symmetry properties of elementary particles are essentially local in character. It follows from this statement that finite gauge symmetry groups must be replaced by corresponding local groups, the parameters of whose transformations vary from point to point. This makes it possible to endow the theory with a new physical object — a gauge field, the interaction with which ensures invariance of the theory with respect to the local symmetry group. Thus, the principle of local gauge invariance is a deep physical principle, which permits the introduction of an interaction purely axiomatically, its form being determined in accordance with the symmetry properties of the theory. Therefore, the properties of gauge fields can be studied even independently of experiment. The problem of the realization of the theoretical concepts in observable phenomena is in itself quite complicated and is thereby distinguished from the mathematical apparatus of the theory. We note that local invariance was used for the first time as a fundamental physical principle in Einstein's general theory of relativity.[4] This idea was subsequently developed by Weyl, who introduced the electromagnetic field through the requirement of invariance of the theory with respect to local, i.e., point-dependent, expansions of the interval: $ds^{2\prime} = \lambda(x)\, ds^2$.[5] But the principle of local gauge invariance took its final form as a physical principle in the above-mentioned work of Yang, Mills, Utiyama, and Sakurai (see §2).

The gravitational and electromagnetic fields, with which the idea of gauge invariance was first associated, refer to universal interactions. The gravitational field interacts universally with all massive particles, and the electromagnetic field with all charged particles. Local gauge invariance led to the discovery of universal nuclear interactions mediated by unstable vector particles — resonances, which interact identically with all particles that carry isospin. Universality of certain weak interactions was also observed, and in this connection attempts were also made to apply the method of gauge fields to this case.[6] For a number of years these attempts had no success, but the final result exceeded all expectations. After the discovery of the mechanism of spontaneous generation of the masses of vector mesons (the Higgs mechanism[7] (1964)) and the formulation of a renormalization procedure for gauge models with spontaneous symmetry breaking ('t Hooft[8] (1971), Slavnov[9] (1972), and Taylor[10] (1971)), it became possible to construct a unified renormalizable theory of the weak and electromagnetic interactions of elementary particles, the simplest variant of which was the Weinberg—Salam model[11] (1967). This model predicted that neutral currents necessarily exist, and until they were discovered experimentally in 1973 this was regarded as an argument against the theory. Subsequent experiments confirmed more complicated quark gauge models which afford a unified description of the strong, weak, and electromagnetic interactions of hadrons.[12] At the present time, the incorporation of gravity into the general scheme of renormalizable interactions is under consideration.

The basis of the theory of gauge fields comprises symmetry principles and the hypothesis of locality of the fields, which converts global symmetries into local symmetries.

The principle of local gauge invariance reflects a deep relationship between the universality of the various interactions, conservation of the vector currents, and the existence of the interactions themselves. This principle determines the form of all interactions, irrespective of their physical nature, and thereby opens the way to the construction of a unified and consistent theory of the interactions of elementary particles. At the same time, the principle of local gauge invariance, like Einstein's general principle of relativity, gives the theory a form which admits a purely geometrical interpretation. As a result, it becomes possible to develop and generalize Einstein's idea that the geometry of space is not specified ad hoc, but is determined by the interaction of physical bodies.[13] In other words, geometry acquires a dynamical character and effectively reflects the influence on a distinguished test particle (or field) of all the remaining matter in the world.

The geometrization of gauge fields shows that 4-dimensional space-time is merely a particular case of possible dynamical geometries. An arbitrary gauge field corresponds to the geometry of a fiber space obtained from ordinary space-time by replacing its points by "internal" spaces in which the gauge group acts. Thus, the classical theory of gauge fields, like general relativity, becomes a purely geometrical theory. The resulting unified theory of the various interactions (strong, weak, electromagnetic, and gravitational) is also a geometrical theory. Its unity consists in the existence of a *general principle* according to which a geometry corresponding to each of the interactions is constructed.[14] In terms of the geometry of a fiber space, the motion of particles interacting with any gauge field becomes free (forceless). As in general relativity, this eliminates the distinction between inertial (or free) motions and noninertial motions (which take place under the action of external forces). This makes it possible to describe gauge fields by means of simple geometrical concepts (connection coefficients and curvature tensors) and renders geometry experimentally testable. The transition from 4-dimensional space-time to a fiber space implies the recognition of an astonishing possibility: the physical space determined by the interactions may be multidimensional or even infinite-dimensional. From this point of view, however, the description of microprocesses in ordinary space-time terms implies a certain projection of the "true" physical geometry of the interactions into a geometry produced by our macroscopic instruments. Therefore it would be very useful to know what we lose when this projection is made.

Local Symmetries and Geometrization of Interactions. *Local Spatial Symmetries and the Gravitational Field.* Suppose that we have a square plate of thin glass and a sphere. The flat uniform glass plate will represent flat (Euclidean) space, and the surface of the sphere will represent curved (Riemannian) space. Suppose now that we must "wrap" the glass plate around the sphere.

Let us cut our large glass square into a set of tiny squares and "cover" the sphere with them. This operation is a model of the process of covering a curved surface (or space) by local maps (or coordinate grids). It is easy to see that the whole flat plate can be covered by a single map, while the sphere cannot. It is for this reason that we had to take a set of tiny squares (local maps), in order to fit them as closely as possible to the points of the sphere. Proceeding in this way, we replaced the sphere by a set of small flat surfaces, which are interrelated in a definite manner, for example, rotated with respect to one another by a fixed angle. In other words, we can say that the difference between the set of tiny flat squares assembled into a single flat plate and the same set of squares assembled into a sphere is that the angle of rotation between their planes is zero in the first case but nonzero in the second. Translated into geometrical language, this means that a curved space can be represented as a set of flat spaces "joined" by *connection coefficients*. The connection coefficients determine the magnitude of the mutual "rotation" or "displacement" of neighboring local flat spaces (Fig. 1). Therefore the connection coefficients are zero when the small squares are stuck together into a

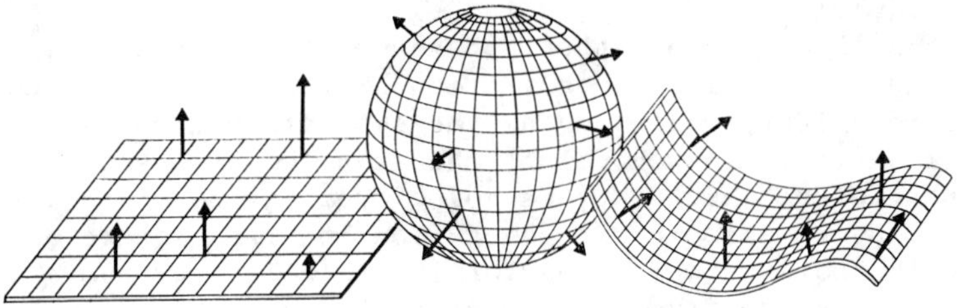

FIGURE 1

plane, but nonzero when they are stuck together into a sphere. Thus, the *sphere* is equivalent to the *set of planes + connection coefficients*.

Let us now compare the symmetry groups of the plane and the sphere, or more precisely, the groups of motions for which these objects are, as one says, transformed into themselves.

If the large square about which we spoke at the beginning is rotated through a right angle around an axis passing through its center and perpendicular to its surface, it will occupy the same position as it did before the rotation (see Fig. 1). Since for a person who did not observe the process of rotation itself this state is no different from the original one, we say that after such a rotation the square has been transformed into itself. Note that after this transformation *all the points* of the plate have moved in one and the same plane and have rotated through one and the same angle, i.e., have undergone *one and the same motion*. Now if we select some tiny square attached to the sphere and rotate the sphere through a right angle around an axis passing through the center of this square and perpendicular to its surface, the sphere will also coincide with itself. But in this case the only points which undergo the previous motion are those belonging to the chosen tiny square. Points of neighboring squares turned slightly on the sphere with respect to one another undergo rotations in different planes, i.e., *different motions*. This means that while the *flat plate* was *as a whole symmetric* with respect to the considered rotations, *on the sphere the previous symmetry has become only local*, i.e., it exists for each tiny square individually, but not for all of them collectively. Note that this fact does not rule out the possibility that the sphere as a whole has an intrinsic symmetry different from that of the flat plate.

Thus, locally the sphere possesses the same symmetry as the plane as a whole. Localization of the symmetry reduces to the fact that, while having the same structure (i.e., type of motion) at each point, the transformations must have parameters which vary in going from one point to another. In our example, in going from one tiny square on the sphere to another, we are performing rotations around a new axis, whereas the flat plate was rotated as a whole in one plane around a single axis.

Now let us imagine that both the plane and the sphere are very large and that the observer is very small. Suppose that the observer has the possibility of learning something about the space in which he finds himself, but that all his observations are "tied" to the point at which he is situated and to the instant of time at which he makes the measurements. Obviously, all results of the measurements will reflect only local properties of the space surrounding the observer. Can he establish the nature of the space as a whole? Can he, while situated at a point, distinguish a sphere from a plane? This is precisely the question that was raised for the first time in physics by Einstein.[13] Einstein's answer is contained in his *principle of equivalence*. Usually, this principle is formulated as the principle of (local) equality of the inertial and gravitational masses. But the principle of equivalence can also be given another form, namely, *flat space + a gravitational field is locally equivalent to a curved Riemannian space* (i.e., is indistinguishable from it[15]).

It is easy to see that the principle of equivalence in this form is very similar to the local equivalence of a sphere and a plane established in the example given above. For complete correspondence, it is sufficient to identify the connection coefficients (a geometrical concept) with the gravitational field (a physical concept). We then obtain a geometrical interpretation of gravitation.

What is the geometry of the world about us? In a certain sense, the principle of equivalence implies that there can be no unique answer to this question. We can suppose that space is flat and that all bodies are subject to the influence of a universal field that penetrates all matter, or that there is no field but that space is curved. In this case, the question of the geometry of space as a whole is equivalent to the question of the behavior of physical fields at arbitrarily large distances from the source. The symmetry properties of space become symmetry properties of interactions. The topology of space as a whole is reflected in the properties of the interactions. Thus, geometry and physics are interlocked.

Note that the geometrical interpretation of the gravitational field became possible as a result of localization of the space-time symmetry, i.e., the transition from flat space-time to a Riemannian space which is curved but which locally possesses the same symmetries. Other forms of interaction, namely, those mediated by gauge fields, also admit a purely geometrical interpretation. It is only in this case that the local symmetries are internal symmetries of elementary particles.

Local Internal Symmetries and Gauge Fields. To illustrate clearly what local symmetries are, consider the following example. Suppose that a ping-pong ball is moving along some trajectory and that we do not see whether it is rotating around its center of mass, although we know that the law of conservation of angular momentum is satisfied. How can we describe the positions of points of the surface of the ball at an arbitrary instant of time if the angular velocity of its intrinsic rotation can vary?

As is well known from mechanics, the free flight of a ball is determined only by the motion of its center of mass. The free motion of the center of mass is independent of whether the ball is rotating and whether the speed and direction of the axis of rotation are constant. Rotation around the intrinsic center of mass is an additional (internal) degree of freedom which is present for every body (more precisely, there are three degrees of freedom, since rotation is possible in any plane). If the character of rotation changes, to maintain the law of conservation of angular momentum we must assume that during its flight the ball is acted upon by some force field which twists it or brakes its rotation. This force field is an analog of a gauge field.

Gauge transformations is the name given to those transformations of the functions describing the motion of a particle which are not reflected in the observable characteristics of the motion, i.e., do not alter its physical state. In this sense, rotations of a ball around its center of mass are an analog of gauge transformations of an internal symmetry if we are interested only in the trajectories of the motion of the ball. Localization of this internal symmetry leads to a change of the angular-velocity vector of the intrinsic rotation of the ball. Disappearance of the localization, i.e., the presence of symmetry transformations with constant parameters, corresponds here to the establishment of a constant velocity of rotation along the entire trajectory. It is obvious that the gauge field also vanishes in this case. The constant angular-velocity vector of the ball corresponds in the theory of condensed media to constant properties of a medium throughout its volume, which manifests itself as a constant order parameter (for example, magnetization vector). In the theory of superconductivity, the global symmetry is described by a constant phase of the wave function of an electron.

The intrinsic rotations of a ball are unobservable unless some mark is made on the ball, for example, a stripe is painted, making it possible to observed its rotation. But the intrinsic rotations can be made observable only by breaking the internal symmetry, since the stripe renders different rotations of the ball inequivalent. This simple example illustrates another important fact: whatever symmetry is present, it implies the existence of identical, i.e., indistinguishable, states, whereas observation and measurement presuppose a distinction between the states, i.e., symmetry breaking. This symmetry breaking is always associated with an influence on the system, i.e., with the appearance of some force field.[16] In other words, *to make a symmetry observable, it must be broken.* An internal microscopic symmetry can become macroscopic and, in principle, observable if in a macroscopically large space-time region the local internal symmetry becomes global (order develops). It then becomes possible to observe macroscopic quantum phenomena. Examples of this kind are provided by the quantization of magnetic flux in superconductors and the appearance of coherent emission (lasers). The classical theory of gauge fields describes microscopically disordered systems and, as a rule, its predictions become experimentally observable on macroscopic scales under special conditions (phase transitions).

Invariance with respect to local gauge transformations means that it is impossible to measure the relative phase of the wave function of a particle at two different world points. This assertion is illustrated by means of the following example involving balls. Suppose that a rotating ball is placed at each point of the Universe. If two such balls are situated at points separated from one another by a spacelike interval, it is impossible to establish their angle of rotation with respect to one another simply because the velocity of light is finite. This is true in any space-time V_4.

Each local internal symmetry can be associated with its own gauge field, whose source in the case of invariance with respect to an ordinary (i.e., global) gauge group is a conserved quantity — a vector or tensor current density. In the example involving balls, the source of the gauge field is the intrinsic angular-momentum density of the balls.

Internal Spaces and a Fiber Space over V_4. Internal symmetries can be understood as symmetries of some internal space whose points correspond to different states of a particle, which are not associated with its position in space. An example of an internal symmetry is isospin invariance, or the fact that the nuclear forces are independent of the charges of the particles. As a result of isospin invariance of the nuclear forces, the proton and neutron are indistinguishable in the absence of an electromagnetic field. Two indistinguishable particles can be regarded as two states of the same particle. We label these states by the values of an internal quantum number — the isospin: $\frac{1}{2}$ (the proton) or $-\frac{1}{2}$ (the neutron). This gives an isotopic doublet. It is also possible to have richer isospin multiplets containing three or more particles. The influence of the electromagnetic field on an isospin multiplet leads to breaking of the isospin symmetry and a decomposition of the multiplet into individual components (particles), which behave differently with respect to the electromagnetic field.

Localization of internal symmetries, like localization of space-time symmetries, makes it necessary to introduce a new physical object — a gauge field. The concept of a gauge field was first introduced by Yang and Mills in connection with an attempt to construct a theory of the strong interactions on the basis of the requirement of invariance with respect to the local group of isospin transformations. In 1954 they proposed a method of introducing a vector field which is responsible for the strong interactions between nucleons and which is related to a conserved isospin current. The idea of the method was as follows.[1] Conservation of isospin is identical to the requirement of invariance of all interactions with respect to rotations of the isospin. This means that the orientation of the isospin has no physical significance when electromagnetic interactions can be neglected. In this case, the distinction between the proton and the neutron becomes completely arbitrary. However, it is usually understood that this arbitrariness is limited by the following condition: as soon as one chooses what to call the proton and what to call the neutron at one point of space-time, the freedom of choice disappears at other space-time points, even at points separated from the first point by a spacelike interval.

This situation is incompatible with the hypotheses of short range and locality of the fields, on which ordinary physical theories are based. In fact, suppose that there is no electromagnetic field and that the proton and neutron are indistinguishable. Let us now assume that an electromagnetic field is switched on at one of the points or in some region of space, thereby establishing what particle is a proton and what particle is a neutron. This distinction will become established in other regions of space only to the extent that the electromagnetic field reaches these regions. Obviously, this cannot occur instantaneously at all points of space, since the velocity of propagation of light (the electromagnetic field) is finite. Therefore Yang and Mills proposed to introduce the requirement of invariance of all interactions with respect to independent rotations of the isospin at all points of space-time, so that the relative orientation of the isospin at different points of space-time becomes meaningless (if the electromagnetic field is neglected). Thus, one requires invariance with respect to the isotopic gauge transformation $\psi' = S\psi$, where S is a rotation of the isospin depending on the choice of the point.

Invariance of the theory with respect to local isotopic rotations is ensured by the introduction of a triplet of vector fields, whose quanta are identified with the triplet of ρ mesons. The multiplets of vector mesons correspond in the geometrical interpretation to the concept of connection coefficients of a fiber space, which play the role of "forces."

A *fiber space* is obtained from ordinary space-time if its points are replaced by new spaces (fibers), i.e., it is assumed that the "points" have "internal structure." Internal symmetries of elementary particles then become symmetries acting within the fibers (internal spaces), and space-time symmetries transform the fibers belonging to different space-time points into one another. The geometry of a fiber space generalizes Riemannian geometry and includes it as a particular case.

An internal space cannot be identified with ordinary space-time (which we shall call the "world space"), even when the coordinates of its points are also space-time points. The simplest example is the same rotating ball. Identification of the internal and space-time symmetries of the motions of the ball would imply identification of its intrinsic and orbital rotations. In mechanics, a coupling between the intrinsic rotations of a ball and its motions through space can occur in the presence of friction in the surrounding medium. In the general case, one must consider independently *two spaces*: one to describe the motion of the center of mass (ordinary space), and the other — an "internal" space — to describe the rotations around the center of mass. As was observed by Cartan, the groups of symmetry transformations with respect to points lying inside or outside a body are isomorphic to one another but not identical. For example, if we are considering rotations, the operators of the corresponding groups are inverses of one another. In the example involving balls, we can assume that there is a single space which is counted twice, since it plays a double role. But in elementary-particle theory these are two different spaces.

It was noticed fairly quickly that the transition from a symmetry which is given over all space to a local symmetry which exists only in the neighborhood of a point is reminiscent of the transition from the flat absolute space-time of Minkowski* to the Riemannian space of the general theory of relativity, which locally possesses the same properties as Minkowski space. In fact, a Riemannian space can be represented as a manifold whose "points" are flat Minkowski spaces, these being "interlinked" by Ricci or Christoffel connection coefficients. The geometrical concept of a connection coefficient in a Riemannian 4-dimensional space corresponds in physics to the gravitational interaction. If, in a similar way, we consider a 4-dimensional manifold whose "points" are spaces of the representations of an internal symmetry group, we obtain an example of a fiber space. The connection coefficients introduced in it correspond to vector potentials of gauge fields, or multiplets of vector mesons. This geometrical interpretation of gauge fields allows us to consider the trajectories of particles interacting with a gauge field as free trajectories in a fiber space. Thus, in the description of any interactions mediated by some gauge field, we can get rid of the concept of force and make the theory of such interactions purely geometrical, as in general relativity.

Interaction and Geometry. Principles of Relativity, Geometry, and the Concept of Force. Every contemporary physical theory is based on some principle of relativity. A principle of relativity is formulated as a requirement of invariance of the theory with respect to some symmetry group. As a rule, it is assumed that this

---

*I.e., a flat 4-dimensional space in which the time plays the role of the fourth coordinate. The interval (or length) in this space is defined as the square root of the sum of the squares of the spatial coordinates minus the square of the time interval, multiplied by the square of the velocity of light. Owing to the presence of the minus sign in the expression for the 4-dimensional interval, the geometry of Minkowski space is said to be not Euclidean, but pseudo-Euclidean.

group reflects the symmetry properties of space-time as a whole. Before the creation of Einstein's general theory of relativity, the role of such groups was played by finite Lie groups (the Galilean group, which includes rotations and displacements in 3-dimensional space and, separately, displacements in time; the Lorentz group, which includes rotations in 4-dimensional Minkowski space, where the three spatial coordinates and the time are regarded as coordinates on an equal footing; and the Poincaré group — the group of motions, i.e., rotations and displacements of Minkowski space). In Einstein's theory, the principle of relativity was based on an infinite group for the first time.

The principle of relativity singles out a definite class of coordinate systems known as *inertial systems* (in the sense of this principle of relativity), in which, by definition, the motion of particles is assumed to be rectilinear and the particles themselves are free. The observed deviation of the trajectories from inertial trajectories and the interaction between the particles are described by means of the concept of a force field.

Einstein's conception of motion makes all trajectories inertial. This corresponds to invariance of the theory with respect to arbitrary continuous transformations of the spatial coordinates and the time, which are also called general covariant transformations. In Newtonian mechanics all inertial trajectories are interrelated by Galilean transformations, and in the special theory of relativity by Lorentz transformations. The groups of Galilean, Lorentz, and general covariant transformations specify the principles of relativity of the corresponding theories. They determine the degree of symmetry of the possible force-free motions. Violation of this symmetry is identified with the action of a force. Force-free inertial motions provide a realization of space-time geometry, which is introduced axiomatically.

Thus, every physical theory is based on some postulate about the geometrical properties of space-time, and this postulate finds its expression in the principle of relativity of the theory.[17-21] In this sense, geometry logically precedes experiment. The laws of physics cannot be expressed without the use of geometry, although geometry, taken by itself, does not correspond to any experiments or any experimental science.[22] The geometrical postulates of a physical theory reflect the choice of the means of measurement by which this theory can be tested.[23]

Absolute Space and Dynamical Geometry. According to the law of inertia in classical mechanics, a body maintains its state of rest or uniform rectilinear motion until it is acted upon by a force. Consequently, in

Newtonian mechanics we learn of the existence of a force acting on a body through the deviation of its trajectory from a rectilinear trajectory. But how is a rectilinear trajectory defined? For this purpose, we must compare it with a standard straight line. In other words, we need a trajectory which is rectilinear by definition, and we need a fixed procedure for comparing trajectories. As a model of a straight line in mechanics, one often uses light rays. This choice is equivalent to the assumption that the quanta of light — photons — are not subject to mechanical forces, i.e., that their mass is zero. In principle, the straight line can be taken to be the trajectory of any particle undergoing inertial motion, i.e., a free particle. But do free particles and straight lines exist? For theories like Newtonian mechanics, this is a fundamental question, since the deviation of the studied trajectories from inertial trajectories serves as a measure of the intensity of the interaction or force acting on a particle.

If a particle does not interact with anything, it is unobservable, since an observation (measurement) presupposes an influence on the particle. Absolutely free particles and absolutely straight lines are completely unobservable objects — mythical entities. The concept of inertial or free motion is relative. It depends on the class of possible interactions which is chosen. Absolute Newtonian space is the space of a static coordinate system associated with the source of the gravitational field. Inertial motion in this space is observable, for example, by means of light, i.e., the electromagnetic field. Thus, mechanics is in principle not a closed theory. It presupposes the existence of nonmechanical interactions, which make it possible to observe inertial mechanical motions and to perform measurements. It is understood here that the influence on the system under investigation and on the standard is sufficiently small during the measurement. When it is not possible to satisfy this condition, one goes over to the quantum picture, which explicitly takes into account the influence of the process of measurement on the system under investigation. Quantization removes the inconsistency of the procedure of comparison with unobservable standards.

We turn now to the general theory of relativity. What is the difference between the points of view of Newton and Einstein? Einstein's theory does not simply generalize the classical (Newtonian) theory of gravitation. It is constructed on totally different principles and is in essence a new type of physical theory. The main special features of this theory are the absence of a concept of interaction (or force) and a new conception of space-time. These features follow from a new understanding of the role of the principles of relativity and symmetry in a physical theory.

Following Poincaré,[17] Einstein[13] assumed that physics and geometry are not independent in experiment; experimental tests can be made only for the combination: geometry + physical laws. Experiment does not provide a proof of the existence of any particular geometrical space and, correspondingly, geometry irrespective of the physical laws on which it is based. Indeed, before an experimenter performs an experiment, he makes a number of explicit or implicit assumptions about the conditions of the experiment. For example, he assumes that the result is independent of the position on the Earth at which the experiment is performed and the orientation of the apparatus. Thus, it is postulated beforehand that the space in which the experiment is performed is homogeneous and isotropic.

If a space is homogeneous, i.e., as a whole it possesses some symmetry, then all geometrical objects in it are characterized by a set of numbers — the invariants of the various representations of the symmetry group of this space. These numbers correspond to the properties of the geometrical objects which remain unchanged under transformations that take the space under consideration into itself. For example, for a circle in the plane, the radius is such a number. The radius (a distance) is an invariant of the group of motions of the plane. The properties of the circle are determined by the fact that it is the geometrical locus of points equidistant from a given point. In a homogeneous space, it is sufficient to know its group of motions to describe everything that can occur in it: the properties of geometrical objects and the relations between them. This is the meaning of the famous "Erlanger Programm" of F. Klein. The use of homogeneous spaces in physics means that the properties of the studied objects (for example, particles) are formulated in terms of the invariants which characterize the representations of the symmetry group of space-time and the internal symmetries. The numerical values of the invariants correspond to integral conserved quantities: energy, momentum, angular momentum, spin, isospin, and so forth.

Einstein was the first to introduce into physics the Riemannian geometrical point of view, according to which space as a whole and nonlocal characteristics (for example, length) are determined only step by step. In general, a Riemannian space does not admit any motions, i.e., it does not possess any degree of homogeneity. The invariants of Riemannian geometry are the differential invariants of the group of arbitrary continuous transformations of the coordinates (general covariant transformations). In that sense, like finite Lie groups corresponding to Klein spaces, the group of general covariant transformations has no invariants. In exactly the same way, they do not lead to ordinary conservation laws and, consequently, invariants of the group of local internal symmetries. Their subgroups corresponding to transformations with constant parameters give ordinary (weak) conservation laws. The local symmetries themselves determine the form of the differential invariants, which become Lagrangians of interacting fields. It is for this reason that they are dynamical symmetries. Dynamical symmetries make it possible to formulate the properties of interactions.

An important property of local symmetries is the existence of identity relations between the extremals and their derivatives. These identities can be expressed in the form of conservation laws which are strong, i.e., are satisfied independently of the specific form of the Lagrangian and the equations of motion. Integration of such conservation laws yields invariants having a topological meaning.

Thus, we have before us two completely different types of physical theories. In theories of the first type, the structure of space-time is rigid and absolute, with some degree of symmetry. It is assumed that there can exist free particles, which move along geodesic (rectilinear) trajectories. The fact that curved trajectories are actually observed is described by the concept of interaction (a force field). All fields (including the gravitational field) are on an equal footing and differ only by the law of propagation and interaction with the currents that produce them.

In theories of the second kind, the interaction is completely eliminated. It is regarded as a manifestation of the dynamical nature of geometry. All the results of measurements then refer directly to the geometrical properties of space-time. In this case, it is no longer necessary to distinguish inertial trajectories as standards. It is sufficient to compare the trajectories of two arbitrary particles or bodies. The distance between them, called the geodesic interval, is proportional to the curvature tensor of space-time (or the stress tensor of a gauge field). Geometrization of interactions rids physical theory of the division into two dissimilar parts: an unobservable (geometrical) part and an observable (physical) part. Since experiment reflects only the sum of these parts, the division which is usually encountered always admits an arbitrariness and serves as a source of nonuniqueness in the correspondence between theory and experiment (the conventionalism of physical theory).[17]

If interaction is geometrized, the geometry of physical space becomes experimentally testable. In this case, it is possible to answer the question of the nature of the geometry of the physical world. It is sufficient to indicate what physical bodies or processes provide realizations of the basic geometrical concepts: a point, straight line, sphere, vector, and so forth. Such realizations are always approximate, since they involve idealizations. The answer to our question depends on this correspondence. It is well known that Euclidean geometry can be realized by means of solid and immutable macroscopic bodies (as far as this is possible). Among classical theories, this corresponds to classical Newtonian mechanics. The electrodynamics of photons realizes the geometry of Minkowski space, the gravitational field outside the sources realizes Riemannian geometry, and free gauge fields realize the geometry of a fiber space.

The Role of Geometrical Theories. Geometrical field theories are frequently regarded as pure mathematics, bearing no direct relation to experiment. This is due, in particular, to the fact that one of the basic concepts of any geometrical theory is the concept of a test body. A geometrical theory is a theory of the motion of test bodies. At the same time, attempts to find a physical model of a test body often encounter difficulties, since the basic property of a test body is its ability to feel the influence of an external field without exerting an inverse effect on the field. We note that one of Einstein's objections to Newtonian mechanics was the fact that in this theory space-time is absolute, "having a physical effect, but not itself influenced by physical conditions." Einstein considered that "it is contrary to the mode of thinking in science to conceive of a thing ... which acts itself, but which cannot be acted upon."[13] But the idea of something which experiences an influence from without, while not acting itself, is equally contrary to the mode of scientific thinking. However, a test body, by definition, must have precisely this property. Is this possible, and under what conditions? An affirmative answer to this question would mean that it is possible to specify a class of real physical objects which under certain conditions can play the role of test bodies, i.e., can move along geodesic trajectories. This would specify a class of physical objects whose motion is described by a geometrical theory, i.e., it would specify its domain of applicability.

Obviously, an experimental test of any theory makes sense only within its domain of applicability. This was understood particularly clearly in the study of quantum phenomena. Nobody is surprised when it is said that before studying the properties of some state by the methods of quantum theory, this state must be "prepared." But it turns out that before studying geodesic motion by the methods of the general theory of relativity, this motion must also be "prepared." The degree of accuracy with which geodesic motion occurs is the same for all spheres satisfying the condition $\rho R = \text{const}$, where ρ is the density of matter in the sphere and R is its radius. Estimates show that a high degree of accuracy is achieved for sufficiently massive bodies, such as planets, which are very remote from one another. Therefore the natural domain of applicability of general relativity is the motion of celestial bodies and cosmology, as well as the motion of artificial test bodies, such as drag-free space probes.[23]

The condition $\rho R = \text{const}$ determines the degree of universality of general relativity under realistic conditions. Geometrized electrodynamics also contains a parameter which determines the degree of its universality: $e/m = \text{const}$, where e is the charge of a particle (i.e., a test body) and m is its mass. It is obvious that in general relativity ρ plays the role of the charge, while $1/R$ corresponds to the mass of a test body.

A physical model of a fiber space is provided by any set of elements (atoms, molecules, spins, and so forth) which are identical, but oriented differently at different space-time points. For this reason, the classical theory of gauge fields has proved to be very useful in the physics of condensed and ordered media. It is possible that solid-state physics, phase transitions, and the properties of continuous media will become a major field of application of the geometrical theory of gauge fields. Geometrization of the interaction in this case implies the transition to the idea that all the particles of the medium are free (i.e., do not interact with one another, or are not situated in some external field). This point of view is similar to the quasiparticle formalism, which is being successfully applied in solid-state physics.

The geometrical description of interactions makes it possible to axiomatize physics. This means that: 1) internal and space-time symmetries are unified in a natural way; 2) there is a natural criterion for choosing the form of Lagrangians of interacting fields; 3) the solutions of the equations of classical gauge fields are classified according to their algebraic and topological properties; 4) there is a basis for the algebra of fields.

However, the main advantage of a geometrical theory of

interactions is its sovereignty. In his polemics with Bohr, Einstein did not simply oppose an ideology that was foreign to him. The basis of his objections to the quantum approach, to whose development he himself contributed greatly, was a vision of an ideal theory in which all representations of physical reality appear in terms of their theoretical analogs. A theory of this kind would not contain any elements adopted from experiment and not having a purely theoretical interpretation. Only such a "nondualistic" theory can be compared with experiment as a whole. In Einstein's opinion, quantum mechanics did not satisfy this requirement, nor did the theory of the gravitational field with sources, which was developed by Einstein himself. The goal of eliminating this dualism was in fact the motive for Einstein's long search for a unified geometrical theory of interactions.

As is well known, the goal of finding a general way of looking at various physical phenomena or a general mathematical apparatus is as old as physics itself. But until the present century, physics has generally striven to find a universal substance filling all space, whose properties could account for all physical phenomena. This led to hypotheses about universal fluids, such as phlogiston or the ether. From a modern point of view, the ether is a mechanical model of the geometrical properties of space-time. Therefore, without prejudice to the theory, it can be eliminated from it and replaced by axioms about space-time relationships. As is well known, an understanding of this circumstance led to the creation of the special theory, and subsequently to the general theory of relativity. In general relativity, the universal gravitational interaction was identified with a curvature of space-time. In geometrical unified theories of gravitation and electromagnetism (Einstein, Weyl, Eddington, Rainich, Kaluza, Fock, and others), the gravitational and electromagnetic fields were unified by the assumption that these two interactions are manifestations of a non-Euclidean character of space-time, but of a more complicated nature than the curvature corresponding to only the gravitational interaction.

In contemporary group-theoretic approaches, one sometimes seeks the largest group of transformations which includes both the space-time and internal symmetries of elementary particles. It has been found that such a group, if it is nontrivial, must be infinite. In contrast to these approaches, in geometrical unification of symmetries, one constructs a space which is naturally endowed with both forms of symmetry, specified by arbitrary groups. The geometry in the "base" and in the "fiber," i.e., the space-time and internal symmetries, are in general not related to one another. This reflects the independence of the quantum numbers corresponding to internal symmetries and determining, for example, the selection rules in elementary-particle reactions of their space-time characteristics. Thus, the construction of a fiber space makes it possible to unify any symmetries. The relation between the internal and spatial symmetries manifests itself in the violation of local gauge invariance. It is also important to note that instead of a single substance, or a single function with a large number of components, or a single large group, in the geometrical theory of gauge fields one specifies a general principle according to which the geometry of the space corresponding to each interaction is constructed.

Use of the geometry of a fiber space makes it possible to treat not only each gauge field individually, varying only the structure of the "fiber" in accordance with the gauge group, but also several fields simultaneously, if several fibers are introduced over each point of space-time. The interaction of different fields with each other or with the gravitational field can then be determined by means of projections of the corresponding fibers onto one another or onto the tangent space to the base, i.e., a projection onto space-time. All the results of geometrical unified theories of gravitation and electrodynamics using 4-dimensional or 5-dimensional geometry can be obtained in this way.

Thus, the scientist concerned with contemporary elementary-particle theory is reminiscent of those who sit in the Platonic cave with their backs to the fire and attempt to determine from the dancing shadows on the wall what is happening to the objects moving nearby behind their backs and casting these shadows. We do not know the nature of the "internal world" of elementary particles or of the internal symmetries. Nevertheless, from the images of these internal properties caught by our macroscopic and 3-dimensional measuring instruments, we are attempting to reconstruct what is going on in this mysterious and inaccessible world which is known as the "elementary particle." But while we cannot "turn around" and "see the essence," we can attempt to understand how the "shadow" is obtained and what the "fire" is. By seeing the image and knowing how it is obtained, we might "construct the essence."

§2. GAUGE FIELDS AND ELEMENTARY-PARTICLE PHYSICS

The past fifteen to twenty years of elementary-particle physics have been like a kaleidoscope. New trends, theories, models, and techniques have suddenly arisen, have yielded several striking results, and have faded away just as suddenly. Dispersion relations, Reggeism, higher symmetries, quarks and partons, current algebra, black holes, fiber spaces, charmed particles, strings, solitons, Wilson expansions, asymptotic freedom and confinement for quarks, "conspiracy of poles," and hydrodynamics — what have these in common? Is there a "wood" behind these "trees," or do all these paths lead in different directions? For a long time the answer was not clear. Now, however, we can say that there is at least one theory in which all the paths meet. This is the theory of gauge fields. It makes use of new physical ideas and mathematical apparatus to give a unified description of all forms of interactions of elementary particles: strong, weak, electromagnetic, and gravitational. At the same time, it admits a purely geometrical interpretation, and in its classical aspect it can be regarded as a theory of a continuous medium. Therefore, it is linked in a natural way to statistical physics and solid-state physics, borrowing some methods from them and giving them some of its own.

The experimental and theoretical discoveries of recent years show that the traditional classification of the forms of interactions according to their strength may be meaningful only at low energies. With increasing energy, the weak interactions grow and may become stronger than the electromagnetic interactions, while the strong interactions become weaker and may approach the weak and electromagnetic interactions in order of magnitude.[24] In that case, the strong, weak, and electromagnetic interactions, which we are accustomed to distinguishing, become manifestations of a single universal interaction. If universality is interpreted as the presence of a single coupling constant, it appears that this may happen at energies 10^{16}—10^{18} GeV, when gravitation must also be taken into account. But in any case, a truly unified theory of all interactions must incorporate a theory of gravity. Thus, it is possible that in the near future a step will be taken in elementary-particle physics analogous to the creation of Maxwell's electrodynamics in the last century, when it was understood that light, electric, and magnetic phenomena are different forms of one and the same electromagnetic interaction. Maxwell's theory made it possible to predict and discover radio waves and the mechanical action of light, and as a result of this theory the mechanical picture of the world was replaced by an electromagnetic picture. Contemporary unified renormalizable models of the strong, weak, and electromagnetic interactions are the result of almost 40 years of attempts to construct a Lagrangian theory of the weak and strong interactions in the form and image of electrodynamics.[25] It is possible that these will also change our picture of the world.

The theory of gauge fields is based on symmetry principles, the most important of which is the principle of local gauge invariance. This principle was used for the first time by Weyl,[5] who showed that Dirac's theory can be made invariant with respect to the local group of phase transformations of the wave functions (gauge transformations of the second kind) if in the Lagrangian the ordinary derivatives are replaced by covariant derivatives according to the rule $\partial_\mu \to \nabla_\mu = \partial_\mu - ieA_\mu$, where A_μ is the vector potential of the electromagnetic field, and e is the electric charge. The possibility of such a substitution points to the universality of the electromagnetic field and the corresponding coupling constant e. At the same time, it means that the interaction introduced in this way admits a purely geometrical treatment in the language of the connection coefficients of space-time. The latter was demonstrated by Weyl himself, who showed that the electromagnetic field can be identified with the additional nonmetric connection coefficients of space-time, and local gauge transformations with point-dependent expansions of the 4-dimensional interval. In Weyl space, there is an interesting geometrical analog of the Bohr quantization of orbits. In particular, if we consider the Coulomb radially symmetric electromagnetic field (the hydrogen atom) in Weyl space, it turns out that the Bohr orbits (electron orbits) are those trajectories along which parallel transport of a vector does not change its length.[26,27] No other trajectories in Weyl space satisfy this requirement. The main features of Weyl's treatment of the electromagnetic field (universality, geometrizability, and the presence of classical analogs of quantization) are preserved in the modern theory of gauge fields.

Yang—Mills Fields. The requirement of local isospin invariance is analogous to the requirement of gauge invariance of charged fields in electrodynamics, $\psi' = \exp[i\alpha(x)]\,\psi$, where such invariance is ensured by the introduction of the electromagnetic field A_μ, which transforms according to the law

$$A'_\mu = A_\mu + (1/e)\,\partial\alpha/\partial x^\mu, \tag{2.1}$$

and by the replacement of the ordinary derivative in the Dirac equation by the "covariant" derivative:

$$\partial_\mu \to \nabla_\mu = \partial_\mu - ieA_\mu.$$

Similarly, local isospin invariance is ensured by the substitution

$$\partial_\mu \to \nabla_\mu = \partial_\mu - igB_\mu, \tag{2.2}$$

where the B_μ are 2 × 2 matrices ($\hbar = c = 1$; $x_4 = it$), of which three are Hermitian ($\mu = 1, 2, 3$), while B_4 is anti-Hermitian; g is the isotopic charge.

From the invariance requirement

$$S\,(\partial_\mu - igB'_\mu)\,\psi' = (\partial_\mu - igB_\mu)\,\psi, \tag{2.3}$$

we obtain the law of transformation for B_μ (B are the quanta of the Yang—Mills field):

$$B'_\mu = S^{-1}B_\mu S + (i/g)\,S^{-1}\,\partial S/\partial x^\mu.$$

The last term is similar to the gauge term in a gauge transformation of the electromagnetic potentials.

By analogy with the stress tensor $F_{\mu\nu}$ of the electromagnetic field, we construct the tensor

$$\mathcal{F}_{\mu\nu} = \partial B_\mu/\partial x^\nu - \partial B_\nu/\partial x^\mu + ig\,(B_\mu B_\nu - B_\nu B_\mu),$$

which transforms according to the law

$$\mathcal{F}'_{\mu\nu} = S^{-1}\,\mathcal{F}_{\mu\nu}S. \tag{2.4}$$

The line of reasoning outlined above can be applied to a field ψ with arbitrary isospin. In this case, we have different matrices S, i.e., representations of the group of rotations in 3-dimensional space. Different fields with the same total isospin, i.e., belonging to the same representation S, interact with one and the same matrix field B_μ. The product of representations $S = S^{(a)}\,S^{(b)}$ produces the sum of the B_μ fields corresponding to each representation: $B_\mu = B_\mu^{(a)} + B_\mu^{(b)}$. The field B_μ can be represented in the form

$$B_\mu = 2\mathbf{A}_\mu \cdot \mathbf{T}, \tag{2.5}$$

where $\mathbf{T}$ is the matrix of the representation of the group of isotopic rotations O_3, and the fields $\mathbf{A}_\mu$ are the same for all representations. In isospin space, $\mathbf{T}$ and $\mathbf{A}_\mu$ are 3-dimensional vectors. Then

$$\mathcal{F}_{\mu\nu} = 2\mathbf{F}_{\mu\nu}\cdot\mathbf{T};\quad \mathbf{F}_{\mu\nu} = \partial\mathbf{A}_\mu/\partial x^\nu - \partial\mathbf{A}_\nu/\partial x^\mu - 2\,g\mathbf{A}_\mu \times \mathbf{A}_\nu.$$

The same $\mathbf{F}_{\mu\nu}$ interact with all fields ψ, irrespective of the representation S to which ψ belongs. If we consider only infinitesimal isotopic gauge transformations $S = 1 - 2\,i\mathbf{T}\cdot\delta\omega$, the transformations $\mathbf{A}_\mu$ take the form

$$\mathbf{A}'_\mu = \mathbf{A}_\mu + 2\,\mathbf{A}_\mu \times \delta\omega + (1/g)\,\partial_\mu\delta\omega.$$

Here $\delta\omega$ denotes the set of parameters of the transformations. Note that infinitesimal transformations can be distinguished only if the local group is a Lie group.

We choose a Lagrangian density invariant with respect to local isotopic gauges in the form

$$L = -(1/4)\,\mathbf{F}^{\mu\nu}\cdot\mathbf{F}_{\mu\nu} - \bar{\psi}\gamma^\mu\,(\partial_\mu - ig\tau\cdot\mathbf{A}_\mu)\,\psi - m\bar{\psi}\psi.$$

Variation of this Lagrangian with respect to $\mathbf{A}_\mu$ and $\bar{\psi}$ leads to the equations

$$\left.\begin{aligned} &\partial_\nu \mathbf{F}^{\mu\nu} + 2g(\mathbf{A}_\nu \times \mathbf{F}^{\mu\nu}) + \mathbf{J}^\mu = 0; \\ &\gamma^\mu(\partial_\mu - ig\boldsymbol{\tau}\cdot\mathbf{A}_\mu)\psi + m\psi = 0, \end{aligned}\right\} \tag{2.6}$$

where $\mathbf{J}^\mu = ig\bar{\psi}\gamma^\mu\tau\psi$. The divergence of $\mathbf{J}_\mu$ is nonzero:

$$\partial\mathbf{J}^\mu/\partial x^\mu = -2\,g\mathbf{A}_\mu \times \mathbf{J}^\mu.$$

If, however, we introduce the quantity

$$\tilde{\mathbf{J}}^\mu = \mathbf{J}^\mu + 2\,g\mathbf{A}_\nu \times \mathbf{F}^{\mu\nu},$$

we obtain a conservation law in the form

$$\partial\tilde{\mathbf{J}}^\mu/\partial x^\mu = 0. \tag{2.7}$$

The field equations can be supplemented by a condition on $\mathbf{A}_\mu$ analogous to the Lorentz gauge for the electromagnetic potentials:

$$\partial\mathbf{A}_\mu/\partial x^\mu = 0.$$

This condition eliminates the scalar part in the field $\mathbf{A}_\mu$, leaving the components corresponding to spin 1, and imposes a restriction on the possible isotopic gauge transformations. In particular, the transformation $S = 1 - i\boldsymbol{\tau}\cdot\delta\boldsymbol{\omega}$ must satisfy the condition

$$\partial^2\delta\boldsymbol{\omega}/\partial x^{\mu 2} + 2\,g\mathbf{A}_\mu \times \partial\delta\boldsymbol{\omega}/\partial x^\mu = 0,$$

which, using (2.2), can be represented in the form

$$\nabla_\mu\nabla_\mu\delta\boldsymbol{\omega} = 0.$$

This is an analog of the condition on the parameters of the gauge transformation (2.1) in electrodynamics:

$$\partial^2\alpha/\partial x^{\mu 2} = 0. \tag{2.8}$$

As is well known, the condition (2.8) distinguishes the wave solutions in Maxwell's equations.

Written in terms of the vector potential in the Lorentz gauge, the Yang–Mills equations take the form

$$\Box\mathbf{A}_\mu + 2\,g\mathbf{A}_\nu \times (2\,\partial_\nu\mathbf{A}_\mu - \partial_\mu\mathbf{A}_\nu - 2\,g\mathbf{A}_\mu \times \mathbf{A}_\nu) = 0. \tag{2.9}$$

Spherically Symmetric Solutions of the Classical Yang–Mills Equations. Free Fields. An analogy between classical gauge fields and the electromagnetic field was demonstrated by Ikeda and Miyachi[28] in 1962, and subsequently by Loos[29] in 1965. These authors obtained static spherically symmetric solutions of the source-free Yang–Mills equations in the Lorentz gauge for the case of isospin and SU(3) symmetries. In addition, Loos showed that the spherically symmetric solutions of the equations for a gauge field with any internal symmetry always lead to the Coulomb form. In fact, in the general case the spherically symmetric solution of the Yang—Mills equations has the form

$$\mathrm{A}_i = (x^i/r)\, \mathrm{f}\, (r, t),\ i = 1, 2, 3;$$

$$\mathrm{A}_4 = \mathrm{i}\varphi\, (r, t),$$

or, in polar coordinates, $\mathrm{A}_r = \mathrm{f}\,(r, t)$, $\mathrm{A}_\theta = \mathrm{A}_\varphi = 0$, $\mathrm{A}_4 = \mathrm{i}\,\varphi\,(r, t)$.

In the static case, the equations (2.9) take the form

$$\Delta \mathrm{f} - 2\,\mathrm{f}/r^2 + 2\,g\,[\mathrm{f} \times \mathrm{f}' + \varphi\times\varphi' + 2\,g\varphi\times(\mathrm{f}\times\varphi)] = 0; \qquad (2.10)$$

$$\Delta\varphi + 4\ g\mathrm{f} \times [\varphi' + g\mathrm{f} \times \varphi] = 0. \qquad (2.11)$$

The gauge conditions $\partial_i \mathrm{A}_i = 0$ reduce to the conditions

$$\mathrm{f}' + 2\mathrm{f}/r = 0. \qquad (2.12)$$

Here a prime denotes differentiation with respect to r, and Δ is the Laplacian. Integrating (2.12), we obtain $\mathrm{f} = \mathrm{a}/r^2$ (where a is a constant of integration). Then (2.10) and (2.11) give

$$\varphi \times [\varphi' + 2\ g\mathrm{a} \times \varphi/r^2] = 0; \qquad (2.13)$$

$$(r^2\varphi')' + 4\ g\mathrm{a} \times [\varphi' + g\mathrm{a} \times \varphi/r^2] = 0; \qquad (2.14)$$

here $\varphi = 0$ is a trivial solution of these equations. Suppose that $\varphi \neq 0$. Then it follows from (2.13) that there exists a scalar function $\lambda(r)$, such that

$$r^2\varphi' = -2g\mathrm{a} \times \varphi + \lambda\varphi. \qquad (2.15)$$

Substituting (2.15) into (2.14), we obtain

$$(\lambda' + \lambda^2/r^2)\ \varphi = 0,$$

from which $\lambda = k(1 - k/r)^{-1}$, where k is a constant of integration. When $\lambda \neq 0$, (2.15) reduces to the equation

$$r^2\,(\lambda\varphi)' = -\,2\ g\mathrm{a} \times (\lambda\varphi).$$

The general solution of this equation is

$$\varphi = (1 - k/r)\ [\bar{\mathrm{a}} + \bar{\mathrm{b}} \cos\,(2g\,|\mathrm{a}|/r) + \bar{\mathrm{c}} \sin\,(2\,g\,|\mathrm{a}|/r)], \qquad (2.16)$$

where $\bar{\mathrm{a}}$, $\bar{\mathrm{b}}$, and $\bar{\mathrm{c}}$ are constants of integration satisfying the conditions $\mathrm{a}\times\bar{\mathrm{a}} = 0$, $|\mathrm{a}|\,\bar{\mathrm{c}} = \mathrm{a} \times \bar{\mathrm{b}}$, and $|\mathrm{a}|\,\bar{\mathrm{b}} = -\,\mathrm{a} \times \bar{\mathrm{c}}$. This means that: 1) $\bar{\mathrm{a}} \parallel \mathrm{a}$; 2) $\bar{\mathrm{a}}$, $\bar{\mathrm{b}}$, and $\bar{\mathrm{c}}$ form a right-handed orthogonal system; and 3) $|\bar{\mathrm{b}}| = |\bar{\mathrm{c}}|$. When $\lambda = 0$ (i.e., $k = 0$), the general solution of (2.15) is given by Eq. (2.16), in which we must put $k = 0$.

In the case $\mathrm{a} \times \varphi \neq 0$, the solution of (2.15) has the following properties. In going from one point of space to another, the vector φ "rotates" in group space around an "axis" a, the angle α between a and φ remaining constant. The "angular velocity" ω of this rotation can be obtained as follows. The component of "velocity" of the end of the vector φ in the direction $a \times \varphi$, determined from Eq. (2.15), is

$$\frac{(a\times\varphi)\cdot\varphi'}{|a\times\varphi|} = -2g\frac{|a\times\varphi|}{r^2} = -2g|a|\cdot|\varphi|\frac{\sin\alpha}{r^2}. \tag{2.17}$$

Dividing (2.17) by the distance of the end of the vector φ from the "axis" a (i.e., by $|\varphi|\sin\alpha$), we obtain an expression for the "angular velocity" ω, namely, ω: $\omega = -2g|a|/r^2$.

The particular solution $A_i = 0$, $A_4 = i(c_1/r + c_2)$, $c_1 = (0, 0, c_1)$, $c_2 = (0, 0, c_2)$, and $c_1, c_2 = \text{const}$ will be called the canonical solution. It can be interpreted as a neutral B field.

Thus, for the B quanta of the Yang—Mills field we have the canonical solution

$$B_i = 0;\ B_4 = 2i\,(c_1/r + c_2)\cdot T. \tag{2.18}$$

It corresponds to a stress tensor of the field with components

$$\mathcal{F}_{ij} = 0;\ \mathcal{F}_{i4} = 2i\,(x^i/r^2)\ c_1\cdot T. \tag{2.19}$$

The general solution is

$$B_i = 2\,(x^i/r^3)\ a\cdot T;\ B_4 = 2\,i\varphi\cdot T, \tag{2.20}$$

where φ is determined by Eqs. (2.16) and (2.17). In this case,

$$\mathcal{F}_{ij} = 0;\ \mathcal{F}_{i4} = -2\,i\lambda\,(x^i/r^3)\ \varphi\cdot T. \tag{2.21}$$

By choice of the gauge, the general solution can be brought to canonical form.

The conserved integral quantities satisfy the relations

$$4\pi c_1 = -i\int J_4\, d^3x;\quad 4\pi k(\bar{a} + \bar{b}) = i\int J_4 d^3x.$$

The spherically symmetric solution for a gauge field with SU(3) symmetry found by Loos is analogous in its properties to the solution of Ikeda and Miyachi for isospin symmetry given above. The components of the vector potential in Loos's solution have the form

$$\left.\begin{aligned} &A_r = ic^j H_j/r^2,\ j = 1, 2;\ A_0 = A_\varphi = 1;\\ &A_t = (ia/r)\,[E_3 \exp(i\beta) + E_{-3}\exp(-i\beta)],\end{aligned}\right\} \tag{2.22}$$

where $\beta = \beta_0 - c^j a_j(3)/r$; H_1, H_2, E_3, E_{-3} are the generators of SU(3); here $H_1 = (\sqrt{3}/3)\ I_3$ and $H_2 = (1/2)\ Y$. The root vector $a^j(3) = (-\sqrt{3}/6, 1/2)$ belongs to E_3; c^j and a are real constants; and

$$F_{rt} = -(ia/r^2)\,[E_3 \exp(i\beta) + E_{-3} \exp(-i\beta)]. \qquad (2.23)$$

As in the case of isospin symmetry, the solution (2.22) and (2.23) leads to the Coulomb form through the transformation $S = \exp(ic^j H_j/r)$ with the condition $c^j a_j(3) = 0$, which means that $c^j H_j$ is proportional to $\mathbf{Q} = \mathbf{I}_3 + {}^1/_2\mathbf{Y}$, i.e., $\mathbf{A}_r$ is proportional to the electric charge $\mathbf{Q}$.

As r varies, as in the solution of Ikeda and Miyachi, the vector $r\mathbf{A}_t$ rotates in group space. As a result, for large r we have $\beta \to \beta_0$, and for $r \gg c^j a_j$ the solution is practically of the Coulomb type.

If we put the velocity of light equal to unity $(c = 1)$, the coefficients in the equations can be chosen in such a way that the matrix elements of $\mathbf{A}_r$ have dimensions $1/l$, where l is a length, and the generators of the group, the root vectors, and a are dimensionless, $c^j \sim l$. Then $c^j a_j \sim l$, where l determines the dimensions of the region inside which the field is of short range, and outside which it tends rapidly to the Coulomb form. Thus, if we consider a gauge field of a single point particle, it does not differ fundamentally from the Coulomb field, and the nonlinearity of the field equations does not manifest itself. However, for two point particles the solution can no longer be reduced to the Coulomb form.

M o n o p o l e s a n d I n s t a n t o n s. In 1969 Wu and Yang, who used the gauge $\partial^0 A_i^a = 0$, found a static solution of the Yang—Mills equations which couples the internal and space-time indices. This solution had the form[30]

$$A_j^a = \varepsilon_{iaj}\, x^j f(r)/r; \quad A_0^a = 0; \quad r = (\mathbf{x}^2)^{1/2}$$

and was singular at $r = 0$. However, if one seeks a similar solution not for free SU(2) gauge fields, but for fields interacting with scalar Higgs fields φ^a, it is possible to construct a static spherically symmetric solution describing a magnetic monopole without string singularities. Such a solution was obtained for the first time by 't Hooft and Polyakov[31] in 1974:

$$A_i^a = -e^{-1}\varepsilon_{iaj}\, x^j f(r)/r; \quad A_0^a = 0; \quad \varphi^a = x^a u(r)/r.$$

To construct this solution, it was necessary to assume that the local symmetry is spontaneously broken, i.e., that there is a distinguished direction in isospace determined by the triplet of fields φ^a or by a single chiral vector field n^a, and to identify the direction of φ^a at each point of V_4 with the direction of the radius vector in ordinary space. Therefore the monopole mass obtained in this way is proportional to the mass of the vector mesons associated with the SU(2) gauge invariance of the theory: $M \sim m_V/g^2$.

In what sense does the solution of 't Hooft and Polyakov describe a magnetic monopole? Let us assume that the electromagnetic group U(1) is a subgroup of a larger gauge group with a compact covering group (in our case, a subgroup of SU(2)) and define the physically observable electromagnetic field as the contraction $F_{\mu\nu} = n_a F^a_{\mu\nu} - e^{-1}\varepsilon_{abc}\, n^a (\nabla_\mu n^b)(\nabla_\nu n^c)$. In the static spherically symmetric case, we have $F_{\mu\nu} = -\varepsilon_{\mu\nu a}\, r^a/er^3$. Consequently, the solution obtained above describes the magnetic field of a point source, $H^a = -r^a/er^3$, situated at the origin $r = 0$. The total magnetic flux is $-4\pi/e$. It satisfies the Schwinger quantization condition $eg = 1$. The tensor $F_{\mu\nu}$ satisfies the equations

$$\partial_\nu F^{\mu\nu} = J^\mu_e; \quad (1/2)\,\varepsilon^{\mu\nu\alpha\beta}\partial_\nu F_{\alpha\beta} = J^\mu_m,$$

from which we see that the currents are conserved. The magnetic charge is defined as

$$Q_m = (1/4\pi)\int J^0_m d^3r = (-1/8\pi e)\int \varepsilon^{ijk}\,\varepsilon^{abc}\,(\partial_i n^a)(\partial_j n^b)(\partial_k n^c)\, d^3r.$$

From the requirement of single-valuedness of n^a, we obtain $Q_m = (-1/4\pi e)\int_{n^2=1} k d^2 n = -k/e$, where k is the number of turns of the sphere in isospace in the integration over the sphere in ordinary space. For $n^a =$ const or r^a/r, $k = 0$ or 1. The quantity k is called the *degree of the mapping* of S^2 into S^2 (a topological invariant). Conservation of magnetic charge is not related to dynamics, since it follows from the existence of a constant scalar isovector field n^a and the identities $\varepsilon_{\mu\nu\alpha\beta}\partial^\nu\partial^\alpha (n^a A^\beta) \equiv 0$, which are valid in the absence of linear singularities in A^a_μ.

A generalization of this approach to groups of higher dimensions than SU(2) leads to SO(4) and SO(3, 1) gauge groups when the internal space and the base have the same dimensions. The first solution of this type was constructed by A. A. Belavin, A. M. Polyakov, A. S. Schwarz, and Yu. S. Tyupkin in 1975 and has become known as an *instanton*. Instanton solutions are regular, localized in space and time ($S < \infty$), satisfy the condition $F^a_{\mu\nu} = \pm {}^*F^a_{\mu\nu} = (\pm 1/2)\,\varepsilon_{\mu\nu\tau\lambda} F^{a\tau\lambda}$ of self-duality of the field tensor, and minimize the action integral in the Euclidean V_4, since $\int (F^a_{\mu\nu} \mp {}^*F^a_{\mu\nu})^2\, dV \geqslant 0$ and hence $S = \int (F^a_{\mu\nu})^2\, dV \geqslant \pm \int (F^a_{\mu\nu}\, {}^*F^a_{\mu\nu})\, dV$. The action plays the same role in the description of instantons as the energy in the description of particles. Instantons are also known as pseudoparticles. There exist only two types of instanton solutions of the equations of an SO(4) gauge field in the Euclidean V_4:

$$1)\ A^{\alpha\beta}_\mu = 2Y^{\alpha\beta}_\mu (x^2 + \lambda^2)^{-1}; \quad F^{\alpha\beta}_{\mu\nu} = 4\delta^{\alpha\beta}_{\mu\nu}\lambda^2 (x^2 + \lambda^2)^{-2};$$

$$2)\ A^{\alpha\beta}_{\mu} = (Y^{\alpha\beta}_{\mu} - Z^{\alpha\beta}_{\mu})(x^2 + \lambda^2_+)^{-1} + (Y^{\alpha\beta}_{\mu} + Z^{\alpha\beta}_{\mu})(x^2 + \lambda^2_-)^{-1};$$
$$F^{\alpha\beta}_{\mu\nu} = 2\delta^{\alpha\beta}_{\mu\nu}[\lambda^2_+(x^2 + \lambda^2_+)^{-2} + \lambda^2_-(x^2 + \lambda^2_-)^{-2}] + 2\varepsilon_{\mu\nu\alpha\beta}[\lambda^2_+(x^2 + \lambda_+)^{-2} - \lambda^2_-(x^2 + \lambda^2_-)^{-2}],$$

where $Y^{\alpha\beta}_{\mu} = x^{\alpha}\delta^{\beta}_{\mu} - x^{\beta}\delta^{\alpha}_{\mu}$ and $Z^{\alpha\beta}_{\mu} = \varepsilon_{\mu\alpha'\beta'\gamma}\, x^{\gamma}\, g^{\alpha\alpha'}\, g^{\beta\beta'}$; λ, λ_+, λ_- are scale factors, and $\lambda_+ \neq \lambda_-$. Solution 1, obtained by A. A. Belavin et al., is real, everywhere regular and vanishing as $x^2 \to \infty$ in the Euclidean V_4, and becomes singular in going over to Minkowski space, since SO(4) is replaced by SO(3, 1) and it is possible to have $x^2 + \lambda^2 = 0$ as a consequence of the change of sign in the metric. Solution 2 describes a pair of pseudoparticles: an instanton and an anti-instanton.

An instanton, like a monopole, is characterized by a conserved current, which is not related to dynamics and which gives as the charge the topological invariant $q = \int F^{\alpha}_{\mu\nu}(\pm)\, {}^*F^{\alpha}_{\mu\nu}(\pm)\, d^4x$, where $F^{0a}_{\mu\nu}(\pm) = (1/2)(F^{0a}_{\mu\nu} \pm (1/2)\,\varepsilon_{abc}\, F^{bc}_{\mu\nu})$. For solution 1, we have $q = 1$. The anti-instanton has $q = -1$. Instantons are long-range fields and must be taken into account in the infrared problem. They describe quantum fluctuations of the vacuum. The SO(4) gauge field in flat V_4 is not the gravitational field.

Utiyama's Theory and its Development. Gauge Fields of General Form. In 1956 Utiyama[2] showed that the requirement of local invariance of the action integral

$$S = \int \bar{\psi}\Gamma^{\mu}\partial_{\mu}\psi d^4x \qquad (2.24)$$

with respect to an arbitrary semisimple Lie group can be used to obtain a vector field corresponding to this group, as well as ordinary conservation laws. In Utiyama's work, the postulate of local gauge invariance of a Lagrangian theory was regarded as a general principle which makes it possible to introduce new vector fields and to determine the form of the interaction of these fields with the original fields, relating the new fields to conservation laws. The gauge group was taken to be an arbitrary semisimple Lie group. The theory was constructed by means of the variational formalism. The new postulate of invariance made it possible to answer the following questions:

1. What will be the nature of the field $A_{\mu}(x)$ which is introduced on the basis of the requirement of invariance?

2. How does this new field transform under transformations of the group $G_{\infty r}$?

3. What is the form of interaction between the field $A_{\mu}(x)$ and the original field ψ?

4. Can we determine the new Lagrangian $L'(\psi, A)$ from the

original one $L(\psi)$?

5. What types of equations for the field $A_\mu(x)$ will be admissible?

Indeed, suppose that ψ is subjected to a gauge transformation of the form

$$\psi' = S\psi, \tag{2.25}$$

where $S = 1 + \underset{a}{I}\,\varepsilon^a(x)$, and $\underset{a}{I}$ is the generator of some representation of the Lie group G_r, with respect to which the action integral (2.24) is invariant. It is easy to see that (2.25) is obtained from the infinitesimal transformations of G_r by replacing the parameters by arbitrary functions of the coordinates. Thus, at each point we have a repetition of the algebra of the finite Lie group G_r, but the parameters of the transformations vary from point to point. Since the transformations depend on r functions, and not numbers, we shall denote the group of local gauges by the symbol $G_{\infty r}$ (an infinite group).[32] The action integral (2.24) is not invariant with respect to $G_{\infty r}$ because its variation contains nonvanishing derivatives of the parametric functions. To restore invariance of the Lagrangian, it is sufficient to introduce in it an interaction with some vector field by replacing the ordinary derivative by the "covariant" (or compensating) derivative according to the rule $\partial_\mu \to \nabla_\mu = \partial_\mu - A^a_\mu \underset{a}{I}$. In Utiyama's theory, this rule is no longer mnemonic, as in the theory of Yang and Mills, but is obtained as a result of solution of the conditions of invariance of the Lagrangian. These conditions are given in §5, where we construct a complete Lagrangian theory of gauge fields.

The conditions of invariance analogous to (2.3) determine the transformation properties of the vector potential of the gauge field with respect to the group of local gauges:

$$\delta A^a_\mu = f^a_{bc} A^c_\mu \varepsilon^b + \partial_\mu \varepsilon^a,$$

where f^a_{bc} are the structure constants of the group G_r, and ε^a are parameters. This law of transformation of A^a_μ ensures identical (i.e., covariant) transformation properties with respect to $G_{\infty r}$ of the wave function ψ and its covariant derivative $\delta(\nabla_\mu \psi) = \varepsilon^a \underset{a}{I}\, \nabla_\mu \psi$.

Knowing the law of transformation of A^a_μ, we can use the requirement of invariance with respect to local gauges to find a Lagrangian for the free field A^a_μ, depending on A^a_μ and its first derivatives. The simplest relativistically invariant and gauge-invariant Lagrangian has the form

$L_0 = -(1/4)\, F^{\mu\nu}_{a} F^{a}_{\mu\nu}$. An arbitrary Lagrangian for the free field A^a_μ reduces to an arbitrary function of L_0.[39] In the general case, the stress tensor of the gauge field, sometimes called simply the f i e l d t e n s o r, has the form $F^a_{\mu\nu} = \partial_{[\mu} A^a_{\nu]} - {}^1\!/_2\, f^a_{bc} A^b_{[\mu} A^c_{\nu]}$. It transforms with respect to $G_{\infty r}$ homogeneously: $\delta F^a_{\mu\nu} = f^a_{bc} F^c_{\mu\nu}\, \varepsilon^b(x)$. Lowering the parametric index in the expression for $F^a_{\mu\nu}$ by means of the "group metric" $g_{ab} = f^{\,l}_{am} f^{m}_{l\,b}$, we obtain the field tensor $F_{a\mu\nu} = g_{ab} F^b_{\mu\nu}$. Its law of transformation is

$$\delta F_{a\mu\nu} = -\, f^b_{ac} F_{b\mu\nu}\varepsilon^c(x).$$

Clearly, since g_{ab} does not become degenerate only for semisimple groups, the gauge fields obtained by Utiyama correspond only to semisimple groups.

Assuming that the total Lagrangian has the form $L_T = L_0(F) + L(\psi, \nabla_\mu\psi)$, Utiyama wrote the equations for the gauge field in the form $\delta L_T/\delta A^a_\mu = 0$, and also the conservation law

$$\partial_\mu \tilde{J}^\mu_a = 0 \qquad (2.26)$$

for the current

$$\tilde{J}^\mu_a = -\big(\underset{a}{I}\,\psi \partial L/\partial\nabla_\mu\psi + f^b_{ac} A^c_\nu\, \partial L_0/\partial F^b_{\mu\nu}\big). \qquad (2.27)$$

Here the Lagrangian L_0 of the free field A^a_μ is assumed to be an arbitrary function of $F^a_{\mu\nu}$ satisfying the condition $f^a_{bc} F^b_{\mu\nu}\, \partial L_0/\partial F^c_{\mu\nu} = 0$. The first term in (2.27) is the ordinary current J^μ_a, which is conserved in the case of a global symmetry of L. The second term appears because of the locality of the symmetry. In the particular cases of Abelian and isotopic groups, (2.26) implies conservation laws for the electromagnetic and isospin currents which are identical to the conservation laws in electrodynamics and in the Yang–Mills theory. But Utiyama introduced these laws not by differentiation of the field equations, but by analyzing the conditions of local invariance of the Lagrangian, assuming, first, that the field equations are satisfied, i.e., that the conservation law is weak, and secondly, that the current can be defined as

$$J^\mu_a = \partial L_T/\partial A^a_\mu. \qquad (2.28)$$

This definition of the current, taken together with the conditions of local invariance of the Lagrangian, leads to (2.27), but is not always equivalent to the ordinary definition. For massive gauge fields, the two definitions of the current, (2.27) and (2.28), lead to different expressions because the definition (2.28) is sensitive to the potential energy, while (2.27) is not. In Utiyama's theory, the conservation law (2.27) is satisfied on the extremals

$\delta L_T/\delta A^a_\mu = 0$, as a consequence of the relation which follows from the conditions of invariance of the Lagrangian:

$$\partial_\mu \tilde{J}^\mu_a \equiv \partial_\mu (\delta L_T/\delta A^a_\mu). \tag{2.29}$$

Apart from the divergence of an antisymmetric tensor, this always implies that $\delta L_T/\delta A^a_\mu = \tilde{J}^\mu_a$. Therefore we can say that the source of any gauge field is a conserved current.

It was shown for the first time in Refs. 32 and 35 that the Lagrangian theory of gauge fields is a particular case of the theories that are invariant with respect to infinite groups. The variational formalism for infinite groups developed there (see Chapter II) made it possible to construct a consistent Lagrangian theory of gauge fields which includes, in particular, the results of Yang and Mills and of Utiyama.

Lorentz Group and the Gravitational Field. Attempts to interpret gravitation as a gauge field occupy a special position, since, first, this is the only physical field which is directly related to the structure of space-time, and secondly, we have here localization of coordinate transformations (and not gauge transformations), so that homogeneous spaces such as Minkowski space are replaced by nonhomogeneous spaces such as Riemannian space, and in the general case by fiber spaces.[14,15,32-36]

As we have already mentioned, if a Lie group is localized, then the algebra is preserved only locally. In exactly the same way, the concept of a representation of a finite Lie group G_r becomes local. Therefore, before considering groups of coordinate transformations, we must introduce, in addition to the world coordinate grids covering all space-time, local coordinate systems of the tangent space to V_4 attached to each world point. The basis vectors of a local orthogonal coordinate system in the general theory of relativity are called t e t r a d s. Thus, by choosing the tangent space as the fiber at each point, we obtain the tangent fiber space over V_4. With localization, the nonhomogeneous group of motions of flat space decomposes into two different groups in the sense of a group of transformations: the homogeneous subgroup goes over into a local symmetry which transforms the tetrad basis at each point, and the displacements go over into a group of continuous automorphisms mapping the local spaces onto one another (the general covariant group). Clearly, the fiber and the base have the same dimensions in this case.

After Weyl,[5] gravitation was first considered as a gauge field by Utiyama.[2] In Utiyama's theory, the gravitational field is introduced as follows. Two groups of transformations are considered: the local Lorentz group acting in the tangent space, and the group of arbitrary point transformations in V_4. In addition to the ordinary field variables, such as the wave functions of particles, tetrads are introduced in the Lagrangian as 16 new independent variables. A definite dependence is postulated between the covariant derivatives of the world and tetrad components of the tensors. This leads to a dependence between the gauge field $A_\mu(kl)$ corresponding to the local Lorentz group and the tetrads. Using the relation between the metric tensor and the tetrads, and assuming that the connection coefficients in the base are symmetric in the lower indices, Utiyama shows that the gauge field $A_\mu(kl)$ can be expressed in terms of the Christoffel symbols and is therefore identical with the gravitational field. Then the stress tensor $F_{\mu\nu}(kl)$ of the field can be expressed in terms of the Riemann curvature tensor of V_4. The tetrads are used to construct a Lagrangian of the gauge field, which is linear in $F_{\mu\nu}(kl)$ and is identical with the ordinary scalar curvature of V_4.

In Utiyama's theory, Einstein's equations are obtained as follows. Two coordinate systems are introduced: a local Lorentz system (the x system) and an arbitrary curvilinear system of world coordinates (the u system). Quantities referring to the x system are denoted by Latin indices, and those referring to the u system by Greek indices.

The square of the invariant length of an infinitesimal interval can be written in the form

$$ds^2 = g_{ik}dx^i\,dx^k = g_{\mu\nu}\,du^\mu du^\nu;\ g_{\mu\nu}\,(u) = (\partial x^i/\partial u^\mu)\,(\partial x^k/\partial u^\nu)\,g_{ik}.$$

The two groups of functions (tetrads) $h^k_\mu(u) = \partial x^k/\partial u^\mu$ and $h^\mu_k(u) = \partial u^\mu/\partial x^k$ relate local quantities to world quantities. These satisfy the relations

$$g_{ik}\,h^i_\mu\,h^k_\nu = g_{\mu\nu};\ g_{\mu\nu}h^\mu_i\,h^\nu_k = g_{ik};\ h^\mu_k h^l_\mu = \delta^l_k;$$

$$g^{ik}\,h^\mu_i h^\nu_k = g^{\mu\nu};\ g^{\mu\nu}h^i_\mu\,h^k_\nu = g^{ik};\ h^\mu_k\,h^k_\nu = \delta^\mu_\nu;$$

$$\det|g_{\mu\nu}| = g = -\,h^2 = -\,|\det(h^k_\mu)|^2.$$

The four vectors h^μ_1, h^μ_2, h^μ_3, h^μ_4 specify a local Lorentz frame of reference (the x system) at each world point. The wave functions Q^A are defined with respect to this x system. In particular, spinors are introduced as functions whose local components transform according to the spinor representation of the group of Lorentz rotations of the local coordinate system. The world components of the wave functions are obtained from the local components by means of the tetrads: $Q^\mu\,(u) = h^\mu_k(u)\,Q^k$. The action integral can be written in the form $I = \int |h|\,L\,(Q^A,\,h^\mu_k,\,Q^A_{,\mu})\,d^4x$. It is invariant with respect to two types of transformations:

1) a Lorentz transformation

$$\delta h^k_\mu = \varepsilon^k{}_l h^l_\mu ; \ \delta Q^A = {}^1/_2 \underset{(kl)}{T^A{}_B} Q^B \varepsilon^{kl}, \tag{2.30}$$

where $\underset{(kl)}{T^A{}_B}$ is an element of the N × N matrix of the representation of the Lorentz group, and the u^μ do not change;

2) a general point transformation

$$u^{\mu'} = u^\mu + \lambda^\mu (u), \tag{2.31}$$

where $\lambda^\mu(u)$ is an arbitrary function of u; here $\delta h^k_\mu = (\partial\lambda^\nu/\partial u^\mu)\, h^k_\nu$, $\delta Q^A(u) = Q^{A'}(u') - Q^A(u) = 0$, and $\delta Q^A_{,\mu} = -(\partial\lambda^\nu/\partial u^\mu)\, Q^A_{,\nu}$.

Localization of the transformations (2.30) leads to the replacement of the ordinary derivative by the covariant derivative:

$$\partial_\mu \to \nabla_\mu = \partial_\mu - {}^1/_2 A_\mu (kl) \underset{(kl)}{T^A_B} Q^B,$$

where $A_\mu(kl)$ is the gauge field. In particular, if Q^B is a spinor field ψ, then

$$\nabla_\mu \psi = \partial\psi/\partial x^\mu - (i/4)\, A_\mu\, (kl)\, [\gamma^k\gamma^l]\, \psi,$$

where γ^k denotes the usual Dirac matrices. This definition of the covariant derivative of spinors in Riemannian space was proposed earlier by Fock and Ivanenko[37] in 1929.

A compensating treatment of gravitation was also considered in 1961 by Brodskiĭ, Ivanenko, and Sokolik,[38] who introduced the Ricci connection coefficients through the requirement of invariance of the first-order equation $\Gamma^\mu\partial_\mu\psi - m\psi = 0$ for particles of arbitrary spin with respect to the localized homogeneous Lorentz group.

A generalization of Utiyama's gravitational theory was proposed in 1961 by Kibble,[39] who took the gauge group to be not the homogeneous Lorentz group, but the Poincaré group, assuming, however, that local displacements do not affect local rotations. With respect to local displacements, tensor transformations of the tetrads can be regarded as gauge transformations. Then the tetrads become a second gauge field and are included automatically. The same dependence between the covariant derivatives of the tensor components in the fiber and in the base as was postulated by Utiyama, but with no restriction on the symmetry of the connection coefficients of the base, enabled Kibble to express the gauge field $A_\mu(kl)$ in terms of the Christoffel symbols and the torsion tensor of space-time. Kibble proposed to relate the torsion tensor to the spin properties of the material sources of the gravitational field. Kibble's theory is not identical with the general theory of relativity and is not a consistent theory.

The interpretation of the gravitational field as a gauge field leads in general to a non-Einsteinian theory of gravitation, since the basic field variables here are the connection coefficients rather than the metric tensor. In the geometrical treatment, such a gravitational field corresponds to spaces of affine connection, which may not be endowed with a metric. Following the gauge ideology, the free-field Lagrangian in this theory must be taken to be a Lagrangian which is quadratic in the curvature tensor, namely, $L = -(1/4)\, R_{\mu\nu}\,(ik)\, R^{\mu\nu}\,(i\, k)$, which was originally proposed in 1918 by Weyl[5] (see also Ref. 33).

The metric tensor can be introduced in a gauge theory of gravitation either as an additional field variable or as a gauge field associated with the general covariant transformations (Konopleva[32] (1968)). In the second case, the simplest local gauge-invariant Lagrangian is the scalar curvature (see §8). It is obvious that by varying it with respect to the field variable, i.e., with respect to the metric, we obtain Einstein's equations.

If the theory possesses two types of local symmetries, namely, the local Lorentz and general covariant symmetries, there must also be two types of gauge fields: connection coefficients and a metric (or tetrads). The Lagrangian in this case can be taken to be a sum of Lagrangians which are linear and quadratic in the curvature tensor.[33,40] Two groups of equations are obtained as the field equations in such a theory. Variation of the Lagrangian with respect to the metric gives a generalization of Einstein's equations. Variation with respect to the connection leads to quasi-Maxwell equations for the curvature tensor (Refs. 40—42 and §13). If the metric is covariantly constant with respect to the connection under consideration, the class of solutions of the generalized gravitational field equations includes all the spaces of general relativity.[40] Renewed interest in the gauge theory of gravitation with a Lagrangian quadratic in the curvature tensor was aroused in 1974 by the work of Yang.[43]

Gauge Theories of the Strong Interactions. Sakurai's Theory and SU(3) Symmetry. In 1960 there appeared an important paper by Sakurai,[3] in which a universal theory of the strong interactions was proposed. Sakurai's theory was based, first, on a result of Pais, according to which there do not exist any other exact internal symmetry properties apart from baryon number, hypercharge, and isospin, and secondly, on local gauge invariance. Through the requirement of local gauge invariance, some vector field can be associated with each of the above-mentioned quantities. Sakurai claimed that these three fields were sufficient to account for the mass spectrum which was known at that time.

In addition, Sakurai's theory predicted the production of resonances and yielded a number of beautiful results and explanations (in particular, the multiplicity of pion production in proton—antiproton annihilation, the effect of a repulsive core, and so forth). According to Sakurai, the strong interactions have a vector character. The objects which mediate them must be vector particles with a mass equal to the mass of several pions and with a small lifetime (resonances). However, Sakurai's theory did not allow the production of single pions and kaons, and from the point of view of this theory Yukawa couplings were purely phenomenological.

On the basis of Sakurai's ideas, Ikeda et al. in 1960 and Ne'eman, Gell-Mann, and Glashow in 1961 proposed an eightfold symmetry for the strong interactions.[44] In contrast with Sakurai's theory, which was based on three independent localized symmetries (two one-parameter symmetries, as in electrodynamics, corresponding to conservation of baryon number and hypercharge, and the localized group of isotopic rotations), in the new theory these conservation laws followed from a single symmetry group, namely, from the group SU(3). This group combined all the baryons (as well as the mesons) into a single 8-dimensional representation. Several new particles and a large number of selection rules were predicted. The forms of the couplings with Yukawa potentials were determined for pions, kaons, and χ mesons. The mass formulas deduced from SU(3) symmetry were in good agreement with experiment. While Sakurai's model led to the discovery of the vector mesons ρ, ω, and φ, SU(3) symmetry was responsible for the discovery of the Ω resonance and the appearance of new models of the hadrons known as *quark models*.

From the point of view of SU(3) symmetry, the experimentally observed particles are not elementary, since they do not form multiplets of the lowest dimension. Therefore there may exist in principle "smaller" structural units, of which the observed particles are composed. Such particles have become known as quarks. Quarks were introduced by Gell-Mann and Zweig in 1964 as hypothetical "constituents" of nucleons and mesons, which form the lowest possible representation of the group SU(3) — a triplet.[45] Quarks were introduced with the aim of reducing the number of necessary "building blocks of the world," since the number of known particles which were traditionally regarded as elementary had become too large. For a long time, quarks were not taken seriously because they seemed excessively exotic objects. It was necessary for them to possess fractional electric charges and move almost freely inside hadrons, but they were never supposed to appear outside hadrons. In spite of the fact that "quark counting" yielded predictions in agreement with experiment, quarks (and, subsequently, their generalization — partons, i.e., any "parts" of hadrons) were until recently regarded as auxiliary, rather than real objects. The situation changed after the discovery in November 1974 of the mysterious ψ (or J) particles, which belong to a new family of particles.[46,47] It was possible to explain the properties of the ψ particles by introducing a fourth "charmed" c quark. This made it possible to represent the ψ particles as c h a r m o n i u m, i.e., a charmed quark—antiquark system analogous to positronium (the e^+e^- system), and to predict its excited levels, which were also observed. The forces which hold the quarks inside hadrons were explained by means of a non-Abelian gauge vector field of gluons, which carries a new quantum number — c o l o r.

The theory of non-Abelian gauge fields (i.e., fields associated with a non-Abelian gauge group) has proved to be the only asymptotically free theory, i.e., a theory which predicts vanishing of the interaction in the region of large momenta (or at small distances), and a growth of the interaction at small momenta (at large distances). Most of the mechanisms of quark retention (q u a r k c o n f i n e m e n t) that have so far been proposed make use of the properties of non-Abelian gauge fields in some way. However, the question of whether quarks can nevertheless escape has not been definitively resolved, since it entails the complicated problem of infrared divergences in gauge models. At the present time, this problem is being intensively investigated.

Reggeization of Gauge Fields and Dual Models. The surprising property of asymptotic freedom of the theory of non-Abelian gauge fields means, among other things, that this theory is free from the celebrated problem of "zero charge," from which quantum electrodynamics suffers. This problem, in its time, aroused such pessimism among theoreticians that some of them resolved to construct a theory of elementary particles without resorting to quantum field theory, but using only the properties of the S matrix. This led to the advent of Reggeism. However, Reggeism could not be developed into a consistent theory of elementary particles, although as a phenomenological model it remains useful for the description of experiments. It has provided an idea which is very valuable for the classification of elementary particles, namely, the concept of R e g g e t r a j e c t o r i e s on which particles "lie." Regge trajectories, like quarks, make it possible to reduce the number of "building blocks of the world," since a single trajectory can embrace an infinite number of particles, in spite of the fact that the trajectory itself is described by only two parameters: its slope and intercept. There is now a strong tendency towards reconciliation of Reggeism and quantum field theory; this was seen, for example, in the creation of the calculus of Regge fields, using the elements of the Lagrangian quantum field-theoretic approach. This tendency is manifested even more strongly in dual models — another variant of the S-matrix approach, which is intimately related to Reggeism.[48,49] In recent years, so-called dual models of "strings" have been intensively investigated. A string is a one-dimensional extended (and in this sense, classical) system with an infinite number of degrees of freedom. If in dual models the slope of the trajectory $\alpha(t)$ tends to zero and certain masses are fixed, it is possible to reproduce the results of certain field-theoretic models: the $\lambda\varphi^3$ model, the massless Yang—Mills theory, and the massive Yang—Mills theory with spontaneous symmetry breaking. Dual and field-theoretic models are usually distinguished by the fact that the former give Regge behavior in the tree approximation, while the latter do not. However, if the series of perturbation theory is summed in field theory, then under certain conditions Regge behavior can occur. As was shown in Refs. 50 and 51, the necessary conditions for Reggeization are fulfilled in theories with non-Abelian gauge fields. It may be that non-Abelian gauge theories are the only field theories which are capable of being "unified" with dual models. In the case of such unification, the theory of strong interactions acquires the qualities of the geometrical electrodynamics of Weyl discussed previously, namely, strings and their excitations can be identified with solutions like vortices, dislocations, solitons, and kinks of the classical nonlinear equations. At the same time, strings can be identified with the neutral gluon field, and their ends can be identified with quarks, which in turn can be regarded as magnetic monopoles.[51] Thus, we obtain geometrical and topological interpretations of certain quantum relations[52] and ensure the universality of the strong interactions.

It is interesting to note that dual models do not contain unphysical states (ghosts) only in spaces of dimension 26 (without spinors) and 10 (with spinors). Therefore the idea of a fiber space arises in these models as a reasonable interpretation of the additional spatial dimensions.[53] Dual models with gauge fields may lead to a unified renormalizable quantum theory of all interactions, including gravitation.

Dynamical Symmetries and Chiral Dynamics. Contemporary elementary-particle physics makes use of two types of symmetries: algebraic and dynamical. Algebraic symmetries correspond to groups of transformations with constant (i.e., independent of the space-time point) parameters. Dynamical symmetries are specified by either local gauge groups or nonlocal and nonhomogeneous realizations of algebraic symmetries. Algebraic symmetries make it possible to classify particles according to their quantum numbers into multiplets which form different representations of the chosen symmetry group. Dynamical symmetries make it possible to classify interactions between particles, imposing restrictions on the form of the Lagrangians and leading to useful sum rules, low-energy theorems, and fortunate cancellations between the divergent Feynman diagrams in perturbation theory. Owing to local gauge invariance and its consequences, unified models of the strong, weak, and electromagnetic interactions have been unitary and renormalizable. The basic tool used to prove renormalizability is the Ward identities,[9,10] which are a quantum analog of the identities that follow in the presence of a local symmetry from Noether's second theorem.[32,35]

Dynamical symmetries not only determine the form of the Lagrangian, but also ensure universality of the interaction, in the sense that the first orders of the expansion in the coupling constant are identical with the first orders of the expansion in powers of the energy.[49] This makes it possible to use the corresponding field theory to obtain reasonable results in the low-energy region, independently of the value of the coupling constant.

Previously, the apparatus of dispersion relations was successfully employed for the description of hadron physics at low energies. However, the need to introduce a number of arbitrary parameters made this approach phenomenological in character. The use of the dynamical principle of SU(3) × SU(3) chiral symmetry allowed a considerable reduction in the number of undetermined parameters, imposing additional boundary conditions on the low-energy solutions of dispersion relations. At the present time, it is generally recognized that chiral symmetry provides a good description of low-energy pion physics. The model which is most frequently used here is the so-called nonlinear σ model, which exhibits many properties of theories with local gauge invariance.

The idea of universality of the strong interactions, which is analogous to universality of the electromagnetic and gravitational interactions, and the idea of a distinguished role of vector conserved currents make it possible to find a general approach to the description of the electromagnetic properties of heavy particles (hadrons) and the deviations from electrodynamics due to the strong interactions. The treatment of gravitation as a gauge field and the application to it of the model of tensor dominance, in their turn, make it possible to take into account the contributions of the gravitational field to the strong interactions.

Universal Strong Interactions and Vector Dominance. In quantum field theory, universality of the electromagnetic interactions (i.e., universality of the electric charge as the coupling constant with the electromagnetic field) is formulated as a relation between the amplitudes characterizing emission of a photon γ of very low frequency. For example, for a proton p and positron e^+, we have

$$\frac{A\,(e^+ \leftarrow\ \rightarrow e^+ + \gamma)}{A\,(p \leftarrow\ \rightarrow p + \gamma)} = \frac{Q\,(e^+)}{Q\,(p)} = 1, \tag{2.32}$$

where Q is the electric charge. This relation is satisfied only in the limit $q^2 \to 0$, when the momentum transferred by the photon tends to zero. At large momentum transfers, the corresponding amplitudes (form factors) differ by several orders of magnitude.

In a completely analogous way, universality of the strong interactions of the ρ meson can be formulated as the equality

$$\frac{A\,(A \leftarrow\ \rightarrow A + \rho^0)}{B\,(B \leftarrow\ \rightarrow B + \rho^0)} = \frac{T_3^{(A)}}{T_3^{(B)}}, \tag{2.33}$$

where $T_3^{(i)}$ are the third components of the isospins, and A and B are arbitrary particles which carry isospin. Exactly as in the case of photons, we must require equality of the matrix elements as $q^2 \to 0$. However, the ρ meson, unlike the photon, possesses mass. Therefore this relation can be tested experimentally only in the case when the corresponding form factors vary very slowly or when this law of variation is known and the necessary extrapolation can be performed.

The relation (2.33) is also a consequence of the postulate of the identity between currents and fields proposed by Kroll, Lee, and Zumino[54]:

$$J_\mu^a\,(x) = (m_\rho^2/f_\rho)\rho_\mu^a\,(x), \tag{2.34}$$

where $J_\mu^a(x)$ is the isovector electromagnetic current density, and $\rho_\mu(x)$ is the ρ-meson field operator.

The original formulation of universality in the quantum theory of the strong interactions was closely connected with the model of vector dominance, i.e., with the assumption that the main contribution to the strong interactions of elementary particles comes from exchanges of a single vector meson. In other words, vector dominance in the strong interactions means dominance of the diagrams of the type shown in

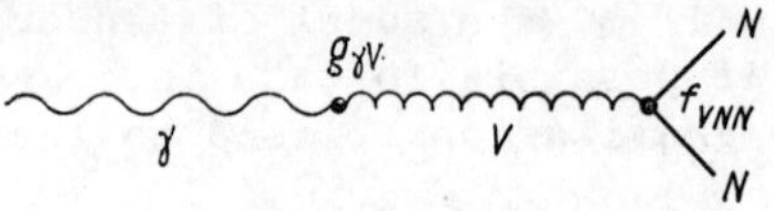

FIGURE 2

Fig. 2. As a consequence, it leads to universality of vector interactions. For example, for the ρ meson, we have $f_{\rho\pi\pi} = f_{\rho NN} = f_\rho$ (which is the coupling constant for the interaction of the ρ meson with any particle that carries isospin). In addition, there is an effective coupling constant for the photon with a vector meson, which in the case of the ρ meson has the form $g_{\gamma\rho} = e\ (m_\rho^2/f_\rho)$, where m_ρ is the mass of the ρ meson.

From the point of view of quantum electrodynamics, the appearance of diagrams such as that of Fig. 2 means that it is possible for a photon to be converted into a vector meson ("the photon acquires mass"), which then decays into an electron—positron (or muon) pair, and also that vector mesons can be produced in electron—positron collisions at very high energy (in colliding beams). In addition to the three ρ mesons (ρ^+, ρ^-, ρ^0), which form an isovector, in accordance with SU(3) symmetry of the strong interactions there are two isoscalar (i.e., isospin-zero) vector mesons ω and φ, for which analogous processes are possible: $\omega \rightleftarrows e^+ + e^-$; $\varphi \rightleftarrows e^+ + e^-$.

The proportionality between the electric charge and the coupling constant for the interaction of vector mesons with other particles makes it possible to find a relation between the matrix elements for the processes $\gamma + A \to B$ and $\rho\ (q^2 = 0) + A \to B$, where A and B are arbitrary particles. In other words, the strong interactions involving vector mesons reproduce, as it were, the analogous electromagnetic processes (the photon is simply "replaced by" the ρ meson). Another example of this kind is provided by the relation between the reactions

$$\omega \to \pi^0 + \gamma \leftarrow\to \omega \to \pi + \rho;$$

$$\gamma + N \to \pi + N \leftarrow\to \pi + N \to \rho\ (\omega, \varphi) + N;$$

$$\gamma + p \to \rho^0 + p \leftarrow\to \rho^0 + p \to \rho^0 + p.$$

The idea of universality is tested by comparing the values of the coupling constants obtained by different methods. The values of the coupling constant for ρ mesons calculated in different processes agree with a high degree of accuracy and give $f_\rho^2/4\pi \sim 2$.[44]

The discrepancies between the predictions of the vector-dominance model and experiment must be associated with the approximate character of this model, rather than with the idea of universality on which it is based. In a comparison with experiment, the choice of a model which gives an adequate realization of some physical principle is always ambiguous and complex.

Strong Gravitation. Here we consider the problem of taking into account the contribution of the gravitational field to the strong interactions of elementary particles. The gravitational field is regarded as a gauge field corresponding to Weyl coordinate transformations (scaling symmetry): $x^{\mu\prime} = \lambda(x) x^{\mu}$. It is easy to see that these transformations form a local Abelian gauge group, and in accordance with the general ideology they lead to the existence of a gauge field. The source of this field is assumed to be the trace of the energy—momentum tensor of the strongly interacting heavy particles (hadrons), which is replaced (dominated) by a scalar particle — the quantum of the gauge field. Another variant of allowance for gravitation in the strong interactions is tensor dominance, or the introduction of a gauge tensor field (whose carriers are the f and f' mesons, which have spin 2). Tensor dominance requires that the gravitational field should interact not directly with the phenomenological fields of the hadrons, but through an intermediate spin-2 hadron field, with which it interacts bilinearly. The source of the tensor meson field is assumed to be the energy—momentum tensor of the hadrons. Therefore this field must correspond to Einsteinian gravitation.

As experiment shows, scaling symmetry is satisfied approximately for interactions at very high energies.

Universal Weak Interactions and Massive Vector Fields. The fact that the weak interactions have a vector character if the nonconservation of parity is disregarded and that they are approximately universal in strength led a number of physicists to attempt to describe them by means of vector fields associated with local gauge invariance. This was facilitated by the existence of Fermi's theory (1934), in which the weak interactions, by analogy with electrodynamics, were described by a Lagrangian which is a product of the particle current J^{μ}_{W} and the vector W_{μ}:

$$L_{em} = eA_{\mu} J^{\mu}_{em} \to L_W = gW_{\mu} J^{\mu}_{W}; \quad \frac{g^2}{m_W^2 - q^2} \xrightarrow[q^2/m_W^2 \to 0]{} = \frac{g^2}{m_W^2} = \frac{G_F}{\sqrt{2}}$$

(where G_F is the Fermi constant).

Universality of the weak interactions manifested itself in the equality of the coupling constants in all weak processes:

$$\mu \to e\nu\tilde{\nu},\ \Sigma \to nl\nu,\ \pi \to e\nu\pi^0;\quad n \to pe\tilde{\nu},\ \Sigma \to \Lambda e\nu,\ K \to l\nu;$$

$$\mu p \to n\nu,\ \Xi \to \Lambda l\nu,\ K \to l\nu\pi;\quad \Lambda \to pl\tilde{\nu},\ \pi \to l\nu,\ K \to l\nu\pi\pi,$$

where $l = e$ or μ.

Nonleptonic weak decays were approximately characterized by the same universal constant G_F.[55]

However, in addition to the similarities between electrodynamics and the weak interactions, there were also important differences. The weak current, unlike the electromagnetic current, is charged: the charge changes in the transition $n \to p$, and the lepton pair $e^-\tilde{\nu}$ is charged. For a long time, experimentalists did not observe neutral currents similar to the electromagnetic current. The weak forces act over a range $<10^{-14}$ cm, whereas the electromagnetic field is long-range. This is so because of the different masses of the carriers of the interaction: the photon is a massless particle, as it must be in a theory with local gauge invariance, while the intermediate W boson, if it exists, must have mass $m_W > 2$ GeV. The existence of long-range weak forces would contradict the principle of equivalence.[56]

Direct application of the gauge ideology to the weak interactions also proved to be difficult because some of the currents appearing in the theory were not conserved. The currents of isospin and strangeness, which entered into the equations of the theory of strong coupling, ceased to be conserved as soon as the electromagnetic and weak interactions were taken into account. Conservation of the weak current was violated not only in the presence of electromagnetic fields, but also (if the axial-vector part and the strangeness-changing part of the Lagrangian were considered) by the mass terms and the strong interactions. Moreover, it proved to be impossible to simply express the weak hadron currents in terms of fundamental fields corresponding to the known stable particles. Therefore it was necessary to seek a formulation of universality of the weak interactions which was independent of conserved currents. Such a formulation was proposed by Gell-Mann. He adopted the postulate that the multiplets of charges of the heavy and light particles have identical algebraic properties. The hypothesis of universality, together with the hypothesis of conservation of the vector current, made it possible to explain the experimentally observed approximate equality of the muon decay constant and the vector coupling constant of the neutron decay.

Thus, to construct a theory of universal weak

interactions, it was necessary to find a means of describing neutral currents with allowance for symmetry breaking in charge space, to learn to assign mass to the vector fields without violating local gauge invariance, to take into account the contribution of the strong interactions, and to formulate a renormalization procedure.

The last requirement is connected with the following circumstance. In first-order perturbation theory, Fermi's theory gives good agreement with experiment. But with increasing energy, this agreement might become much worse. The higher-order corrections have little effect on the results of the first approximation only when the theory is renormalizable, i.e., when the structure of the divergences does not change and their number does not grow with increasing order of perturbation theory. Electrodynamics is a renormalizable theory. The four-fermion theory of the weak interactions is not renormalizable. However, it was found that the theory of weak interactions can be made renormalizable if the electromagnetic and weak interactions are combined into a single scheme on the basis of the theory of gauge fields. This was done for the first time in the Weinberg—Salam model (1967), whose advantages were, it is true, appreciated only later. Subsequently, many other, more complicated models which unify the weak, electromagnetic, and strong interactions were proposed. The next subsection is devoted to these models.

Unified Models of the Strong, Weak, and Electromagnetic Interactions. At contemporary energies, all the interactions of elementary particles can be divided into several basic types, which differ strongly from one another in the size of the coupling constant. If they are arranged in order of decreasing coupling constants, we obtain the well-known hierarchy of interactions: strong, electromagnetic, weak, and gravitational. The origin of this hierarchy is unknown, and it is not clear whether it will be preserved with increasing energy. Each form of interaction is characterized by its own symmetry group, according to whose representations the particles are classified.

Algebraic symmetries predict equality of the masses of particles belonging to the same multiplet. But experiment shows that even the masses of the proton and neutron, which form an isotopic doublet, differ somewhat. In other cases, the mass splittings in multiplets are even larger. Therefore algebraic internal symmetries of elementary particles must, as a rule, be broken by adding to the Lagrangian terms having a lower degree of symmetry than the original Lagrangian. These symmetry-breaking terms must explain the small mass splittings which are observed experimentally in the multiplets of particles.

The idea of a gradual violation of an initial high symmetry down to the complete absence of this symmetry was usually associated with the idea of a hierarchy of types of interactions differing in strength. The strongest interaction must possess the highest symmetry and must conserve the largest possible number of properties of the elementary particles. Therefore the nuclear forces were associated with the group SU(3) or with higher symmetries. Breaking of the symmetry down to SU(2) corresponded to the transition to the weak interactions, and the transition to the symmetry U(1) or O(2) described inclusion of the electromagnetic field. The gravitational interaction destroyed every internal symmetry.

In 1964 Higgs proposed another symmetry-breaking mechanism, which made it possible to obtain arbitrary finite mass splittings in multiplets.[7] This mechanism has become known as spontaneous symmetry breaking. With spontaneous symmetry breaking, there is actually no violation of the symmetry of the original Lagrangian $\mathscr{L}$. One adds to $\mathscr{L}$ a nonlinear Lagrangian of scalar fields (the Higgs Lagrangian) with the same symmetry as $\mathscr{L}$. But certain components of the scalar fields are assigned nonzero vacuum expectation values, as a result of which the symmetry of the vacuum is lower than that of the original Lagrangian. The fields in the Lagrangian are then redefined in such a way that they all have zero vacuum expectation values. This procedure apparently destroys the symmetry between the components of the fields. But in reality it simply becomes a hidden symmetry.

When a continuous algebraic symmetry is spontaneously broken, there appear unphysical massless bosons, in accordance with Goldstone's theorem. When a local gauge symmetry is spontaneously broken, instead of unphysical bosons, there appear masses of the vector gauge fields. The values of the resulting masses can be expressed in terms of the coupling constant λ of the scalar fields and can be made arbitrary, since λ is normally not fixed by the theory. Use of the Higgs mechanism makes it possible to assign masses to the vector fields, while preserving the local gauge invariance of the theory. The local gauge invariance of the theory, in its turn, ensures its renormalizability.

It is difficult to overestimate the importance of this fact. After the creation of quantum electrodynamics, there appeared for the first time a field theory in which calculations to any order of perturbation theory could be carried through to completion (at least in principle), and this theory is the theory of gauge fields.

But for all its attractiveness, the Higgs mechanism has one drawback: it introduces an arbitrary parameter λ into the theory and reduces its predictive power. This parameter can be fixed if we require, for example, that the theory is supersymmetric, i.e., that it possesses symmetry between bosons and fermions. Another way of understanding the meaning of λ is to determine it from some physical considerations — to assign a concrete meaning to the scalar Higgs field. This approach presupposes the usual interpretation of the scalar Higgs field as a field describing a condensate (like the condensate of Cooper pairs in a superconductor). Then the spontaneous symmetry breaking becomes dynamical, and the vector fields acquire mass as a result of a phase transition.[57] This idea can be traced back to the classical work of Bardeen, Cooper, and Schrieffer,[58] Bogolyubov,[59] and Ginzburg and Landau[60] on superconductivity.

Another free parameter which occurs in unified gauge models is the *mixing angle*. This parameter appears because the experimentally observed particles are associated not with the fields that enter into the Lagrangian, but with linear combinations of them. Thus, in the unified model of weak and electromagnetic interactions due to Weinberg and Salam (1967), the physical photon is associated with the combination $(a - b_3)$, where b_3 is the neutral component of an SU(2) triplet of vector fields responsible for the weak interactions, and a is the electromagnetic field, associated with U(1) invariance of the Lagrangian. The combination of fields $(a + b_3)$ is identified with the Z meson responsible for the neutral weak currents. The mixing angle of the fields a and b_3 is called the *Weinberg angle* θ_W. Its value is determined experimentally. According to contemporary data, $\sin^2\theta_W \approx 0.23$.

The weak and electromagnetic interactions of hadrons are described by unified gauge models using the group SU(3) × SU(2) × U(1) or its generalizations. It is obvious that here there are more possible combinations of fields and that there may be several mixing angles. Contemporary gauge models of the strong, weak, and electromagnetic interactions can be divided into three groups,[25] depending on the method of mixing the strong and weak interactions. Usually, the purpose of this mixing is to suppress the neutral strangeness-changing currents. The first group comprises models in which the neutral component of a multiplet of spinor fields is split. For example, instead of the doublet of quarks $\binom{\mathscr{P}}{\mathscr{N}}$, one considers the doublet $\binom{\mathscr{P}}{\mathscr{N}\cos\theta_C + \lambda\sin\theta_C}$, where λ is the strange quark and θ_C is the Cabibbo angle. This necessitates the introduction of an additional doublet with a new quark $\mathcal{P}'$ (or c): $\overset{\mathcal{P}'}{\left(-\mathcal{N}\sin\theta_C + \lambda\cos\theta_C\right)}$. Thus, the theory involves a new quark with a new quantum number — "charm."

Another group of models comprises the "pseudo-Cabibbo" models. In these models, the doublets $\binom{\mathcal{P}}{\mathcal{N}}$ and $\binom{\mathcal{P}}{\lambda}$ are taken to be representations of two different, mutually commuting SU(2) groups. In each of the SU(2) groups, the intermediate vector boson interacts with either of the currents $\bar{\mathcal{N}}\mathcal{N}$ or $\bar{\lambda}\lambda$, but there are no particles interacting with $\bar{\mathcal{N}}\lambda$. Such a scheme cannot be realized if it is assumed that the $\mathcal{P}$ quarks in the two doublets are the same. Therefore Georgi and Glashow proposed to assume that one of the doublets is left-handed, and the other right-handed. The semileptonic weak interactions with $\Delta S = 0$ then have the usual $(V - A)$ form, while the interactions with $\Delta S = 1$ have a $(V + A)$ form.

In the third group of models, the strong and weak interactions are distinguished in the sense that the intermediate boson does not interact directly with the quarks, but is first converted into some strongly interacting vector meson. In work with models of this type, it is useful to adopt the hypothesis of vector dominance, which had been used successfully in the earliest gauge models of strong interactions.[44] The structure of such models leaves sufficient freedom for a successful description of experiment.

It is necessary to say a few words about the quest for the high symmetry which must characterize the hypothetical universal interaction that manifests itself at low energies as the strong, weak, electromagnetic, and gravitational forces. There are two main motivations for this quest. First, it is necessary to understand the origin of the mixing angles and to fix them theoretically. Secondly, it is desirable to have a single coupling constant for all forms of interactions and thereby reduce the number of fundamental physical constants. However, the enlargement of the symmetry brings about a catastrophic growth in the number of new predicted particles which are not observed experimentally, as well as an instability of the known stable particles.[12] To avoid these unpleasant features, it is necessary to devise a complex mechanism of confinement of the extra particles and to spontaneously break the high symmetry down to the observed SU(3) × SU(2) × U(1) or SU(2) × U(1) symmetry.

Let us summarize what has been said.

The history of attempts to construct a unified theory of the weak and electromagnetic interactions in the form and image of electrodynamics goes back more than 40 years, but it is only now that we seem to have found an acceptable variant of such a theory.

The theory of gauge fields was formulated mainly during the period 1967—1969. The classical Lagrangian formulation and the geometrical interpretation of gauge fields took their modern form in the works of Konopleva.[14,32] The functional-integral method of quantization of gauge fields and the diagrammatic technique were developed by Faddeev and Popov[61] and DeWitt.[62] However, to construct realistic models of elementary particles and to compare the theory with experiment, it was necessary to solve the problems of the masses of gauge fields and the renormalizability of the theory. The first problem arose because all interactions of elementary particles, apart from the electromagnetic and gravitational interactions, are short-range, i.e., they must be mediated by massive particles. A favorable solution of the second problem made the theory reasonable in any energy region. Both problems were solved successfully during the period 1971—1973. The history of the ideas which made it possible to find the solution and references to them can be found in Ref. 63. Since that time, gauge theories and unified gauge models were at the center of attention of theoreticians and also, for 1.5—2 years, of experimentalists, especially after the discovery at the end of 1973 of the neutral currents predicted by these models.[64] Neutrino processes involving neutral currents may play an important role in the energetics of stars. At the present time, neutral currents are being intensively investigated.

A very interesting possibility for neutrino experiments is provided by gauge models with a small nonzero neutrino mass. It turns out that in this case there can be neutrino oscillations, namely, alternating transformations of electron neutrinos into muon neutrinos and vice versa. This idea was put forward for the first time in 1957 by Pontecorvo.[65] Experiments to observe neutrino oscillations are being planned in several research centers.[66]

In the domain of electromagnetic and weak interactions of hadrons, a particular achievement of gauge models is the explanation of the behavior of the ψ and Υ particles and the prediction of their spectroscopy, which has been confirmed experimentally. Experiments have also confirmed the existence of a heavy lepton, which had been predicted by many unified models of the electromagnetic and weak interactions. Next, there are experiments to observe an intermediate W boson. In the domain of strong interactions, a major success of quantum chromodynamics was the prediction and discovery of charmed particles (mesons and baryons). It would be very interesting to test the prediction of deviations from Bjorken scaling in deep inelastic processes, which follow from asymptotically free quark—gluon models of the strong interactions. But the errors in the corresponding experiments are still too large to distinguish a constant (the prediction of the parton model) from a weakly varying power of a logarithm, obtained in the quark—gluon model.

Thus, from the point of view of the theory of gauge fields, we have the following picture. All elementary particles are divided into two classes: the basic particles (nucleons and leptons) and the mediator particles (mesons). The mesons mediate the interaction between the basic fields. In the quark model, the basic particles consist of three quarks, and the mesons consist of two. The quarks are bound to one another by "glue" (g l u o n s). This "glue" has special properties. It can be drawn together into filaments (strings), whose elasticity under tension grows linearly with the distance. Therefore quarks cannot escape, and only bound states of quarks are observed. This quantum field-theoretic picture has a geometrical analog in the fiber space-time. The fields describing the basic particles correspond to sections of the fiber space-time. The mesons correspond to the connection coefficients of the fiber space. An internal symmetry of elementary particles (isospin or color) is a symmetry of the group space of the gauge group (fiber).

The geometrical picture corresponds to the transmission of an interaction through a medium. Therefore gauge models have naturally led to the notion of the vacuum as a medium, i.e., a system with an infinite number of degrees of freedom. In this vacuum, pairs of particles can be produced in such numbers that a condensate is formed, and polarization effects are so strong that they can completely screen a charge which is introduced from without. The phenomenon of asymptotic freedom can be regarded as antiscreening produced by the dispersion of the vacuum. Of course, the theory of gauge fields contains many Big Questions, but the elegance of the theory itself and the extremely interesting experimental discoveries of recent years offer hope that it will be possible to cope with them. This would provide a bridge between elementary-particle physics, statistical physics, and solid-state physics.

REFERENCES

[1]C. N. Yang and R. L. Mills, Phys. Rev. 96, 191 (1954).
[2]R. Utiyama, Phys. Rev. 101, 1597 (1956).
[3]J. J. Sakurai, Ann. Phys. (N.Y.) 11, 1 (1960).
[4]A. Einstein, Collected Works [Russian translation], Vol. 1, Nauka, Moscow (1966).

[5]H. Weyl, Gravitation und Elektrizität, Sitz. Preuss. Akad. Wiss., Berlin (1918); Z. Phys. 56, 330 (1929).
[6]A. Salam and J. C. Ward, Nuovo Cimento 11, 568 (1959); 19, 165 (1961).
[7]P. W. Higgs, Phys. Lett. 12, 132 (1964).
[8]G. 't Hooft, Nucl. Phys. B33, 173 (1971).
[9]A. A. Slavnov, Teor. Mat. Fiz. 10, 153 (1972).
[10]J. C. Taylor, Nucl. Phys. B33, 436 (1971).
[11]S. Weinberg, Phys. Rev. Lett. 19, 1264 (1967); A. Salam, in: Proc. of the 8th Nobel Symposium, Stockholm (1968), p. 367.
[12]Proc. of the 18th Intern. Conf. on High Energy Physics, Tbilisi (1976); D1.2-10400, JINR, Dubna (1977).
[13]A. Einstein, Collected Works [Russian translation], Vol. 2, Nauka, Moscow (1966). The quotation given in the text is from: A. Einstein, The Meaning of Relativity, 6th Ed., Methuen, London (1956), p. 54.
[14]N. P. Konopleva, in: Proc. of the Intern. Seminar on Vector Mesons and Electromagnetic Interactions, JINR, Dubna (1969).
[15]N. P. Konopleva, in: Lokale Symmetrien Eichfelder und Geschichtete Räume: Ideen des Exakten Wissens, No. 10, Deutsche Verlags-Anstalt, Stuttgart (1971), p. 671.
[16]N. P. Konopleva and G. A. Sokolik, "The problem of identity and the principle of relativity" [in Russian], in: Éinshteĭnovskiĭ sbornik 1967 (Einstein Anthology 1967), Nauka, Moscow (1967).
[17]H. Poincaré, Dernières Pensées, Paris (1913).
[18]V. A. Fock, Primenenie ideĭ Lobachevskogo v fizike (Application of Lobachevskii's Ideas in Physics), GTTI, Moscow (1950).
[19]N. P. Konopleva, in: Éinshteĭnovskiĭ sbornik 1975—1976 (Einstein Anthology, 1975—1976), Nauka, Moscow (1978).
[20]N. P. Konopleva and G. A. Sokolik, "The possible and the actual in field theory and their relation to the general principle of relativity" [in Russian], in: Prostranstvo i vremya v sovremennoĭ fizike (Space and Time in Modern Physics), Naukova Dumka, Kiev (1968).
[21]N. P. Konopleva and G. A. Sokolik, Vopr. Filos. No. 1, 118 (1972).
[22]I. Kant, Kritik der Reinen Vernunft (1781) [English translation: Critique of Pure Reason, 2nd Ed., Dutton (1979)].
[23]N. P. Konopleva, Usp. Fiz. Nauk 123, 537 (1977) [Sov. Phys. Usp. 20, 973 (1977)].
[24]D. B. Cline, A. K. Mann, and C. Rubbia, Sci. Am. 234, No. 1, 44 (1976).
[25]S. Weinberg, Rev. Mod. Phys. 46, 255 (1974).
[26]E. Schrödinger, Z. Phys. 12, 13 (1922).

[27]F. London, Z. Phys. 42, 375 (1927).
[28]M. Ikeda and Y. Miyachi, Prog. Theor. Phys. 27, 474 (1962).
[29]H. G. Loos, Nucl. Phys. 72, 677 (1965).
[30]T. T. Wu and C. N. Yang, in: Properties of Matter under Unusual Conditions (eds. H. Mark and S. Fernbach), Interscience, New York (1969).
[31]G. 't Hooft, Nucl. Phys. B79, 276 (1974); A. M. Polyakov, Pis'ma Zh. Eksp. Teor. Fiz. 20, 430 (1974) [JETP Lett. 20, 194 (1974)].
[32]N. P. Konopleva, in: Gravitatsiya i teoriya otnositel'nosti (Gravitation and the Theory of Relativity), Nos. 4—5, Kazan State University (1968).
[33]N. P. Konopleva, Vestn. Mosk. Univ., Ser. Fiz., No. 3, 73 (1965); in: Problemy gravitatsii. Tezisy dokladov II sovetskoĭ gravitatsionnoĭ konferentsii (Problems of Gravitation. Abstracts of Contributions to the 2nd Soviet Conf. on Gravitation), Tbilisi (1965).
[34]N. P. Konopleva and H. A. Sokolik, Nucl. Phys. 72, 667 (1965).
[35]N. P. Konopleva and G. A. Sokolik, in: Problemy teorii gravitatsii i élementarnykh chastits (Problems of the Theory of Gravitation and Elementary Particles), Atomizdat, Moscow (1966).
[36]N. P. Konopleva, in: Tezisy IV vsesoyuznoi mezhvuzovskoĭ konferentsii po problemam geometrii (Abstracts of the 4th All-Union Inter-University Conf. on Problems of Geometry), Tbilisi (1969).
[37]V. Fock and D. Ivanenko, C. R. Acad. Sci. 188, 1470 (1929).
[38]A. M. Brodskiĭ, D. D. Ivanenko, and G. A. Sokolik, Zh. Eksp. Teor. Fiz. 41, 1307 (1961) [Sov. Phys. JETP 14, 930 (1962)].
[39]T. W. B. Kibble, J. Math. Phys. 2, 212 (1961).
[40]N. P. Konopleva, in: Tezisy 5-ĭ mezhdunarodnoĭ gravitatsionnoĭ konferentsii (Abstracts of the 5th Intern. Conf. on Gravitation), Tbilisi (1968), p. 27.
[41]N. P. Konopleva, in: Problemy teorii gravitatsii i élementarnykh chastits (Problems of the Theory of Gravitation and Elementary Particles), No. 3, Atomizdat, Moscow (1970).
[42]A. G. Iosif'yan, K. P. Stanyukovich, and G. A. Sokolik, Dokl. Akad. Nauk SSSR 159, 1261 (1964) [Sov. Phys. Dokl. 9, 1108 (1965)].
[43]C. N. Yang, Phys. Rev. Lett. 33, 445 (1974).
[44]M. Ikeda, S. Ogawa, and Y. Ohnuki, Prog. Theor. Phys. 23, 1073 (1960). M. Gell-Mann, Report CTSL-20, California Institute of Technology (1961) [reprinted in: The Eightfold Way (eds. M. Gell-Mann and Y. Ne'eman), Benjamin, New York (1964), p. 11]. S. L. Glashow and M. Gell-Mann, Ann. Phys. (N.Y.) 15, 437 (1961).
[45]M. Gell-Mann, Phys. Lett. 8, 214 (1964).

[46]J. J. Aubert et al., Phys. Rev. Lett. 33, 1404 (1974). J.-E. Augustin et al., Phys. Rev. Lett. 33, 1406 (1974).
[47]G. S. Abrams et al., Phys. Rev. Lett. 33, 1453 (1974).
[48]V. de Alfaro, S. Fubini, G. Furlan, and C. Rossetti, Currents in Hadron Physics, North-Holland, Amsterdam (1973).
[49]"Nonlocal, nonlinear, and nonrenormalizable field theories" [in Russian], Preprint D2-9788, JINR, Dubna (1976).
[50]J. M. Cornwall, D. N. Levin, and G. Tiktopoulos, Phys. Rev. Lett. 30, 1268 (1973).
[51]M. T. Grisaru, H. J. Schnitzer, and H.-S. Tsao, Phys. Rev. Lett. 30, 811 (1973).
[52]L. D. Faddeev and L. A. Takhtadzhyan, Usp. Mat. Nauk 29, 245 (1974); Teor. Mat. Fiz. 21, 169 (1974).
[53]E. Cremmer and J. Scherk, Nucl. Phys. B108, 409 (1976).
[54]N. M. Kroll, T. D. Lee, and B. Zumino, Phys. Rev. 157, 1376 (1967).
[55]L. B. Okun', in: Élementarnye chastitsy. Tr. 1-ǐ shkoly fiziki ITÉF (Elementary Particles. Proc. of the 1st School of Physics at the Institute of Theoretical and Experimental Physics, Moscow), No. 1, Atomizdat, Moscow (1973).
[56]T. D. Lee and C. N. Yang, Phys. Rev. 98, 1501 (1955).
[57]D. A. Kirzhnits and A. D. Linde, Usp. Fiz. Nauk 115, 534 (1975) [Sov. Phys. Usp. 18, 259 (1975)].
[58]J. Bardeen, L. N. Cooper, and J. R. Schrieffer, Phys. Rev. 108, 1175 (1957).
[59]N. N. Bogolyubov, "Quasiaverages in problems of statistical mechanics" [in Russian], Preprint D-781, JINR, Dubna (1961).
[60]V. L. Ginzburg and L. D. Landau, Zh. Eksp. Teor. Fiz. 20, 1064 (1950).
[61]L. D. Faddeev and V. N. Popov, Phys. Lett. 25B, 29 (1967).
[62]B. S. DeWitt, Phys. Rev. 162, 1195, 1239 (1967).
[63]S. Coleman, in: Laws of Hadronic Matter: Proc. of the 11th Course of the "Ettore Majorana" International School of Subnuclear Physics (ed. A. Zichichi), Academic Press, New York (1975). J. Bernstein, Rev. Mod. Phys. 46, 7 (1974). E. S. Abers and B. W. Lee, Phys. Rep. 9C, 1 (1973). A. A. Slavnov and L. D. Faddeev, Vvedenie v kvantovuyu teoriyu kalibrovochnykh poleǐ (Introduction to the Quantum Theory of Gauge Fields), Nauka, Moscow (1978) [English translation, Benjamin/Cummings (1980)].
[64]F. J. Hasert et al., Phys. Lett. 46B, 138 (1973); 73B, 487 (1978).
[65]B. Pontecorvo, Zh. Eksp. Teor. Fiz. 34, 247 (1958) [Sov. Phys. JETP 7, 172 (1958)].
[66]Neutrino 77: Proc. of the Intern. Conf. on Neutrino Physics and Neutrino Astrophysics, Baksan Valley, 18—24 June

1977, Vol. 1, Nauka, Moscow (1978).
[67]A. A. Belavin, A. M. Polyakov, A. S. Schwartz, and Yu. S. Tyupkin, Phys. Lett. 59B, 85 (1975).
[68]E. S. Fradkin, in: Problemy teoreticheskoĭ fiziki (Problems of Theoretical Physics), Nauka, Moscow (1969), p. 386.

CHAPTER II. LAGRANGIAN THEORY OF GAUGE FIELDS

§3. INTRODUCTION

A distinguishing characteristic of all gauge theories is invariance with respect to an infinite group (pseudo-group), whose transformations depend on r arbitrary functions and their derivatives. Owing to this symmetry, gauge fields always describe a system with extra degrees of freedom, and if the parametric functions are replaced by constant parameters, this leads to systems with constraints. Usually, when a theory is invariant with respect to a finite group and contains extra variables, these variables are eliminated by imposing supplementary conditions (gauge conditions). The form of these gauge conditions is not related to the field equations or the equations of motion. In theories of gauge fields, invariance with respect to an infinite group leads to the appearance of identity relations between the extremals of the corresponding Lagrangians and their derivatives (the Noether identities). In quantum theory, the Noether identities correspond to the Ward identities.

When a local gauge invariance is broken and reduced to an r-parameter finite group (for example, when mass terms are introduced in the Lagrangian), the original invariance with respect to an infinite group manifests itself in the presence of supplementary conditions on the field variables, which follow from the field equations. The form of these supplementary conditions is determined by the Noether identities for that part of the field equations which is invariant with respect to the unbroken infinite symmetry group. Fulfillment of these supplementary conditions in the presence of broken symmetry ensures the preservation of local gauge invariance of the action integral on the extremals, even though this symmetry is absent on arbitrary trajectories. Moreover, invariance with respect to an infinite group leads to a unique situation, which is very important from a physical point of view: there occur so-called *strong conservation laws*, whose form is not related to the specific structure of the Lagrangian and the field equations. Currents corresponding to strong conservation laws are called *improper*. Improper currents can always be represented in the form of the divergence of an antisymmetric tensor. Therefore the charges corresponding to them are determined by the boundary values of the fields and by the topology of V_4. The structure of the strong conservation laws is determined only by the structure of the local gauge group and by the Noether identities. Breaking of the local gauge invariance of the Lagrangian by the addition of terms not containing derivatives of the field variables, for example, mass terms, does not alter the expressions for the improper currents in terms of the field variables. The conservation laws also remain unchanged if the supplementary conditions mentioned above are satisfied. It should be noted only that, with the exception of electrodynamics, the strong conservation laws always have the form of covariant, and not ordinary conservation laws, like the conservation laws in Einstein's general theory of relativity. In electrodynamics, the strong conservation laws are identical with the ordinary conservation laws as a consequence of the Abelian nature of the gauge group.

Integration of improper conserved currents leads to multiplets of charges forming the regular representation* of a finite gauge group. The algebra of these charges is not related to the specific structure of the Lagrangian. It is related only to the presence of symmetry with respect to the infinite group, even if this is somewhat broken. The existence of conservation laws which are to some extent independent of the field equations and the Lagrangians makes it possible to obtain useful and simple relations characterizing the interactions of elementary particles without particularizing the details of these interactions, but merely postulating symmetry properties for the conserved currents or charges,[1] as is done in current algebra.

The Lagrangian formulation of the classical theory of gauge fields is the most convenient way of discussing gauge invariance from the point of view of choosing the form of the invariant Lagrangians and analyzing the structure of the conservation laws and their possible violations, and also when supplementary conditions on the field variables are chosen. The use of the variational formalism and the two theorems of Noether then offers definite advantages by tying these questions together into a single whole.

With localization of space-time symmetries, application of Noether's second theorem and the apparatus of Lie derivatives permits explicit construction of locally invariant Lagrangians for arbitrary tensor fields without introducing any additional (gauge) fields. The role of the connection coefficients here is played by combinations of the

---

*A regular representation of a Lie algebra is a representation by matrices whose matrix elements are the structure constants of this algebra.

derivatives of the original field, determined by the Noether identities. The resulting Lagrangians are nonlinear and have a perspicuous geometrical meaning. Evidently, chiral dynamics refers to this type of theory.

Thus, the presence of local gauge invariance (and, in the more general case, invariance with respect to an arbitrary infinite group) leads to the possibility of solving the following problems:

1) given the form of the transformations of the field variables, which form an infinite group, to construct the invariant Lagrangian, to find the field equations and supplementary conditions for them (using here the direct form of Noether's second theorem), and to find the conservation laws (following Noether's first and second theorems);

2) given the form of the field equations and the supplementary conditions on the field variables which follow from them, to establish the forms of the Lagrangian and the group transformations of the field variables (using the inverse form of Noether's second theorem).

The material of this chapter is organized as follows. In §4 we give an account of Noether's first and second theorems,[2] which form the basis of all group and variational approaches, and we formulate the corresponding conservation laws. Noether's first theorem is used in the case of symmetry of the theory with respect to arbitrary finite Lie groups, and the second theorem in the case of symmetry with respect to infinite groups. In §5 we show how a knowledge of the form of the transformations of the field variables generated by a local gauge group can be used to construct the invariant Lagrangian and field equations,[3,4] to determine the forms of the Noether identities and the conservation laws, and to find the supplementary conditions on the field variables that follow from the field equations and the Noether identities when mass terms are present.[5] In other words, Noether's theorems are applied to the construction of a Lagrangian theory of an arbitrary gauge field. The inverse problems (the determination of the group transformations and the establishment of the form of the Lagrangian from the field equations and the supplementary conditions that follow from them[5-7]) are considered in §6. The importance of these problems is connected with the fact that the choice of the supplementary conditions can be dictated by physical considerations, and this entails restrictions on the form of the possible interactions. In §7 Noether's second theorem and its consequences are generalized to isoperimetric problems involving symmetry with respect to an infinite group, in particular, local gauge invariance.[8] We discuss a mechanism of generation of mass of a gauge field, using not Higgs fields, but breaking of local gauge invariance in the isoperimetric problem. It is shown that if an integral supplementary condition possesses only global symmetry, then the action integral remains invariant with respect to the local gauge group on the extremals of the isoperimetric problem, provided that the supplementary conditions which follow from the generalized Noether identities are satisfied. The form of the conservation laws in the presence of integral supplementary conditions is discussed. It is demonstrated that there are proportionality relations between the currents and fields when an infinite symmetry is broken by mass terms.

The interpretation of the general theory of relativity as a theory of a gauge field of a symmetric second-rank tensor[5,9] is discussed in §8. The gauge group is taken to be the group of general covariant (arbitrary continuous) coordinate transformations of 4-dimensional space-time. Lie derivatives are used as the group variations of the field variables.[5,10] This makes it possible to show that in going over to nonhomogeneous space-time there is no need to localize the group of motions of flat space-time, as is done when the ideas of Yang and Mills are naively transferred to space-time symmetries, but it is sufficient to use the apparatus of Lie derivatives,[11] whose form and definition are not related to any specific geometry of space-time. For this reason, the results obtained by this method are also independent of the geometry of space-time.

In §8 we show that the Lagrangian for a symmetric second-rank tensor field $g_{\mu\nu}$, regarded as a gauge field in the sense indicated above, is identical in its structure with the scalar curvature of Riemannian space-time V_4, while the field equations are identical with Einstein's equations. It is therefore possible to identify $g_{\mu\nu}$ with the metric tensor. But we need not make this identification. In that case, $g_{\mu\nu}$ can be regarded as a tensor field in a flat space V_4, describing, for example, the properties of a continuous medium.[5] In §8 we also make use of Lie derivatives to obtain the Noether identities and conservation laws for a gauge vector field and an antisymmetric second-rank tensor field, and we show how they are related to Maxwell's equations.[12] We discuss the general form of the integral conserved quantities corresponding to covariant differential conservation laws, and we construct the integral conservation laws in the general theory of relativity.[13,14] It is shown that the possibility of independent treatment of the 4-momentum and angular momentum in flat space-time is related to the presence of a 4-dimensional normal divisor in the corresponding group of motions. This same circumstance is related to the possibility of identifying P^μ with a space-time vector, although the index μ actually refers not to Minkowski space itself, but to a fiber space over it. The simplest case of a fiber space is a tangent space, which in the case of a flat space is usually identified with the initial space. But in the general case, all integral conserved quantities are vectors in a Lie algebra, i.e., in a space tangent to the fiber.

§4. NOETHER'S THEOREMS

Formulation and Proof of the Theorems. As is well known, the conservation laws in modern field theory are obtained by the variational method by means of Noether's first theorem. This theorem can be formulated as follows.

First Theorem. *If the action integral S is invariant with respect to some group G_r (an r-parameter Lie group), then r linearly independent combinations of Lagrangian derivatives* reduce to divergences; conversely, the latter condition implies invariance of S with respect to some group G_r.*

The expressions of which the divergence is taken in this theorem are called *currents*. If the Lagrangian derivatives are equal to zero (Euler's equations are satisfied), the divergences of the currents vanish. In this way, differential conservation laws are obtained. Integral conservation laws (like the law of conservation of electric charge, or the law of conservation of energy) are obtained by integration of the differential conservation laws over a specially chosen 3-dimensional hypersurface with certain boundary conditions.

Local gauge groups and the group of general covariant coordinate transformations of space-time belong to the class of infinite groups. Local gauge groups are obtained from finite Lie groups of gauge transformations of the wave functions, $\delta\psi = \varepsilon^a \underset{a}{I}\psi$, by replacing the parameters ε^a by functions of the coordinates: $\varepsilon^a(x)$. Since, by definition, a gauge group describes fixed-coordinate transformations, the algebra of a finite Lie group is preserved if such a replacement is made at each fixed point of space-time. This distinguishes local gauge groups from other infinite groups, which in general do not correspond to any algebra. The group of general covariant coordinate transformations of

---

*I.e., the left-hand sides of Euler's equations, whose solutions are called extremals.

general relativity, $x^{\mu'} = f^{\mu}(x)$, can be regarded as an infinite group $G_{\infty 4}$.

The properties of functionals invariant with respect to arbitrary infinite groups $G_{\infty r}$ were investigated by Noether, and her results were formulated in the form of the following theorem (Noether's second theorem).

Second Theorem. If the action integral S is invariant with respect to a group $G_{\infty r}$ involving derivatives up to order k inclusive, then there exist r identity relations between the Lagrangian derivatives and the derivatives of them up to order k. The converse is also true.

Thus, in the case of $G_{\infty r}$ invariance, we have identity relations between Euler's equations and their derivatives, and this leads to a reduction in the number of linearly independent field equations. In other words, in a $G_{\infty r}$-invariant theory, there are always r arbitrary transformations of the field variables, which can be fixed by supplementary gauge conditions. In the case of G_r invariance, Euler's equations are in general linearly independent, and identities connecting them do not exist (the number of variables is equal to the number of equations).

A theory which is invariant with respect to an infinite group $G_{\infty r}$ is also invariant with respect to its subgroup G_r obtained from $G_{\infty r}$ when $\varepsilon^a(x) = \text{const}$. Therefore the divergence relations with which Noether's first theorem is concerned also occur in the case of $G_{\infty r}$ invariance. As a result of the application of both of Noether's theorems, it turns out that when the finite group G_r is a subgroup of the infinite group, and only in this case, the divergence relations corresponding to G_r invariance are linear combinations of the identity relations between the Lagrangian derivatives and the currents become linear combinations of the Lagrangian derivatives. Noether called such currents improper currents. The conservation laws for improper currents are satisfied identically on the extremals. An improper conserved current can be reduced to the divergence of some antisymmetric tensor and therefore possesses a superpotential.

Proof of Noether's Theorems. We shall compare the proofs and results of the two theorems of Noether.[2,5,15] Let

$$S = \int L(x, u, u', u'')\, dx,$$

where u denotes the arbitrary functions describing the system (field variables), x denotes the coordinates (space and time variables), both primes and commas signify ordinary differentiation, and L is the Lagrangian density.

We make the transformation $y = x + \Delta x$, $v(y) = u + \Delta u$. Now the variation of u has the form $\bar{\delta} u = \Delta u - (\partial u/\partial x)\,\Delta x$. Then the condition of invariance of the action takes the form

$$\int [\bar{\delta} L + \operatorname{div}(L\Delta x)]\,dx = 0. \tag{4.1}$$

Since the relation (4.1) is satisfied when an integration is made over any region, the integrand vanishes identically and we obtain a differential condition for $\delta S = 0$:

$$\bar{\delta} L + \operatorname{div}(L\Delta x) = 0, \tag{4.2}$$

where

$$\bar{\delta} L = \frac{\delta L}{\delta u}\,\bar{\delta} u + \partial_\mu \left[\left(\frac{\partial L}{\partial u_{,\mu}} - \partial_\nu \frac{\partial L}{\partial u_{,\nu\mu}} \right) \bar{\delta} u + \frac{\partial L}{\partial u_{,\mu\nu}}\,\bar{\delta} u_{,\nu} \right];$$

here $\delta L/\delta u = \partial L/\partial u - \partial_\mu\,\partial L/\partial u_{,\mu} + \partial^2_{\nu\mu}\,\partial L/\partial u_{,\nu\mu}$ is the Lagrangian derivative.

Equation (4.2) is Lie's differential equation, from which we can find the explicit form of the Lagrangian if the variations Δx and Δu or $\bar{\delta} u$ are known. We represent Eq. (4.2) in the form

$$\frac{\delta L}{\delta u}\,\bar{\delta} u = \partial_\mu \left[\left(-\frac{\partial L}{\partial u_{,\mu}} + \partial_\nu \frac{\partial L}{\partial u_{,\nu\mu}} \right) \bar{\delta} u - \frac{\partial L}{\partial u_{,\mu\nu}}\,\bar{\delta} u_{,\nu} - L\Delta x^\mu \right]. \tag{4.3}$$

Equation (4.3) is an identity with respect to all the arguments that appear in it if S is an invariant. Note also that it holds for any functions u, which are not necessarily solutions of Euler's equations. The behavior of u on the boundary is also immaterial for the fulfillment of (4.3), since the divergence term is not discarded. In what follows, we shall be interested in two cases: 1) u is an extremal; 2) u is a solution of generalized Euler equations with right-hand side of a special type.

Let us consider special cases of invariance.

Invariance with Respect to a Finite Group. Let S be invariant with respect to a finite group G_r, and suppose that the variations have the form $\delta x^\mu = \varepsilon^a \underset{a}{X} x^\mu$, $\delta u = \varepsilon^a \underset{a}{I} u$, and $\bar{\delta} u = (\underset{a}{I} u - \underset{a}{X} x^\mu \partial_\mu u)\,\varepsilon^a$. If $\underset{a}{X} = \xi^\mu_a \partial_\mu$ are the generators of G_r in differential form, then

$$\delta x^\mu = \xi^\mu_a\,\varepsilon^a; \quad \bar{\delta} u = \left(\underset{a}{I} u - \xi^\mu_a\,\partial_\mu u \right) \varepsilon^a. \tag{4.4}$$

The expression (4.4) shows that the variation of the form of a function has a structure analogous to a Lie derivative (see §8). Substituting (4.4) into (4.3), removing the parameters ε^a from the divergence operation, and collecting the terms with the same ε^a, we obtain

$$\frac{\delta L}{\delta u}\left(\underset{a}{I}u-\xi_a^\mu\partial_\mu u\right)=-\partial_\mu\left[\left(\frac{\partial L}{\partial u_{,\mu}}-\partial_\nu\frac{\partial L}{\partial u_{,\nu\mu}}\right)\left(\underset{a}{I}u-\xi_a^\nu\partial_\nu u\right)+\right.$$
$$\left.L\xi_a^\mu+\frac{\partial L}{\partial u_{,\mu\nu}}\left(\underset{a}{I}u_{,\nu}-\xi_a^\lambda u_{,\nu\lambda}\right)\right] \tag{4.5}$$

or

$$\left(\underset{a}{I}u-\xi_a^\mu\partial_\mu u\right)\delta L/\delta u=\partial_\mu J_a^\mu. \tag{4.6}$$

Thus, r linearly independent combinations of Lagrangian derivatives $\delta L/\delta u$ reduce to divergences in the case of invariance of S with respect to G_r, independently of whether u is a solution of Euler's equations.

In the ordinary variational problem of finding the extremum of S, the integral of the right-hand side of (4.5) is transformed into a surface integral and reduces to zero because the variations of the functions and all their derivatives are assumed to vanish on the boundary (the fixed end-point problem). Then the condition $\delta S = 0$ and the arbitrariness of δu imply Euler's equation $\delta L/\delta u = 0$.

In the case of group variations, the δu do not in general vanish on the boundary (the end points are to some extent free), the variational principle gives only the relations (4.5), and Euler's equation is not assumed to hold.

If u is an extremal, i.e., $\delta L/\delta u = 0$, then (4.5) implies the well-known differential conservation law $\partial_\mu J_a^\mu = 0$, where

$$J_a^\mu=-\left[\left(\frac{\partial L}{\partial u_{,\mu}}-\partial_\nu\frac{\partial L}{\partial u_{,\nu\mu}}\right)\left(\underset{a}{I}u-\xi_a^\lambda\partial_\lambda u\right)+L\xi_a^\mu+\right.$$
$$\left.(\partial L/\partial u_{,\mu\nu})\left(\underset{a}{I}u_{,\nu}-\xi_a^\lambda u_{,\nu\lambda}\right)\right] \tag{4.7}$$

is the conserved current. If the Lagrangian does not contain second derivatives of the field variables, it is convenient to represent the conserved current in the form

$$J_a^\mu=-\frac{\partial L}{\partial u_{,\mu}}\underset{a}{I}u+\left(\frac{\partial L}{\partial u_{,\mu}}u_{,\nu}-L\delta_\nu^\mu\right)\xi_a^\nu=-\frac{\partial L}{\partial u_{,\mu}}\underset{a}{I}u+T_\nu^\mu\xi_a^\nu, \tag{4.8}$$

where $T_\nu^\mu = u_{,\nu}\,\partial L/\partial u_{,\mu} - L\delta_\nu^\mu$ is the canonical energy—momentum tensor.

If u is a solution of a "generalized" Euler equation of the form $\delta L/\delta u = \Theta$, where Θ denotes new functions (for example, sources not included in the Lagrangian), then from

(4.5) we obtain "generalized" nonhomogeneous conservation laws like partial conservation of the axial current (the PCAC hypothesis):

$$\partial_\mu J^\mu_a + \Theta \, (\underset{a}{I} u - \xi^\mu_a \partial_\mu u) = 0.$$

It follows from this relation that the difference of the divergence of the current from zero, i.e., the violation of the conservation laws and hence of the symmetry, can be related in a definite way to new sources. The relations (4.5) also show that the expressions for the divergences of the currents can be obtained directly from the field equations by contracting them with the variations of the form of the wave functions produced by the invariance group G_r.[16] The condition $\delta L/\delta u = 0$ implies that $\partial_\mu J^\mu_a = 0$, and in this sense the conservation laws are a consequence of the field equations. Thus, the conservation laws are by no means obtained directly from Noether's first theorem. This theorem merely asserts that if there is a certain symmetry in a variational problem (or in the corresponding physical situation), then it is possible to form expressions (called currents) whose divergences are equal to linear combinations of Lagrangian derivatives. These divergences vanish on extremals, and we then obtain differential conservation laws. However, the divergence relations with which Noether's first theorem is concerned are also satisfied when the field equations coincide only partially with Euler's equations. The choice of the form of the field equations (and hence of the form of the conservation laws) lies beyond the scope of Noether's first theorem.

Invariance with Respect to an Infinite Group. Suppose now that the action integral S is invariant with respect to $G_{\infty r}$. Suppose, as in the case of G_r, that δx and δu are linear in $\varepsilon^a(x)$ and their derivatives. Then the variation of the form can be written as

$$\delta u = a_a (x, u, u', \ldots) \, \varepsilon^a(x) + b^\mu_a(x, u, u', \ldots) \, \partial\varepsilon^a(x)/\partial x^\mu \qquad (4.9)$$

(the first derivatives of $\varepsilon^a(x)$ are sufficient for the existence of improper conservation laws). In this case,

$$(\delta L/\delta u)\bar{\delta} u = (\delta L/\delta u) \, [a_a(x, u, u', \ldots) \, \varepsilon^a(x) + b^\mu_a (x, u, u', \ldots) \, \partial\varepsilon^a/\partial x^\mu].$$

Now on the basis of the identity

$$\varphi \, (x, u, u', \ldots) \, \partial^\tau \varepsilon(x) / \partial x^\tau \equiv (-1)^\tau (\partial^\tau \varphi / \partial x^\tau) \, \varepsilon(x) \mod \operatorname{div}$$

it is possible to replace the derivatives of $\varepsilon^a(x)$ by the functions themselves. Then (4.3) takes the form

$$[a_a \delta L/\delta u - \partial_\mu (b^\mu_a \delta L/\delta u)] \, \varepsilon^a(x) = -\operatorname{div} (B^\mu + b^\mu_a \varepsilon^a \delta L/\delta u). \qquad (4.10)$$

In the absence of second derivatives in the Lagrangian,

$$B^{\mu} = (\partial L/\partial u_{,\mu})\, a_a (x, u, u', \ldots)\, \varepsilon^a + L\Delta x^{\mu} + b^{\nu}_{a}\partial_{\nu}\varepsilon^a \partial L/\partial u_{,\mu}.$$

If we now integrate (4.10) over some region on whose boundary $\varepsilon^a(x)$ and all their derivatives vanish, then the integral of the right-hand side of (4.10) reduces to zero.* Then the condition $\delta S = 0$ implies that

$$\int [a_a\, \delta L/\delta u - \partial_{\mu}(b^{\mu}_{a}\delta L/\delta u)]\varepsilon^a(x)\, dx = 0.$$

Since the functions $\varepsilon^a(x)$ are arbitrary, it follows from this result that the expression in the square brackets is equal to zero:

$$a_a\delta L/\delta u \equiv \partial_{\mu}(b^{\mu}_{a}\delta L/\delta u). \tag{4.11}$$

These are the required dependences between the Lagrangian derivatives and the derivatives of them. The relations (4.11) are linearly independent and also hold for the finite transformations of $G_{\infty r}$. These relations can be written in the form of a covariant conservation law $\nabla_{\mu}(b^{\mu}_{a}\, \delta L/\delta u) \equiv 0$ if we introduce the connection $\Gamma^{b}_{\mu a} = a_a h^{b}_{\mu}$, where h^{b}_{μ} is determined by the condition $h^{b}_{\mu}b^{\mu}_{a} = \delta^{b}_{a}$.

To go over from the infinitesimal transformations of $G_{\infty r}$ to the finite transformations, it is important that: a) δu and δx are linear in $\varepsilon^a(x)$ and their derivatives; b) δu and δx do not contain derivatives of u, since otherwise the finite transformations of $G_{\infty r}$ depend on an infinite number of derivatives of u. The relations (4.11), like (4.5) for G_r, are valid for any functions u, irrespective of whether they are solutions of Euler's equations.

As before, we consider two cases:

1. Let u be an extremal, i.e., $\delta L/\delta u = 0$. Then it follows from (4.11) that

$$\partial_{\mu}(b^{\mu}_{a}\delta L/\delta u) = 0.$$

This relation provides a way of obtaining conservation laws from the field equations by means of differentiation.

2. If u satisfies the equation $\delta L/\delta u = \Theta$, where Θ denotes new functions, then from (4.11) we obtain the following supplementary conditions on Θ:

$$\partial_{\mu}(b^{\mu}_{a}\Theta) = a_a\Theta. \tag{4.12}$$

---

*Since the statement of Noether's second theorem is that invariance with respect to an infinite group produces identity relations between the extremals, to prove the theorem it is sufficient to seek, in the class of *arbitrary* functions $\varepsilon^a(x)$, those functions which vanish together with their derivatives on the surface of integration.

The role of Θ can also be played by terms which violate the $G_{\infty r}$ invariance of the field equations but which are invariant with respect to G_r, for example, mass terms in a gauge-invariant theory. Then Noether's identities lead to supplementary conditions on the field variables, which exclude the extra degrees of freedom. Indeed, suppose that $\delta L/\delta u = \lambda u$, where $\lambda = \text{const}$. Then $\partial_\mu(b^\mu_a u) = a_a u$.

Weak and Strong, or Proper and Improper, Conservation Laws. Noether called the conservation laws (4.7) obtained by means of the first theorem *proper conservation laws*. They are also sometimes known as *weak conservation laws*, since they are satisfied only on extremals. In contrast to the weak conservation laws associated with G_r invariance of the Lagrangian, invariance with respect to an infinite group leads to strong conservation laws, which are satisfied not only on extremals, but for arbitrary u. Strong conservation laws are in fact an expression of the Noether identities. It is such strong conservation laws which occur in the theory of gauge fields, as in general relativity. These conservation laws are improper, since the conserved currents in them are linear combinations of extremals. An example of such a strong (improper) conservation law is provided by the identity satisfied by Einstein's equations:

$$(R^{\mu\nu} - {}^1\!/_2 g^{\mu\nu}R)_{;\mu} \equiv 0. \tag{4.13}$$

In §8 we show that this identity is a particular instance of the Noether identities for $G_{\infty 4}$, which in the general case take the form

$$(\delta\hat{L}/\delta g^{\mu\nu})_{;\mu} \equiv 0, \tag{4.14}$$

where $\hat{L}$ is the Lagrangian density function.

Noether's identities for electrodynamics have the form

$$\partial_\mu(\delta L/\delta A_\mu) \equiv 0. \tag{4.15}$$

If there are no sources and if $L = -{}^1\!/_4 F^{\mu\nu}F_{\mu\nu}$, then (4.15) implies that

$$\partial_\mu \partial_\nu F^{\mu\nu} = 0. \tag{4.16}$$

Integration of this conservation law leads to Gauss's theorem for the fluxes of the electric and magnetic fields through a closed two-dimensional surface: $\oint_\sigma H d\sigma = 0$ and $\oint_\sigma E d\sigma = 0$.

If the sources of the field appear as the right-hand sides of Euler's equations, then from (4.14) and (4.15) we obtain the well-known conservation laws for the current of the sources in electrodynamics and for the energy—momentum tensor in gravitation: $\partial_\mu J^\mu = 0$ and $T^{\mu\nu}_{;\nu} = 0$. Both of these conservation laws are consequences of the Noether identities, i.e., of the supplementary conditions on the field equations, and it is therefore natural to interpret them as conditions on the sources. It is interesting that these conditions on the sources are absolutely indifferent to the specific form of the extremals, i.e., to the form of the left-hand sides of the equations in which they occur. It is important only that these left-hand sides are invariant with respect to the infinite group. Thus, the conservation laws (4.14) and (4.15) are insensitive to the specific structure of the part of the Lagrangian that is invariant with respect to $G_{\infty r}$.

If the Noether identities (4.11) are satisfied, it follows from (4.10) that

$$\operatorname{div}\,[J^{\mu}_{a}\varepsilon^{a}(x) - b^{\nu}_{a}\partial_{\nu}\varepsilon^{a}(x)\,\partial L/\partial u_{,\mu} - b^{\mu}_{a}\varepsilon^{a}(x)\,\delta L/\delta u] = 0, \tag{4.17}$$

where $J^{\mu}_{a} = a_{a}\partial L/\partial u_{,\mu}$. Equation (4.17) is satisfied for arbitrary $\varepsilon^{a}(x)$ and ensures invariance of the action integral with respect to $G_{\infty r}$, as well as the validity of the inverse form of Noether's second theorem.

Making use of Euler's equations $\delta L/\delta u = 0$, we have

$$\operatorname{div}\,[J^{\mu}_{a}\varepsilon^{a}(x) - b^{\nu}_{a}\partial_{\nu}\varepsilon^{a}(x)\,\partial L/\partial u_{,\mu}] = 0. \tag{4.18}$$

Apart from the sign, the left-hand side of Eq. (4.18) is nothing but the variation of the Lagrangian after allowance for Euler's equations and the identities between them. Thus,

$$\delta L = -\{\partial_{\mu}J^{\mu}_{a}\varepsilon^{a}(x) + [J^{\mu}_{a} - \partial_{\nu}(b^{\mu}_{a}\partial L/\partial u_{,\nu})]\,\partial_{\mu}\varepsilon^{a}(x) - b^{\nu}_{a}\partial_{\mu\nu}\varepsilon^{a}(x)\,\partial L/\partial u_{,\mu}\}. \tag{4.19}$$

Since the parametric functions are completely arbitrary, it follows from the condition of invariance of the action that the coefficients of $\varepsilon^{a}(x)$ and their derivatives in δL are equal to zero:

$$\partial_{\mu}J^{\mu}_{a} = 0; \tag{4.20}$$

$$J^{\mu}_{a} = \partial_{\nu}(b^{\mu}_{a}\,\partial L/\partial u_{,\nu}); \tag{4.21}$$

$$b^{\nu}_{a}\partial L/\partial u_{,\mu} + b^{\mu}_{a}\partial L/\partial u_{,\nu} = 0. \tag{4.22}$$

Thus, invariance of the action integral with respect to the infinite group leads on extremals to the ordinary conservation law (4.20) for the current $J^{\mu}_{a} = a_{a}\partial L/\partial u_{,\mu}$, corresponding to G_r invariance. But this current is equal to the divergence of some antisymmetric tensor [by virtue of the conditions (4.21) and (4.22)]. Therefore, by virtue of Gauss's theorem, the corresponding charge $Q = \int J^{0}d^{3}x$ can always be represented in the form of an integral over a closed two-dimensional surface: $Q_a = \oint (\partial L/\partial u_{,i})b^{0}_{a}d\sigma^{i}$.

If the Lagrangian is not invariant with respect to $G_{\infty r}$, but Euler's equations and Noether's identities are satisfied, the relations (4.19) lead to the following equations:

$$\partial\delta L/\partial\varepsilon^a(x) = \partial_\mu J^\mu_a; \qquad (4.23)$$

$$\partial\delta L/\partial\partial_\mu\varepsilon^a(x) = J^\mu_a - \partial_\nu\,(b^\mu_a \partial L/\partial u_{,\nu}) = J^\mu_a - b^\mu_a\partial L/\partial u - \partial_\nu b^\mu_a \partial L/\partial u_{,\nu}; \qquad (4.24)$$

$$\partial\delta L/\partial\partial_{\mu\nu}\varepsilon^a(x) = b^\nu_a \partial L/\partial u_{,\mu} = {}^1/_2\,(b^\nu_a\partial L/\partial u_{,\mu} + b^\mu_a \partial L/\partial u_{,\nu}). \qquad (4.25)$$

If the field variables transform homogeneously with respect to $G_{\infty r}$, i.e., $b^\mu_a = 0$, then Eqs. (4.23) and (4.24) reduce to the equations proposed in 1960 by Gell-Mann and Levy[1]:

$$\partial\delta L/\partial\partial_\mu\varepsilon^a(x) = J^\mu_a; \quad \partial\delta L/\partial\varepsilon^a(x) = \partial_\mu J^\mu_a.$$

These equations for a Lagrangian describing a massive gauge field (with broken local gauge invariance) lead to a proportionality between the current and the field: $J^\mu_a \sim m^2 A^\mu_a$.

§5. LOCAL GAUGE INVARIANCE OF THE LAGRANGIAN AND NOETHER'S SECOND THEOREM

Noether's Second Theorem and Gauge Fields of General Form. Construction of Invariant Lagrangians. Let us consider the problem of finding the simplest Lagrangian which contains derivatives of the field variables A^a_μ no higher than the first and is invariant with respect to a local gauge group in the case when the transformations of A^a_μ have the form

$$\delta A^a_\mu = f^a_{bc}\, A^c_\mu \varepsilon^b + \partial_\mu\varepsilon^a, \quad a = 1, \ldots, r;\ \mu = 1, \ldots, 4, \qquad (5.1)$$

where f^a_{bc} are the structure constants of some finite Lie group, and ε^a are parameters. The vector potential of the gauge field A^a_μ can be interpreted as a multiplet of r (depending on the number of parameters of the gauge group) vector fields A_μ, where the gauge group mixes these fields.

The transformations (5.1) form a group $G_{\infty r}$, and the form of the variation of δA^a_μ corresponds to the case $\Delta x = 0$ and $\bar{\delta} A^a_\mu = \delta A^a_\mu$. In other words, the transformations of $G_{\infty r}$ are fixed-coordinate transformations (the coordinates are unaffected, or the point x is fixed).

Using Noether's second theorem, we construct the action integral $S = \int L\,(A^a_\mu;\ A^a_{\mu,\nu})\,dv$, which is invariant with respect to $G_{\infty r}$. The relations (4.3), which are differential conditions for the invariance of S, take the form

$$(\delta L/\delta A^a_\mu)\delta A^a_\mu \equiv \partial_\mu[-(\partial L/\partial A^a_{\nu,\mu})\delta A^a_\nu]. \qquad (5.2)$$

We substitute δA^a_μ into (5.2) and, since (5.2) must be satisfied identically, we equate to zero the coefficients of $\varepsilon^a(x)$, $\partial_\mu\varepsilon^a(x)$, and $\partial_{\mu\nu}\varepsilon^a(x)$ individually. Then the condition of invariance of S can be written as a system of identities which, when solved from bottom to top, give the general form of the dependence of L on A^a_μ and $A^a_{\mu,\nu}$ (Ref. 3):

$$f^b_{ac} A^c_\mu (\partial L / \partial A^b_\mu) + f^b_{ac} A^c_{\nu,\mu} (\partial L / \partial A^b_{\nu,\mu}) \equiv 0; \tag{5.3}$$

$$\partial L / \partial A^a_\mu + f^b_{ac} A^c_\nu (\partial L / \partial A^b_{\nu,\mu}) \equiv 0; \tag{5.4}$$

$$\partial L / \partial A^a_{(\mu,\nu)} \equiv 0. \tag{5.5}$$

It follows from (5.5) that the derivatives of A^a_μ appear in L only through the antisymmetric expressions $A^a_{[\mu,\nu]}$. Using (5.5), we find from (5.4) that

$$L = L(F^a_{\mu\nu}), \tag{5.6}$$

where

$$F^a_{\mu\nu} = A^a_{\nu,\mu} - A^a_{\mu,\nu} - {}^1/_2 f^a_{bc} \left(A^b_\mu A^c_\nu - A^b_\nu A^c_\mu \right).$$

It is easy to see that $F^a_{\mu\nu}$ transforms homogeneously with respect to (5.1):

$$\delta F^a_{\mu\nu} = f^a_{bc} F^c_{\mu\nu} \varepsilon^b. \tag{5.7}$$

Since the structure of $F^a_{\mu\nu}$ is reminiscent of the Maxwell field tensor, we shall call $F^a_{\mu\nu}$ the stress tensor of the field A^a_μ. Making use of (5.7), we find from (5.3) that

$$L = L\left(F^a_{\mu\nu} f^m_{ac} f^c_{mb} F^b_{\tau\lambda} \right). \tag{5.8}$$

We introduce the quantity

$$F_{a\mu\nu} = f^m_{ac} f^c_{mb} F^b_{\mu\nu} = g_{ab} F^b_{\mu\nu}. \tag{5.9}$$

Its law of transformation is $\delta F_{a\mu\nu} = f^b_{ac} F_{b\mu\nu} \varepsilon^c$. In this case, (5.8) reduces to

$$L = L(F^a_{\mu\nu} F_{a\tau\lambda}). \tag{5.10}$$

By virtue of the Jacobi identity, the quantity $g_{ab} = f^m_{ac} f^c_{mb}$ is symmetric in a and b. It will be used in what follows as an intrinsic metric. The matrix g_{ab} is not degenerate only for semisimple groups, for which the constants f^b_{ac} are antisymmetric in all indices.

The simplest gauge-invariant and also relativistically invariant Lagrangian has the form

$$L_0 = {}^1/_4 F^a_{\mu\nu} F^{\mu\nu}_a. \tag{5.11}$$

An arbitrary invariant Lagrangian is an arbitrary function of L_0; we call L_0 the *Lagrangian of a free gauge field*.

We now determine the general form of an interaction Lagrangian $L_{\text{int}}(A^a_\mu; \psi; \psi_{,\mu})$ which is invariant with respect to $G_{\infty r}$ and depends on two types of functions and their derivatives: on the vector potentials of the gauge fields A^a_μ, which contain in the transformations the first derivatives of the parameters, and on the wave functions of the fields ψ, which transform homogeneously and correspond to particles interacting with the gauge field. Let

$$\left.\begin{aligned} \delta A_\mu^a &= f_{bc}^a A_\mu^c \varepsilon^b + \partial_\mu \varepsilon^a; \\ \delta\psi &= \underset{a}{I} \varepsilon^a \psi; \; \delta\overline{\psi} = -\varepsilon^a \overline{\psi} \underset{a}{I}, \end{aligned}\right\} \tag{5.12}$$

where $\underset{a}{I}$ is the generator of some representation of the group G_r obtained from $G_{\infty r}$ when $\varepsilon^a(x) = \text{const}$.

Suppose that the interaction (the mixed terms in L_{int}) do not depend on derivatives of A_μ^a. To determine L_{int} in this case, we obtain the system of equations

$$f_{ac}^b (\partial L_{\text{int}}/\partial A_\mu^b) A_\mu^c = -(\partial L_{\text{int}}/\partial\psi) \underset{a}{I} \psi - (\partial L_{\text{int}}/\partial\psi_{,\mu}) \underset{a}{I} \psi_{,\mu} + \text{h. c.}; \tag{5.13}$$

$$\partial L_{\text{int}}/\partial A_\mu^a = -(\partial L_{\text{int}}/\partial\psi_{,\mu}) \underset{a}{I} \psi + \overline{\psi} \underset{a}{I} \partial L_{\text{int}}/\partial\overline{\psi}_{,\mu}, \tag{5.14}$$

where h.c. denotes the Hermitian-conjugate terms. It follows from (5.14) that L_{int} depends on arguments which have the form of a covariant derivative:

$$\left.\begin{aligned} \nabla_\mu \psi &= \partial_\mu \psi - {}^1/_2 \underset{a}{I} A_\mu^a \psi; \\ \nabla_\mu \overline{\psi} &= \partial_\mu \overline{\psi} + {}^1/_2 \overline{\psi} \underset{a}{I} A_\mu^a. \end{aligned}\right\} \tag{5.15}$$

It is easy to see that Eq. (5.13) is satisfied by the interaction Lagrangian

$$L_{\text{int}} = \overline{\psi}\gamma^\mu \nabla_\mu \psi - \nabla_\mu \overline{\psi}\gamma^\mu \psi - m\overline{\psi}\psi, \tag{5.16}$$

where m is arbitrary. Thus, the simplest Lagrangian which is invariant with respect to $G_{\infty r}$ and which describes a gauge field and the interaction of this field, for example, with a spinor field ψ, has the form

$$L = -{}^1/_4 F_{\mu\nu}^a \underset{a}{F^{\mu\nu}} + \overline{\psi}\gamma^\mu \nabla_\mu \psi - \nabla_\mu \overline{\psi}\gamma^\mu \psi - m\overline{\psi}\psi, \tag{5.17}$$

where γ^μ denotes the Dirac matrices.

F i e l d E q u a t i o n s. The equations of a gauge field for the Lagrangian (5.10) have the quasi-Maxwell form[17]

$$\left.\begin{aligned} &\partial_\nu \underset{a}{F^{\mu\nu}} - f_{ac}^b \underset{b}{F^{\mu\nu}} A_\nu^c = 0, \\ &\text{or} \;\; \partial_\nu \underset{a}{F^{\mu\nu}} = \overset{0}{J}{}_a^\mu, \;\; \text{or} \;\; \nabla_\mu \underset{a}{F^{\mu\nu}} = 0. \end{aligned}\right\} \tag{5.18}$$

If f_{ac}^b are the structure constants of the group of isotopic rotations, (5.18) are the equations for Yang—Mills bosons. In electrodynamics, we have a single-parameter Abelian group of gauge transformations, for which $f_{ac}^b = 0$.

Therefore (5.18) reduce to Maxwell's equations $\partial_\nu F^{\mu\nu} = 0$.

The equations of motion of particles described by a field ψ and interacting with a gauge field can be written in the form[18]

$$\delta L/\delta\psi = \partial L/\partial\psi - \nabla_\mu \partial L/\partial\nabla_\mu\psi = 0,$$

and for the Lagrangian (5.17) this gives

$$\gamma^\mu \nabla_\mu \psi - m\psi = 0.$$

In curved space-time, the variational equations of a gauge field take the form[19]

$$\partial_\nu (|h|\partial L/\partial \underset{a}{F^{\mu\nu}}) = |h| (\underset{a}{J^\mu} + \underset{a}{\overset{0}{J^\mu}}). \tag{5.19}$$

The equations of motion of particles in gauge and gravitational fields are

$$|h|\partial L/\partial\psi - \nabla^*_\mu (|h|\partial L/\partial\nabla_\mu \psi) = 0,$$

where

$$\nabla^*_\mu (|h|\partial L/\partial\nabla_\mu\psi) = \partial_\mu (|h|\partial L/\partial\nabla_\mu\psi) + |h|(\partial L/\partial\nabla_\mu\psi)\underset{a}{I} A^a_\mu.$$

The weak conservation law for the total current is

$$\partial_\mu \left[|h|\left(\underset{a}{J^\mu} + \underset{a}{\overset{0}{J^\mu}}\right)\right] = 0.$$

The generalized Einstein equations, or the field equations of the tetrads h^μ_ν, are

$$T^\nu_\mu(\text{all fields}) + T^\nu_\mu(\text{all particles}) = 0,$$

i.e.,

$$\Sigma\,(L^\nu_{0\mu} - {}^1\!/_4\,\delta^\nu_\mu L_0) + \Sigma\,((\partial L/\partial\nabla_\nu\psi)\nabla_\mu\psi - L\delta^\nu_\mu) = 0.$$

The physical meaning of these equations is that the sum of the energy—momentum tensors of all the fields entering into the Lagrangian is equal to zero.

If the gravitational field is regarded as a gauge field associated with the homogeneous local Lorentz group, for a Lagrangian quadratic in the Riemann tensor Eq. (5.19) reduces to $R^{\mu\nu}_{ik;\nu} = J^\mu_{(ik)}$.

Noether Identities and Conservation Laws. Replacing a_a by $f^b_{ac}A^c_\mu$ and $\underset{a}{b^\mu}$ by 1 in the identities (4.11), we obtain[4,5]

$$f^b_{ac}A^c_\mu \delta L/\delta A^b_\mu \equiv \partial_\mu \delta L/\delta A^a_\mu, \tag{5.20}$$

where $\delta L/\delta A^a_\mu = \partial L/\partial A^a_\mu - \partial_\nu \partial L/\partial A^a_{\mu,\nu}$.

Since A^a_μ and $A^a_{\mu,\nu}$ appear in L only through $F^a_{\mu\nu}$, it is more convenient to rewrite (5.20) in terms of derivatives with respect to $F^a_{\mu\nu}$. Using the relations

$$\partial L/\partial A^a_\mu |_{A^a_{\mu,\nu}=\text{const}} = (-\partial L/\partial F^b_{\mu\nu}) f^b_{ac} A^c_\nu;$$

$$\partial L/\partial A^a_{\mu,\nu}\big|_{A^a_\mu=\text{const}} = -\partial L/\partial F^a_{\mu\nu} = \partial L/\partial F^a_{\nu\mu},$$

we obtain

$$\frac{\delta L}{\delta A^a_\mu} = \frac{\partial L}{\partial F^b_{\nu\mu}} f^b_{ac} A^c_\nu - \partial_\nu \frac{\partial L}{\partial F^a_{\nu\mu}} = \partial_\nu \frac{\partial L}{\partial F^a_{\mu\nu}} - \frac{\partial L}{\partial F^b_{\mu\nu}} f^b_{ac} A^c_\nu;$$
$$\left(\partial_\nu \frac{\partial L}{\partial F^a_{\mu\nu}} - \frac{\partial L}{\partial F^b_{\mu\nu}} f^b_{ac} A^c_\nu\right) f^a_{kl} A^l_\mu \equiv -\partial_\mu \left(\frac{\partial L}{\partial F^b_{\mu\nu}} f^b_{kc} A^c_\nu\right). \tag{5.21}$$

Making use of the transformation law for $F^a_{\mu\nu}$, the identities (5.20) can readily be written in the form of a covariant divergence of a Lagrangian derivative:

$$\nabla_\mu \; (\partial_\nu \partial L/\partial F^a_{\mu\nu} - f^b_{ac} A^c_\nu \partial L/\partial F^b_{\mu\nu}) \equiv 0, \tag{5.22}$$

or, returning from derivatives with respect to $F^a_{\mu\nu}$ to derivatives with respect to A^a_μ,

$$\nabla_\mu (\delta L/\delta A^a_\mu) \equiv 0.$$

We have obtained a strong conservation law, which holds independently of the form of the field equations for A^a_μ.

The identities (5.21) on the extremals lead to the ordinary (weak) conservation law

$$\partial_\mu \left(f^b_{ac} A^c_\nu \partial L/\partial F^b_{\mu\nu}\right) = \partial_\mu \overset{0}{J}{}^\mu_a = 0 \tag{5.23}$$

for the "self-action" current of the gauge field $\overset{0}{J}{}^\mu_a = f^b_{ac} A^c_\nu \partial L/\partial F^b_{\mu\nu}$.

The conservation law (5.23) is noncovariant with respect to local gauge transformations. The improper current $\overset{0}{J}{}^\mu_a$ is analogous in its meaning to the energy—momentum pseudotensor of the gravitational field in general relativity and might be called a pseudocurrent. It is equal to the divergence of an antisymmetric tensor. In electrodynamics, we have $\overset{0}{J}{}^\mu_a = 0$, owing to the Abelian nature of the gauge group.

Let us now consider the Lagrangian (5.17). In this case, the identities (4.11) take the form

$$f^b_{ac} A^c_\mu \, \delta L/\delta A^b_\mu + (\delta L/\delta\psi) \underset{a}{I} \psi - \bar\psi \underset{a}{I} \delta L/\delta\bar\psi \equiv \partial_\mu \left(\delta L/\delta A^a_\mu\right), \tag{5.24}$$

where

$$\partial L/\partial A^a_\mu = \partial L_0/\partial A^a_\mu + \partial L_{\text{int}}/\partial A^a_\mu = \overset{0}{J}{}^\mu_a + J^\mu_a,$$
$$J^\mu_a = -\; (\partial L_{\text{int}}/\partial\nabla_\mu\psi) \underset{a}{I} \psi + \bar\psi \underset{a}{I} \partial L_{\text{int}}/\partial\nabla_\mu\bar\psi. \tag{5.25}$$

If A^a_μ, ψ , and $\bar\psi$ are solutions of Euler's equations, then from (5.24) we obtain the conservation law

$$\partial_\mu (\overset{0}{J}{}^\mu_a + J^\mu_a) = 0. \tag{5.26}$$

The conservation law (5.26) is covariant. This is the ordinary (weak), but improper, current conservation law. It is interesting that its structure is similar to that of the law of conservation of energy in general relativity: what is conserved in the usual sense is the "matter" current + field pseudocurrent. The current of "matter," i.e., particles interacting with the gauge field, is conserved only in the sense of the covariant divergence. It is easy to see this by writing the Noether identities for the Lagrangian (5.16):

$$(\delta L/\delta\psi) \underset{a}{I} \psi - \bar{\psi} \underset{a}{I} (\delta L/\delta\bar{\psi}) \equiv 0.$$

Exposing the Lagrangian derivatives $\delta L/\delta\psi$ and $\delta L/\delta\bar{\psi}$, we obtain $\partial_\mu J^\mu_a - f^b_{ac} A^c_\mu J^\mu_b = 0$.

Making use of the fact that J^μ_a transforms according to the regular representation of G_r, i.e., $\delta J^\mu_a = f^b_{ac} \varepsilon^c J^\mu_b$, we write the strong conservation law for the particle current in covariant form: $\nabla_\mu J^\mu_a = 0$. For Abelian groups, the conservation law obtained in this way is, as before, homogeneous: $\partial_\mu J^\mu = 0$.

Massive Gauge Fields and Supplementary Conditions. As we have pointed out, the Noether identities, which ensure invariance of the action integral with respect to an infinite group, are satisfied independently of whether the field variables are solutions of Euler's equations for this action. They may also be satisfied when the equations of motion are invariant only with respect to the finite group to which $G_{\infty r}$ reduces when $\varepsilon^a(x) = \text{const}$. But then the Noether identities lead to supplementary conditions on the field variables, which eliminate the extra components.

Let us assume that the vector potentials of a gauge field satisfy not Eq. (5.18), but an equation with mass:

$$\nabla_\nu F^{\mu\nu}_a - m^2 A^\mu_a = 0,$$

which can be written as $\delta L_0/\delta A^a_\mu = m^2 A^\mu_a$.

Then the Noether identities lead to the Lorentz gauge conditions for A^μ_a:

$$m^2 \nabla_\mu A^\mu_a = m^2(\partial_\mu A^\mu_a - f^b_{ac} A^c_\mu A^\mu_b) = 0. \tag{5.27}$$

Since the structure constants for simple groups are antisymmetric in all indices, for $m \neq 0$ the gauge conditions become simply $\partial_\mu A^\mu_a = 0$. When the equation of an extremal $\delta L_0/\delta A^\mu_a = 0$ is satisfied ($m^2 = 0$), supplementary conditions on the vector potentials do not arise. Both cases can be written in the form of the relation

$$\partial_\mu A^\mu_a = \begin{cases} 0 & \text{for } m^2 \neq 0, \\ \text{arbitrary} & \text{for } m^2 = 0. \end{cases} \tag{5.28}$$

By virtue of the inverse form of Noether's second theorem, the existence of identities connecting the field equations and their derivatives indicates the invariance of the corresponding Lagrangian with respect to an infinite group. Indeed, if it has been established that gauge conditions on the vector potentials can be obtained as a consequence of the Noether identities between the field equations, their derivatives, and the ordinary conservation laws, by using the coefficients in the identities as coefficients of the infinite group $G_{\infty r}$ it is possible to reconstruct the Lagrangian from the corresponding variations of the field variables. It is convenient to write the Noether identities in the presence of supplementary conditions in the form[5]

$$a_a\,(\delta L/\delta u - \Theta) \equiv \partial_\mu\,[b^\mu_a(\delta L/\delta u - \Theta)], \tag{5.29}$$

where $(\delta L/\delta u - \Theta)$ corresponds to the part of Euler's equations which is invariant with respect to $G_{\infty r}$, and Θ denotes the sources and all the terms which violate $G_{\infty r}$ invariance. Therefore, if we consider not only gauge fields but also other fields interacting with them (for example, spinor fields), the supplementary conditions for the gauge field associated with the Noether identities can be satisfied only if certain conservation laws for the other fields are satisfied. In fact, it follows from (5.24) that $\nabla_\mu\,(\delta L/\delta A^a_\mu) = \nabla_\mu J^\mu_a$. Consequently,

$$m^2\partial_\mu A^\mu_a \equiv \nabla_\mu J^\mu_a. \tag{5.30}$$

§6. NOETHER'S INVERSE THEOREMS

Noether's First Inverse Theorem. *If r linearly independent combinations of Lagrangian derivatives reduce to divergences (i.e., the relation (4.6) is satisfied), then the action integral is invariant with respect to an r-parameter finite Lie group.* To prove this, we must follow the arguments of §4 in reverse order. Namely, suppose that we have the relation

$$\varphi_a\,(x, u, u_\mu)\,\delta L/\delta u = \partial_\mu J^\mu_a. \tag{6.1}$$

Let us multiply both sides of this equation by ε^a and sum over a. Using the fact that ε^a are numerical parameters, we enter them in the divergence operation on the right-hand side of Eq. (6.1). We introduce the notation

$$\left.\begin{aligned}\bar{\delta}u &= \varphi_a(x, u, u_\mu)\,\varepsilon^a;\ \delta x^\mu = -(\varepsilon^a/L)\left(J^\mu_a + \varphi_a\,\partial L/\partial u_\mu\right);\\ \delta u &= \bar{\delta}u + u_\mu\,\xi^\mu_a\,\varepsilon^a = \bar{\delta}u - (\varepsilon^a/L)\left(J^\mu_a + \varphi_a\,\partial L/\partial u_\mu\right)u_\mu\\ &= -(\varepsilon^a/L)\left(u_\mu J^\mu_a + T\varphi_a + 3L\varphi_a\right),\ \text{where } T \equiv T^\mu_\mu.\end{aligned}\right\} \quad (6.2)$$

Using (4.3), we then find the relation (4.2): $\delta L + \operatorname{div}(L\Delta x) = 0$. Integrating this expression over an arbitrary region dx, we obtain the equality $\delta S = 0$, which implies invariance of the action integral with respect to the infinitesimal transformations determined by the quantities (6.2). The case $\delta x^\mu = 0$, i.e., $J^\mu_a = -(\partial L/\partial u_\mu)\,\varphi_a$, corresponds to fixed-coordinate transformations of an internal symmetry.

The finite transformations can be found by integrating equations of the type $dx^\mu/dt = \xi^\mu$ with appropriate initial conditions. Here t is a running parameter on the one-parameter subgroups.

From the assumption that there must be r and only r independent divergence relations, it follows that the finite transformations corresponding to (6.2) form a group. If this were not so, two transformations performed in succession could lead to a new transformation which did not belong to the class (6.2), and since the action integral would also be invariant with respect to this transformation, we would have new divergence relations that are independent of (6.1), contradicting the hypothesis.

Noether's Second Inverse Theorem. *If there exist r identity relations between the Lagrangian derivatives and the derivatives of them up to order k inclusive, then the action integral is invariant with respect to an infinite group $G_{\infty r}$, whose transformations contain derivatives up to order k.* The proof is analogous to that of the previous case.

If use is made of the identity transformations, the existence of the dependences (4.11) leads, after multiplication by $\varepsilon^a(x)$ and addition, to Eq. (4.10), and from this equation there follow the determinations of δx and δu, and also invariance of the action integral with respect to these transformations. The fact that the resulting transformations form a group follows from the fact that there must exist only r dependences and hence only r arbitrary functions which determine the transformations of $G_{\infty r}$.

Special interest attaches to the case in which the functions u are not solutions of Euler's equations for a $G_{\infty r}$-invariant Lagrangian and the Lagrangian derivatives are nonzero. In that case, as we have mentioned above, Noether's second theorem leads to supplementary conditions on the sources (conservation laws for the currents). The use of

Noether's second theorem in this case makes it possible to determine the form of the invariant Lagrangian from the field equations and the supplementary conditions that follow from them.

Reconstruction of the Lagrangian from the Field Equations and the Supplementary Conditions that Follow from Them. There can be definite physical reasons for a restriction on the number of degrees of freedom of an interacting field as a result of a special choice of supplementary conditions. For example, the gauge conditions (5.28) for vector fields imply that these fields can carry only spin 1 in an interaction, whereas in the absence of supplementary conditions a vector field can carry two spins: 0 and 1.[6] It turns out that if we impose the supplementary conditions (5.28) on the vector fields, i.e., require that they have definite spin and that this spin should be equal to unity, then the multiplets of interacting vector fields form the *regular representations* of various semisimple Lie algebras. Spin is one of the quantum numbers that classify the representations of the Poincaré group, which is the group of motions of Minkowski space-time. Thus, we have here an example of a situation in which the requirement of definite space-time properties of a vector field leads to restrictions on the possible form of interaction between the fields, i.e., determines the internal symmetry of the interactions. In this way, a relation is established between the conservation laws corresponding to internal symmetries (conservation of baryon number, strangeness, isospin, and electric charge) and the space-time property of vector fields of having definite spin. Thus, the existence of truly neutral fields with spin 1 (for example, the photon or ω meson) entails invariances corresponding to conservation of additive quantum numbers (electric charge or strangeness). The existence of charged fields with spin 1 (for example, the ρ meson) leads to invariances such as isotopic invariance, etc. The converse is also true: if there exists some particular conservation law, there is room for a particle responsible for its origin, and one can consider the problem of searching for such a particle.

In the Lagrangian approach outlined in §5, the relation between the supplementary conditions and the invariance group of the Lagrangian arises as a result of Noether's second inverse theorem when the supplementary conditions that are imposed are such that they can be obtained as a consequence of the identity relations between the equations of motion and their derivatives (when the corresponding conservation laws are satisfied).[5] In what follows, we present a method, first proposed by Gell-Mann and Glashow,[7] for establishing the structure of the Lagrangian describing interacting vector fields from the field equations and the supplementary conditions that follow from them (see also Refs. 6 and 9).

Suppose that there exist a certain number r of vector fields $A^a_\mu\,(a=1,\dots,r)$. The most general Lagrangian describing possible interactions with dimensionless coupling constants can be written in the form

$$L_1(x) = -{}^1/_4 f^a_{\mu\nu} f^a_{\mu\nu} - {}^1/_2 (m^2)_{ab} A^a_\mu A^b_\mu + \alpha_{abc}\,\partial_\nu A^a_\mu A^b_\mu A^c_\nu + \beta_{abcd} A^a_\mu A^b_\mu A^c_\nu A^d_\nu + \varepsilon_{\mu\nu\lambda\rho}\left(\gamma_{abc}\,\partial_\mu A^a_\nu A^b_\lambda A^c_\rho + \delta_{abcd} A^a_\mu A^b_\nu A^c_\lambda A^d_\rho\right), \quad (6.3)$$

where $f^a_{\mu\nu} = \partial_\mu A^a_\nu - \partial_\nu A^a_\mu$, and the quantities α_{ijk}, β_{ijkl}, γ_{ijk}, and δ_{ijkl} are real (to ensure Hermiticity of the Lagrangian) numerical coefficients — coupling constants. It is natural to assume that the symmetric Hermitian matrix $\|m^2_{ab}\|$ is diagonalizable and has non-negative eigenvalues. It is not assumed in advance that it is a multiple of the unit matrix, so that the masses of the various fields can be either equal or unequal. The free part of the Lagrangian (6.3) (i.e., the part in which the fields A^a_μ appear only with one and the same Latin index) is written from the outset in such a form that in the absence of interaction (i.e., when $\alpha=\beta=\gamma=\delta=0$ and $\|(m^2)_{ab}\| = m^2 I$, where I is the unit matrix) each field satisfies the usual equation for a vector field with spin 1:

$$\partial_\nu f^{\mu\nu}_a - m^2 A^\mu_a = 0. \quad (6.4)$$

We note also that the terms with γ and δ take into account the possibility of nonconservation of parity. The very definitions of the nonlinear terms lead to the properties

$$\beta_{abcd} = \beta_{bacd} = \beta_{abdc} = \beta_{cdab}; \quad (6.5)$$

$$\gamma_{abc} = -\gamma_{acb}; \quad (6.6)$$

the quantities δ_{abcd} are completely antisymmetric in all indices, as is the discriminant tensor $\varepsilon_{\mu\nu\tau\lambda}$. Using these properties, from the Lagrangian (6.3) we obtain the equations of motion

$$\Box A^a_\mu - \partial_\mu\partial_\nu A^a_\nu - (m^2)_{ab} A^b_\mu - \alpha_{abc}\,\partial_\nu\left(A^b_\mu A^c_\nu\right) + \alpha_{bac}\,\partial_\nu A^b_\mu A^c_\nu + \alpha_{bca}\,\partial_\mu A^b_\nu A^c_\nu + 4\beta_{abcd} A^b_\mu A^c_\nu A^d_\nu + 2\left(\gamma_{abc} - \gamma_{bca}\right)\varepsilon_{\mu\nu\lambda\rho}\,\partial_\nu A^b_\lambda A^c_\rho + 4\delta_{abcd}\,\varepsilon_{\mu\nu\lambda\rho} A^b_\nu A^c_\lambda A^d_\rho = 0. \quad (6.7)$$

Let us take the 4-dimensional divergence of this equation, eliminate $\Box A^a_\mu$ in the resulting relation by means of Eq. (6.7), and require that the alternative (5.28) is satisfied. A necessary and sufficient condition for this is that each combination of terms with the same structure is identically equal to zero. Hence we find the properties of the coefficients α_{abc}, β_{abcd}, γ_{abc}, and δ_{abcd}. The necessity follows from the fact that it would otherwise be impossible to specify arbitrarily the required initial conditions in the Cauchy problem (and in quantum theory, to specify consistently equal-time commutation relations). The sufficiency is obvious. It turns out that: a) the quantities β_{abcd} are related to the coefficients α_{abc} by the equations $8\beta_{abcd} + \alpha_{mca}\alpha_{mdb} + \alpha_{mcb}\alpha_{mda} = 0$; b) the coefficients α_{abc} are completely antisymmetric and satisfy a structure relation which is identical with the Jacobi identity $[\alpha_a, \alpha_b] = -\alpha_{abc}\alpha_c$ for the structure constants of Lie algebras, where we have denoted the matrices by $(\alpha_a)_{bc} = \alpha_{abc}$; c) the quantities γ_{abc} are completely antisymmetric, as a result of which the terms with γ_{abc} completely drop out of the equations of motion; d) $\delta_{abcd} = 0$, i.e., the parity-nonconserving terms are incompatible with the Lorentz condition, so that parity is conserved in the interaction of vector fields with definite spin; e) the masses are restricted by the condition $[\alpha_a, m^2] = 0$. As a result, the Lagrangian (6.3) takes the form

$$L_1 = -\tfrac{1}{4} F^a_{\mu\nu}F^a_{\mu\nu} - \tfrac{1}{2} (m^2)_{ab}A^a_\mu A^b_\mu, \tag{6.8}$$

where $F^a_{\mu\nu} = \partial_\mu A^a_\nu - \partial_\nu A^a_\mu + \alpha_{abc}A^b_\mu A^c_\nu$. We have obtained the Lagrangian of the so-called massive gauge field. It is invariant with respect to the transformations of a gauge group G_r, whose algebra is given by the matrices α_{abc} (in the previous notation, $f^{\,b}_{ac}$):

$$A^{a\prime}_\mu = A^a_\mu + \alpha_{abc}\varepsilon_b A^c_\mu, \tag{6.9}$$

where $\varepsilon_b = \text{const}$. The vector potential of the gauge field A^a_μ corresponds here to multiplets of vector fields A_μ, which form irreducible representations of G_r. Within a single multiplet whose components transform only into one another, the masses of the vector fields are identical. Therefore m^2 can be interpreted simply as the mass of the gauge field.

If the conditions (5.28) are satisfied, under the action of a local gauge group whose transformation parameters depend on the space-time coordinates of the point x the Lagrangian (6.8) is displaced by a total divergence. But in spite of this, the action integral remains invariant with respect to $G_{\infty r}$ if the boundary conditions are such that the surface integral into which the integral of the divergence is transformed vanishes.

In a similar way, it is possible to find the Lagrangian of a system of interacting fields with spins $\frac{1}{2}$, 0, and 1.[6] In this case, the fields with spins $\frac{1}{2}$ and 0 also decompose into multiplets which transform according to the irreducible representations of the gauge group, but representations which are no longer regular.

§7. ISOPERIMETRIC PROBLEMS IN A THEORY WITH LOCAL SYMMETRY

Generalization of Noether's Second Theorem to Isoperimetric Problems. As we saw in the preceding section, the choice of supplementary conditions on the field variables has deep physical meaning. Great importance attaches here to a relation between the symmetry properties of the Lagrangian and the form of the supplementary conditions. Such a relation arises whenever part of an equation of motion coincides with Euler's equation for an action integral which is invariant with respect to an infinite group $G_{\infty r}$, where the equations of motion admit the invariance group G_r obtained from the infinite group $G_{\infty r}$ when $\varepsilon^a(x) = \text{const}$. Such equations of motion can be regarded as Euler's equations for an action integral which is invariant with respect to the infinite group, but in a problem of finding a conditional extremum (as in isoperimetric problems). In this section, we discuss symmetry properties in problems of finding conditional extrema corresponding to physical systems with constraints. As is well known, isoperimetric problems are problems of determining a conditional extremum of some functional in the case when a supplementary condition on the variation is given by the conservation of another functional S_1 of the same variables, i.e., when the supplementary conditions are integral conditions. Since we are interested in conservation laws and supplementary conditions, we shall consider functionals S and S_1 which are invariant with respect to finite or infinite Lie groups. In such cases, there are analogs of Noether's theorems.

Let us consider the generalization of Noether's second theorem to isoperimetric variational problems.

Third Theorem.[5] Suppose that the action integral S is invariant with respect to an infinite group $G_{\infty r}$ and that the fixed functional S_1 is invariant with respect to a finite group G_r, which is a subgroup of $G_{\infty r}$. Then there exist r differential supplementary conditions on the field variables, which follow from the Noether identities, relating the Lagrangian derivatives of the functionals and the derivatives of them.

As a consequence of the third theorem, the following theorem holds.

Fourth Theorem.[5] Suppose that the action integral $S = \int L dv = \int L(x, u, u')\, dv$ is invariant with respect to an infinite group $G_{\infty r}$, whose transformations contain (for simplicity) only first derivatives of the parametric functions, and suppose that $S_1 = \int L_1 dv = \text{const}$ is invariant with respect to the group G_r obtained from $G_{\infty r}$ when $\varepsilon^a(x) = \text{const}$. Then on its extremals the auxiliary functional $S_2 = S + \lambda S_1 = \int (L + \lambda L_1)\, dv = \int L_2 dv$ is invariant with respect to $G_{\infty r}$. The quantity L_2 is displaced under the action of $G_{\infty r}$ by a total divergence, and G_r leaves L_2 unchanged. (Here λ is a Lagrange multiplier.)

In fact, suppose that $\delta u = a_i(x, u, u', \ldots)\,\varepsilon^i(x) + b_i^{\mu}(x, u, u', \ldots)\,\varepsilon^{i}_{,\mu}(x)$, where $\varepsilon^i(x)$ are arbitrary functions which vanish at infinity together with their derivatives, and

$$\delta S_2 = \delta S + \lambda \delta S_1 = \int [(\delta L_2/\delta u)\,\delta u + \operatorname{div}(\delta u \partial L_2/\partial u_{,\mu})]\, dv =$$
$$\int [\,(a_i(x, u, u', \ldots)\,\varepsilon^i(x) + b_i^{\mu}(x, u, u', \ldots)\,\varepsilon^{i}_{,\mu}(x))\ \delta L_2/\delta u +$$
$$\operatorname{div}(\delta u(\partial L_2/\partial u_{,\mu}))\,]dv = \int \{[a_i(\delta L_2/\delta u) - \partial_\mu (b_i^{\mu}\delta L_2/\delta u)]\,\varepsilon^i(x) +$$
$$\operatorname{div}[\,(b_i^{\mu}\delta L_2/\delta u + a_i \partial L_2/\partial u_{,\mu})\varepsilon^i + b_i^{\nu}\partial_\nu \varepsilon^i \partial L_2/\partial u_{,\mu}]\}\, dv.$$

Here

$$\delta L_2/\delta u = \partial L_2/\partial u - \partial_\nu \partial L_2/\partial u_{,\nu} = \partial L/\partial u - \partial_\nu \partial L/\partial u_{,\nu} + \lambda(\partial L_1/\partial u - \partial_\nu \partial L_1/\partial u_{,\nu}).$$

Since the action integral S and the Lagrangian corresponding to it are invariant with respect to $G_{\infty r}$, by Noether's second theorem we have the identities

$$a_i \delta L/\delta u \equiv \partial_\mu (b_i^{\mu}\delta L/\delta u). \tag{7.1}$$

The identities (7.1) hold independently of whether u is a solution of Euler's equations $\delta L/\delta u = 0$, and this is of great importance in our case. On the extremals of the isoperimetric problem, where $\delta L/\delta u = -\lambda \delta L_1/\delta u$, these identities become supplementary conditions of the form

$$a_i \delta L_1/\delta u = \partial_\mu (b_i^{\mu}\delta L_1/\delta u). \tag{7.2}$$

According to the conditions of the isoperimetric problem, $\delta L_1/\delta u \neq 0$, and therefore the conditions (7.2) have nontrivial meaning. They constitute the supplementary conditions on the field variables whose existence is asserted in Theorem 3. The variation of S_2 reduces [when allowance is made for (7.1), (7.2), and the condition $\delta L_2/\delta u = 0$] to an integral of a total divergence. Consequently, if $\varepsilon^i(x)$ and all its derivatives vanish on the boundary, then

$$\delta S_2 = \int \operatorname{div}[a_i \varepsilon^i \partial L_2/\partial u_{,\mu} + b^{\nu}\partial_\nu \varepsilon^i \partial L_2/\partial u_{,\mu}]\, dv =$$
$$\oint (a_i \varepsilon^i \partial L_2/\partial u_{,\mu} + b^{\nu}\partial_\nu \varepsilon^i \partial L_2/\partial u_{,\mu}) d\sigma_\mu = 0. \tag{7.3}$$

If $\varepsilon^i = \text{const}$ and $G_{\infty r}$ reduces to G_r, then instead of (7.3) we obtain

$$\delta S_2 = \oint a_i \varepsilon^i (\partial L_2/\partial u_{,\mu})\, d\sigma_\mu = \varepsilon^i \oint J_i^{\mu} d\sigma_\mu = 0, \tag{7.4}$$

since we have assumed invariance of S and S_1, and

consequently of S_2, with respect to G_r. If $\varepsilon^i(x)$ and $\partial_\mu \varepsilon^i(x)$ are completely arbitrary, the following relation must hold:

$$J_i^\mu = \partial_\nu (b_i^\mu \partial L_2/\partial u_{,\nu}) = b_i^\mu \partial L_2/\partial u + \partial_\nu b_i^\mu (\partial L_2/\partial u_{,\nu}).$$

This relation is indeed satisfied on the extremals. Consequently, $\delta S_2 = 0$ on the extremals. This proves Theorem 4.

The structure of the equations (7.2) is similar to that of the conditions (4.12) on the external sources Θ not included in the Lagrangian. Therefore (7.2), like (4.12), can be used in choosing the form of the supplementary conditions. Conversely, the existence of supplementary conditions of the form (7.2) makes it possible to reconstruct the Lagrangian.

In a theory of gauge fields, we can take $L_1 = A_\mu^a A_a^\mu$, $a_i \to f_{il}^b A_\mu^l$, and $b_i^\mu \to 1$ (the arrows denote correspondence). Then the conditions (7.2) take the form

$$f_{ac}^b A_\mu^c A_b^\mu = \partial_\mu A_a^\mu = 0 \tag{7.5}$$

as a consequence of the total antisymmetry of the structure constants of the gauge group in all indices.

The Lorentz conditions for electrodynamics are obtained automatically from (7.5), since the corresponding structure constants are $f_{ac}^b = 0$. As a consequence of Theorem 4, the Lagrangian of a massive gauge field in the Lorentz gauge is invariant with respect to the local gauge group on its extremals.

Extremals of an Isoperimetric Problem and Massive Gauge Fields. First of all, we note that the extremals of a functional $S = \int L(x, u, u', \ldots)\, dv$ subject to a supplementary condition $S_1 = \int L_1(x, u, u', \ldots)\, dv = \text{const}$ are simultaneously extremals of the problem of finding an unconstrained extremum for the auxiliary functional

$$S_2 = S + \lambda S_1 = \int [L(x, u, u', \ldots) + \lambda L_1(x, u, u', \ldots)]\, dv, \tag{7.6}$$

where $\lambda = \text{const}$ is a Lagrange multiplier. In other words, if the variation is taken subject to the integral supplementary condition, the extremals take the form

$$\delta L_2/\delta u = \delta L/\delta u + \lambda \delta L_1/\delta u = 0, \tag{7.7}$$

where it is assumed that $\delta L/\delta u \neq 0$ and $\delta L_1/\delta u \neq 0$.

Isoperimetric problems can be inverted in the sense that the system of extremals is unchanged if the roles of the varied functional and the supplementary condition are interchanged, i.e., if S_1 is varied subject to the condition $S = \text{const}$. The role of the Lagrange multiplier in this case is played by $1/\lambda$, and this is of no significance for homogeneous extremals: $\delta L_2/\delta u = (1/\lambda)\delta L/\delta u + \delta L_1/\delta u = 0$.

A conditional extremum can be used to change the

symmetry of a problem, for example, to go over from a $G_{\infty r}$-invariant theory to a G_r-invariant theory. For this purpose, it is sufficient to take S_1 to be invariant with respect to G_r, if S is invariant with respect to $G_{\infty r}$. In this way, it is possible to introduce the mass of a gauge field and thereby remove the restriction $m = 0$. Indeed, consider the variational problem of finding the extremum of the functional $S = -\int_{V_4} F^a_{\mu\nu} F^{\mu\nu}_a \, dv$, where $F^a_{\mu\nu} = \partial_{[\mu} A^a_{\nu]} - {}^1/_2 f^a_{bc} A^b_{[\mu} A^c_{\nu]}$ is the stress tensor of a gauge field and A^a_μ is the vector potential of this field, subject to the following condition on another functional:

$$S_1 = \int_{V_4} A^a_\mu A^\mu_a \, dv = b = \text{const} \neq 0. \qquad (7.8)$$

Here $a = 1, \ldots, r$, where r is the number of parameters of the gauge group, $\mu = 1, \ldots, 4$ is a space-time index, and V_4 will be assumed to be a finite region.

According to the general rule, we form the auxiliary functional

$$S_2 = S + \lambda S_1 = \int_{V_4} \left(-F^a_{\mu\nu} F^{\mu\nu}_a + \lambda A^a_\mu A^\mu_a\right) dv \qquad (7.9)$$

and find its extremum. The resulting extremal will also be an extremal for S satisfying the condition (7.8). The constants of integration and the constant λ are then determined from the boundary conditions and the supplementary condition (7.8). If the constant λ is identified with the square of the mass of the gauge field, $2m^2$, then S_2 is the action for the massive gauge field, and the mass has the meaning of the Lagrange multiplier. A similar interpretation of a mass is sometimes used in classical mechanics.

Euler's equations corresponding to $\delta S_2 = 0$ have the form

$$F^{\mu\nu}_{a;\nu} - m^2 A^\mu_a = 0. \qquad (7.10)$$

The system of extremals is unchanged if Euler's equations are represented in the form

$$F^{\mu\nu}_{a;\nu}/m^2 - A^\mu_a = 0. \qquad (7.11)$$

The equations (7.11) can now be regarded as Euler's equations corresponding to variation of the functional $S_1 = \int_{V_4} A^a_\mu A^\mu_a dv$ subject to the supplementary condition

$$S = -\int_{V_4} F^a_{\mu\nu} F^{\mu\nu}_a \, dv = a = \text{const}. \qquad (7.12)$$

The quantity S determines $1/m^2$. The integral S is an integral of the field invariants over the 4-dimensional region in which the field exists. It is determined not only by the invariants themselves, but also by the topology of V_4.

If the gravitational field is regarded as a gauge field corresponding to the local Lorentz group, Eq. (7.12) becomes

$$S = \int_{V_4} R^{\mu\nu\tau\lambda} R_{\mu\nu\tau\lambda}\, dv = \text{const}, \tag{7.13}$$

where $R_{\mu\nu\tau\lambda}$ is the Riemann curvature tensor. For self-dual fields, $F^a_{\mu\nu} = \pm {}^*F^a_{\mu\nu}$, and doubly self-dual fields, $R_{\mu\nu\tau\lambda} = \pm\widetilde{R}_{\mu\nu\tau\lambda}$, the integrals (7.12) and (7.13) are topological invariants.

An analogous situation occurs in relation to the constants for the interaction of particles with a gauge field. Namely, the appearance of an interaction with a gauge field can be treated from the same standpoint as the appearance of a mass term, i.e., as the replacement of the problem of finding an unconstrained extremum for the free (or massless) Lagrangian by the problem of finding a conditional extremum subject to integral supplementary conditions. Thus, the Dirac equations with interaction

$$\gamma^\mu (\partial_\mu \psi - i\varepsilon \underset{a}{I} A^a_\mu \psi) + m\psi = 0 \tag{7.14}$$

provide an extremum for the functional $S_3 = \int (\bar{\psi}\gamma^\mu\partial_\mu\psi + m\bar{\psi}\psi)dv$ subject to the supplementary condition

$$S_4 = \int J^\mu_a A^a_\mu dv = \text{const} \neq 0. \tag{7.15}$$

The coupling constant (the charge) plays the role of a Lagrange multiplier and is determined by the value of S_4. At the same time, Eq. (7.14) provides an extremum for the functional $S_4 = \int J^\mu_a A^a_\mu dv = \int \bar{\psi}\gamma^\mu \underset{a}{I}\psi A^a_\mu dv$ subject to the supplementary condition

$$S_3 = \int (\bar{\psi}\gamma^\mu \partial_\mu \psi + m\bar{\psi}\psi)\, dv = \text{const}. \tag{7.16}$$

In this case, the coupling constant is determined in terms of the constant value of S_3, and renormalization of the coupling constant corresponds to a modification in the value of the action integral for free ("bare") particles.

Thus, the transition from a massless to a massive gauge field, like the transition from free to interacting Dirac particles, can be represented as a transition from a variational problem of finding an unconstrained extremum to the requirement of extremality of the same action integral subject to supplementary integral conditions of the type (7.8), (7.12), (7.15), or (7.16).

Spontaneous Breakdown of Local Gauge Symmetry. In contemporary gauge models of the interactions of elementary particles, the Higgs mechanism is used to obtain the masses of vector gauge fields. In order to obtain the masses of vector gauge fields by means of this mechanism, it is usual to add to the Lagrangian $L = -{}^1/_4 F^a_{\mu\nu} F^{\mu\nu}_a$ of a massless vector field the Lagrangian of a scalar Higgs field (L_X). The components of the vector field are then replaced by linear combinations of the vector and scalar fields such that the resulting vector field is massive, i.e., the new Lagrangian contains terms proportional to the squares of the redefined components of the vector field. The massless scalar fields are no longer present in the new Lagrangian, and we therefore say that the scalar particles have "disappeared" and that the vector field has acquired mass.

Perturbation theory is usually constructed in the neighborhood of the minimum of the potential energy. This minimum can be found by varying the potential $V(\varphi)$, which in general can be highly nonlinear in the fields, i.e., can describe "self-action" of the system. The second derivative of the potential energy with respect to the field, evaluated at the position of the minimum of the potential energy, is called the *square of the mass of the field*, and the fourth derivative at the same point is called the *coupling constant*.

The condition $V = V_{min}$, i.e., $\delta V/\delta\varphi = 0$, can be replaced by the condition $S_1 = S_{1min}$, i.e., $\delta S_1/\delta\varphi = 0$, where $S_1 = \int V(\varphi)dv$. Then it is possible to determine the mass of the field by means of the method of the isoperimetric problem, by studying the extremals of S_1. The mass obtained by differentiation of the potential energy with respect to the field is the mass of fluctuations around a fixed solution and naturally depends on this solution.

For simplicity, consider a single scalar field φ.[20] Its Lagrangian has the form

$$L_\varphi = {}^1/_2 (\partial_\mu\varphi)^2 - V(\varphi) = {}^1/_2 (\partial_\mu\varphi)^2 + L_1(\varphi).$$

Euler's equation for this Lagrangian takes the form

$$\Box\varphi = -\partial V(\varphi)/\partial\varphi.$$

The character of the solutions of Euler's equations depends on the form of the function $V(\varphi)$. To determine the mass of the field φ, we must consider the behavior of the solutions of the equations for the field φ near the solution which gives the minimum of V, i.e., near the extremals $-\delta L_1/\delta\varphi = \partial V(\varphi)/\partial\varphi = 0$. If $V(\varphi)$ has the form $V(\varphi) = (\mu/2)\varphi^2 + (\beta/4)\varphi^4$, the minimum of the potential energy is attained when $\varphi = 0$. The leading term of the expansion of the potential at this point is $V(\varphi) = (\mu/2)\varphi^2$, where $\mu = (\partial^2 V/\partial\varphi^2)_{\varphi=0} > 0$. Euler's equation for L_φ near $\varphi = 0$ takes the form $\Box\varphi - \mu\varphi = 0$, or $\ddot{\varphi} + \mu\varphi - \Delta\varphi = 0$. The solutions of this equation have the form of plane waves $\varphi \sim \exp(\pm ik_\mu x^\mu)$, where the wave vector $k_\mu = (k_0, \mathbf{k})$ satisfies the relation $k_\mu^2 = \mu$, or $k_0^2 = \mathbf{k}^2 + \mu$.

If $V(\varphi)$ has the form of the Higgs Lagrangian

$$L_X = V(\varphi) = -(\mu/2)\varphi^2 + (\beta/4)\varphi^4,$$

its extrema are determined by the solutions of Euler's equation

$$\delta L_X/\delta\varphi = -\mu\varphi + \beta\varphi^3 = +\beta\varphi(\varphi^2 - \mu/\beta) = 0,$$

$\varphi_1 = 0$ and $\varphi_2 = \pm\sqrt{\mu/\beta}$. Of these, φ_1 is a local maximum, and the theory in the neighborhood of $\varphi = 0$ is unstable. Two local minima are attained when $\varphi = \eta = \pm\sqrt{\mu/\beta}$. Near these solutions, $V(\varphi) = -\mu\eta^2/4 + \mu(\varphi - \eta)^2$, and $m^2 = (\partial^2 V/\partial\varphi^2)_{\varphi=\eta} = 2\mu > 0$, where m is the mass of the quanta of the field $\Delta\varphi = (\varphi - \eta)$.

In going over to the quantum picture, the values of the classical field which give the minimum of the potential energy become the vacuum expectation values of this field. Thus, different vacua are possible in nonlinear theories.

In theories which are invariant with respect to a local gauge group, the set of vacua is determined by the factor group G/H, where H is the stability group of the vacuum, which takes a given vacuum into itself. The classical trajectories connecting different vacua and satisfying the condition $S = \text{const}$ play a major role in quantum field theory and in the theory of phase transitions. The fluctuations of the field in the neighborhood of topologically nontrivial solutions lead to a spectrum of anomalously large masses.

We shall apply the method of the isoperimetric problem to obtain the masses of gauge fields without using additional scalar Higgs fields. In mechanics, as is well known, the symmetry group of a theory is a group which leaves the potential unchanged, and the mass is an invariant of this group.

Suppose that a gauge field is described by the action integral $S_1 = \int A_\mu^a A_a^\mu dv$ and, in addition, that the integral supplementary condition $S = -{}^1\!/_2 \int F_{\mu\nu}^a F_a^{\mu\nu} dv = \text{const}$ is imposed. We shall consider the theory in the neighborhood of the extremum of S_1. Euler's equations corresponding to S_1 have the form $A_\mu^a = 0$.

We now vary the integral

$$S_1 = \int (f_{bc}^a f_{akl} A_\mu^b A_\nu^c A_\mu^k A_\nu^l + m^2 g_{ab} A_\mu^a A_\mu^b)\, dv.$$

Euler's equations for S_1 take the form

$$2 f_{bc}^a f_{akl} A_\nu^c A_\mu^k A_\nu^l + m^2 g_{ab} A_\mu^a = 0$$

or

$$A^a_\mu\,(m^2 g_{ab} - 2f_{ald} f^d_{cb} A^c_\nu A^l_\nu) = 0. \tag{7.17}$$

The solution $A^a_\mu = 0$ of Eq. (7.17) gives a local maximum of S_1. We obtain the second solution from the condition $2f_{ald}\, f_c{}^d{}_b A^c_\nu A^l_\nu = m^2 g_{ab}$, from which, by contracting with g^{ab}, we find $4\delta_{lc} A^c_\nu A^l_\nu = rm^2$. This relation is satisfied for the solution corresponding to spontaneous symmetry breaking. For semisimple groups, $g_{ab} = -2\delta_{ab}$.

If we now choose a coordinate system such that $A^a_i = 0$, where $i = 1, 2, 3$ and $A^a_0 = \varphi^a$ is a multiplet of scalar fields, we obtain

$$\delta_{ab}\varphi^a\varphi^b = rm^2/4. \tag{7.18}$$

For a one-parameter Abelian group, $\varphi^2 = m^2/4$, as in other cases of spontaneous symmetry breaking, if we put $S_1 = \int (A_\mu A_\mu A_\nu A_\nu - m^2 A_\mu A_\mu) dv$. For the group SU(2), the relation (7.18) becomes $(\varphi^a)^2 = 3m^2/4$.

The search for solutions of field equations near some specific solution is equivalent to the transition from the problem of motion of a single test body in an external field to the problem of relative motion of two bodies, one of which is regarded as constrained. A condition of the type $S = \text{const}$ makes it possible to fix the constrained trajectory.

Conservation Laws in Isoperimetric Problems. The relation (7.4) yields conservation laws for an isoperimetric problem in the form $\oint J^\mu_i d\sigma_\mu = 0$. Here

$$J^\mu_i = a_i\, \partial L_2/\partial u_{,\mu} = \underset{0}{J^\mu_i} + \lambda a_i\, \partial L_1/\partial u_{,\mu}, \tag{7.19}$$

where

$$\underset{0}{J^\mu_i} = a_i\, \partial L/\partial u_{,\mu}.$$

Symmetry Breaking and Conservation of the Currents. The relation (7.19) shows how the form of the conserved currents is changed when supplementary conditions of the type occurring in the isoperimetric problem are imposed. Namely, the currents are unchanged if L_1 does not contain derivatives of u (the currents "do not feel" the potential energy). In particular, it follows from (7.19) that the transition to massive gauge fields [conditions of the type (7.8)] does not alter the form of the conserved currents. However, allowance for the Noether identities leads to a new relation between the currents and fields. In fact, in this case, as before,

$$\underset{0}{J^\mu_i} = \underset{0}{J^\mu_{\infty i}} = b^\mu_i (\delta L/\delta u) = J^\mu_i - \lambda a_i (\partial L_1/\partial u_{,\mu}).$$

On the extremals of the isoperimetric problem, $\delta L_2/\delta u = 0$, we obtain

$$J_i^\mu - \lambda a_i\,(\partial L_1/\partial u_{,\mu}) = -\lambda b_i^\mu\,(\delta L_1/\delta u). \qquad (7.20)$$

For massive gauge fields, the relation (7.20) leads to proportionality of the current and field: $J_i^\mu = m^2 A_i^\mu$. Thus, proportionality of the current and field (like the appearance of the mass of a gauge field) can be understood as a consequence of the implicit introduction of the supplementary condition (7.8) and the transformation of the ordinary variational problem for gauge fields into an isoperimetric problem.

In conclusion, we note that the supplementary conditions (7.2), like the invariance of S_2 with respect to $G_{\infty r}$, are satisfied only on the extremals $\delta L_2/\delta u = 0$, i.e., they follow from the equations for the field, whereas the invariance of S is determined by its form and is independent of the field equations. This circumstance is perhaps an argument in favor of adoption of an action principle in a form such that the part of the Lagrangian which breaks the symmetry (the interaction) is varied, while the invariant part is fixed and determines the value of the coupling constant.

§8. TENSOR GAUGE FIELDS AND LIE DERIVATIVES

Nonhomogeneous Spaces and Lie Derivatives. In this section, general relativity is regarded as a theory of a gauge field of a symmetric second-rank tensor $h_{\mu\nu}$ in space-time with arbitrary geometry. The gauge group is the group of arbitrary continuous coordinate transformations of general relativity. The form of the variations is defined by means of Lie derivatives. The field $h_{\mu\nu}$ is then identified with the field of the metric tensor $g_{\mu\nu}$ and effects the transition to pseudo-Riemannian space-time. In what follows, we discuss the problem of a vector gauge field and a second-rank antisymmetric tensor field.

Groups describing space-time symmetry cannot be localized by the Yang—Mills procedure. In fact, this operation leads to a completely different type of geometry of space-time and to a different concept of space, while localization of internal symmetries does not in general alter the geometry of V_4. An analogy between general relativity and Yang—Mills theories can be developed in the case when the tangent space to V_4 is not identified with space-time itself. In that case, all symmetry properties can be referred to the tangent space, i.e., to Minkowski space-time, without worrying from the outset about whether this symmetry corresponds to local or global properties of V_4. To compare quantities referring to different points of such a manifold stratified into tangent spaces, we introduce the operation of Lie differentiation. Lie derivatives are defined independently of the geometrical structure of the manifold, but they are related to arbitrary continuous coordinate transformations, i.e., to the general covariant group. Thus, if we use Lie derivatives as the variations generated by the infinite group of coordinate transformations of general relativity, we can apply the variational procedure of §5 to obtain the theory of gravitation as a gauge field. The gauge field will then be a second-rank symmetric tensor field $h_{\mu\nu}$, which in general is not necessarily identified with the metric tensor of space-time V_4. If this identification is made, we obtain Einstein's theory. Einstein's equations are obtained automatically as Euler's equations for the simplest gauge-invariant Lagrangian $L = R$. The exposition here follows Ref. 5, in which this approach was first considered.

L i e d e r i v a t i v e s in a Riemannian space (or in an arbitrary manifold) are introduced as follows.[11] Since all quantities are specified in the neighborhood of a point, and the relation between the points is also specified, to compare quantities referring to different points of space it is necessary to introduce two coordinate systems at each of these points. One of these coordinate systems will be the natural system, and the other will be "transported" from the point with which the comparison is made. The transported coordinate system is simply "the same" as at the second point, and the components of the quantities being compared at the first point with respect to the coordinate system transported from the second point are by definition equal to the components of the quantities in the natural coordinate system at the second point. In other words, we simply assign to each point, in addition to the natural coordinate system, the coordinate system and all the values of the components of the quantity under investigation referred to the second point. Once we have the two coordinate systems and the two sets of components at a single point, we can make use of tensor analysis. To do this, it is necessary to carry out an automorphism (mapping) of the space as a whole into itself and to compare the changed values of the components with the transported values. This comparison is independent of the definition of parallel transport, i.e., of the specific properties of space, and leads to the concept of the Lie derivative of a given quantity. It is meaningless to compare quantities referring to different points of a Riemannian space without transferring them in some way to a single point, since such an equality is violated in going over to a new coordinate system. The transition from the natural values of a quantity to the transported values is described by formulas which generalize tensor transformations. For a symmetric second-rank tensor $h_{\mu\nu}$, these give

$$h^*_{\mu\nu}(x) = (\partial x^{*\tau}/\partial x^{\mu})(\partial x^{*\lambda}/\partial x^{\nu})h_{\tau\lambda}(x^*), \tag{8.1}$$

where the asterisks denote the transported values of the coordinates and components of the tensor $h_{\mu\nu}$.

Consider a continuous automorphism of a Riemannian space induced by coordinate transformations of the form

$$x^{*\mu} = x^{\mu} + \xi^{\mu}(x)t, \tag{8.2}$$

where t is a parameter. The operation called Lie differentiation is defined with respect to such transformations. In the case of a symmetric covariant second-rank tensor $h_{\mu\nu}$, the Lie derivative of it is the expression which is readily obtained from (8.1) as the limit $\lim_{\delta t\to 0}\frac{h^*_{\mu\nu}(x^*)-h_{\mu\nu}(x)}{\delta t}$:

$$\delta_L h_{\mu\nu} = \xi^{\tau}(x)\,\partial_{\tau} h_{\mu\nu} + h_{\tau\nu}\,\partial_{\mu}\xi^{\tau}(x) + h_{\mu\tau}\,\partial_{\nu}\xi^{\tau}(x). \tag{8.3}$$

If $h_{\mu\nu} = g_{\mu\nu}$ and the connection coefficients are expressed entirely in terms of derivatives of $g_{\mu\nu}$ (Riemannian connection), the Lie derivative can be rewritten in terms of covariant derivatives: $\delta_L g_{\mu\nu} = \xi_{\nu;\mu} + \xi_{\mu;\nu}$.

Lie groups which are generated by the automorphisms of (8.2) and preserve the metric tensor $g_{\mu\nu}$ (in the sense $\delta_L g_{\mu\nu} = 0$) are called *groups of motions of the Riemannian space*, and the corresponding vector ξ^{τ} is called the *Killing vector*. In a space which admits a group of motions, a knowledge of the components of the metric tensor at a single point can be used to find $g_{\mu\nu}$ at all points by the process of transport. In flat space, the groups of motions become ordinary space-time symmetry groups. For example, in Minkowski space-time the group of motions is the Poincaré group, and in a space-time of constant curvature it is the de Sitter group O(4, 1), from which the Poincaré group is obtained in the limiting case $R \to \infty$ (where R is the radius of curvature of V_4).

Gauge Vector Field and an Antisymmetric Second-Rank Tensor Field. Lie derivatives are defined for quantities of arbitrary dimensions, both integral and half-integral. In fact, this means that an arbitrary manifold V_4 is stratified by means of the various representations of the general linear group GL(4). In other words, over each point of V_4 a fiber is introduced, whose dimension is determined by the dimension of the representation which we require. Since Lie derivatives always contain arbitrary functions ($\sim\Delta x$) and their derivatives, the gauge fields can be taken to be any fields which have only space-time indices and are not related to gauge groups of the type associated with internal symmetries. As an example, consider a vector field A_μ. Its Lie derivative has the form $\delta_L A_\mu = \xi^\tau \partial_\tau A_\mu + A_\tau \partial_\mu \xi^\tau$, and

$$\delta_L A^\mu = \xi^\tau \partial_\tau A^\mu - A^\tau \partial_\tau \xi^\mu.$$

Using (4.11), the Noether identities can be written in the form

$$(\delta\hat{L}/\delta A_\mu)(\partial_\tau A_\mu - \partial_\mu A_\tau) \equiv \partial_\mu (\delta\hat{L}/\delta A_\mu) A_\tau.$$

If this expression is contracted with $\delta\hat{L}/\delta A_\tau$, we obtain

$$A_\tau (\delta\hat{L}/\delta A_\tau) \partial_\mu (\delta\hat{L}/\delta A_\mu) \equiv (\delta\hat{L}/\delta A_\tau) \partial_{[\tau} A_{\mu]} \delta\hat{L}/\delta A_\mu \equiv 0.$$

Therefore, either the conservation law

$$\partial_\mu (\delta\hat{L}/\delta A_\mu) = 0$$

is satisfied, or the vector field satisfies the condition

$$A_\tau (\delta\hat{L}/\delta A_\tau) = 0.$$

When sources are included, we find from the Noether identities that

$$J^\mu F_{\tau\mu} \equiv 2 (\partial_\mu J^\mu) A_\tau, \qquad (8.4)$$

where $F_{\tau\mu} = \partial_\tau A_\mu - \partial_\mu A_\tau$. It follows from this that $\partial_\mu J^\mu (A_\tau J^\tau) = 0$. Therefore, either J^μ is conserved, i.e., $\partial_\mu J^\mu = 0$, or there is no interaction ($A_\tau J^\tau = 0$). Contracting (8.4) with $^*F^{\tau\lambda}$, we obtain

$$J^\lambda (F^*_{\alpha\beta} F^{\alpha\beta}) \equiv 8 \partial_\mu J^\mu (\varepsilon^{\tau\lambda\alpha\beta} A_\tau F_{\alpha\beta}).$$

Antisymmetric Second-Rank Tensor Field. The Lie variation of such a field is given by the formula

$$\delta_L f_{\mu\nu} = \xi^\tau \partial_\tau f_{\mu\nu} + f_{\tau\nu} \partial_\mu \xi^\tau + f_{\mu\tau} \partial_\nu \xi^\tau.$$

The Noether identities have the form

$$\tfrac{1}{2} (\delta L/\delta f_{\mu\nu}) (\partial_\tau f_{\mu\nu} - \partial_\mu f_{\tau\nu} - \partial_\nu f_{\mu\tau}) \equiv \partial_\mu (\delta L/\delta f_{\mu\nu}) f_{\tau\nu}. \qquad (8.5)$$

If $\partial_{[\tau} f_{\mu\nu]} \neq 0$ and $L = \tfrac{1}{4} f_{\mu\nu} f^{\mu\nu}$, then $\delta L/\delta f_{\mu\nu} = \tfrac{1}{2} f^{\mu\nu}$ and the identities (8.5) take the form

$$\tfrac{1}{2} f^{\mu\nu} (\partial_\tau f_{\mu\nu} - \partial_\mu f_{\tau\nu} - \partial_\nu f_{\mu\tau}) = (\partial_\mu f^{\mu\nu}) f_{\tau\nu}$$

or

$$\tfrac{1}{2} f^{\mu\nu} \partial_\tau f_{\mu\nu} = \partial_\mu (f^{\mu\nu} f_{\tau\nu}),$$

from which there follows the "conservation law"

$$\partial_\mu (f^{\mu\nu} f_{\tau\nu} - \tfrac{1}{4} \delta^\mu_\tau f^{\alpha\beta} f_{\alpha\beta}) = 0.$$

In electrodynamics, the quantity in parentheses plays the

role of the energy—momentum tensor of the electromagnetic field. Here the conservation law for this tensor is obtained as a consequence of the Noether identities.

Using the identity ${}^*f^{\mu\nu}f_{\tau\nu} \equiv {}^1\!/_4\,\delta^\mu_\tau\,({}^*f^{\lambda\nu}f_{\lambda\nu})$, where ${}^*f^{\mu\nu} = {}^1\!/_2\varepsilon^{\mu\nu\tau\lambda}f_{\tau\lambda}$ and $\varepsilon^{\mu\nu\tau\lambda}$ is the discriminant tensor, and the antisymmetry of $f_{\mu\nu}$, Eq. (8.5) can be rewritten as

$$({}^*f^{\lambda\nu}f_{\lambda\nu})\partial_\mu(\delta L/\delta f_{\mu\sigma}) \equiv 6{}^*f^{\sigma\tau}\partial_{[\tau}f_{\mu\nu]}\,\delta L/\delta f_{\mu\nu}. \tag{8.6}$$

If $f_{\mu\nu}$ satisfies the equation

$$\partial_{[\tau}f_{\mu\nu]} = 0, \tag{8.7}$$

then (8.6) simplifies:

$$({}^*f^{\lambda\sigma}f_{\lambda\sigma})\partial_\mu\,(\delta L/\delta f_{\mu\nu}) = 0. \tag{8.8}$$

Thus, from any field equation for $f_{\mu\nu}$, we obtain a strong conservation law and supplementary conditions by simple differentiation. If $\delta L/\delta f_{\mu\nu} = f^{\mu\nu}$, it follows from (8.8) that

$$({}^*f^{\lambda\sigma}f_{\lambda\sigma})\partial_\mu f^{\mu\nu} = 0. \tag{8.9}$$

Consequently,

$$\partial_\mu f^{\mu\nu} = \begin{cases} 0, & \text{if } {}^*f^{\lambda\sigma}f_{\lambda\sigma} \neq 0; \\ \text{arbitrary} & \text{if } {}^*f^{\lambda\sigma}f_{\lambda\sigma} = 0. \end{cases}$$

Thus, fulfillment of the conditions ${}^*f^{\lambda\sigma}f_{\lambda\sigma} \neq 0$ and $\partial_{[\tau}f_{\mu\nu]} = 0$ leads to Maxwell's equations

$$\partial_\mu f^{\mu\nu} = 0, \tag{8.10}$$

and, as a consequence of (8.7) and (8.10), to the equations

$$\Box f^{\mu\nu} = 0. \tag{8.11}$$

We note that (8.11), as a consequence of the Noether identities $\partial_\mu\,(\delta L/\delta f_{\mu\nu}) = 0$, can also be derived in the case of a law of transformation of $f_{\mu\nu}$ which differs from (8.4), namely,

$$\delta f_{\mu\nu} = \partial_\mu\xi_\nu - \partial_\nu\xi_\mu. \tag{8.12}$$

This form of transformation of $f_{\mu\nu}$ was used in Ref. 21, where hypothetical particles "notophs" ("backwards" photons) were introduced as quanta of an antisymmetric second-rank tensor field.

Gauge Field of a Symmetric Second-Rank Tensor. If in V_4 we have a symmetric, nondegenerate, second-rank tensor field $h_{\mu\nu}$ and an automorphism of V_4 induced by continuous coordinate transformations of the form (8.2), the conditions of the type (5.3)—(5.5) determine the form of the Lagrangian and field equations which are invariant with respect to the transformations (8.3) of $h_{\mu\nu}$ generated by the automorphism of V_4.[5] The invariant Lagrangian which does not contain derivatives of $h_{\mu\nu}$ has the form $L = 1/\sqrt{-h}$, where $h = \det|h_{\mu\nu}|$. The Lagrangian depending on $h_{\mu\nu}$ and on its first and second derivatives reduces to

$$\hat{L} = \sqrt{-h}\, R, \qquad (8.13)$$

where R is the analog of the scalar curvature of a Riemannian space (in the sense that it is formed from the quantities $h_{\mu\nu}$ and their derivatives in exactly the same way that the scalar curvature is formed from the metric tensor and its derivatives). Raising of indices in the usual sense does not occur here, since the inverse matrix $h^{\mu\nu}$, satisfying the condition $h^{\mu\sigma}h_{\sigma\nu} = \delta^{\mu}_{\nu}$, is used for the contraction of indices. It is not necessary to introduce a metric to construct an invariant Lagrangian. Another consequence which follows automatically from invariance conditions of the type (5.3)—(5.5) is the fact that the derivatives of $h_{\mu\nu}$ appear in the Lagrangian only through combinations $\Gamma^{\lambda}_{\mu\nu}$ which, are formed from $h^{\mu\nu}$ and the derivatives of $h_{\mu\nu}$ as the Christoffel symbols are.

The Euler equations obtained by variation of the Lagrangian (8.13) with respect to $h^{\mu\nu}$ are the Einstein equations[5,9]

$$R_{\mu\nu} - {}^{1}/{}_{2} h_{\mu\nu} R = 0.$$

The use of the concept of Lie derivatives on arbitrary manifolds makes the meaning of the resulting equations very perspicuous. In fact, the form of the invariant Lagrangian is determined only by the form of the variations of the field variables. The structure of the Lie derivative depends only on the tensor dimensions of the functions and is therefore the same for all symmetric second-rank tensors and for all geometries. Lie derivatives are always generated by a group, this being infinite, describing a mapping of an arbitrary manifold onto itself. This group coincides outwardly with the general covariant transformations (or the "local group of displacements") and therefore leads to the same results as Riemannian geometry, in the sense of the structure of the differential invariants.

The Noether identities corresponding to the transformations (8.3) have the form ${}^{1}/{}_{2}\,(\delta\hat{L}/\delta h_{\mu\nu})\,(\partial_{\tau}h_{\mu\nu} - \partial_{\nu}h_{\mu\tau} - \partial_{\mu}h_{\tau\nu}) \equiv \partial_{\nu}\,(\delta\hat{L}/\delta h_{\mu\nu})h_{\mu\tau}$, or

$$\nabla_{\nu}\,(\delta\hat{L}/\delta h_{\mu\nu}) \equiv 0, \qquad (8.14)$$

where $\delta\hat{L}/\delta h_{\mu\nu} = \partial\hat{L}/\partial h_{\mu\nu} - \partial_{\tau}\partial\hat{L}/\partial(\partial_{\tau}h_{\mu\nu})$ and $\hat{L} = \sqrt{-h}L$. The covariant derivative here must be understood formally in the sense of the connection coefficients formed from $h_{\mu\nu}$ and its derivatives.

Equation (8.14) is a strong conservation law, which is valid independently of the specific form of the Lagrangian or of Euler's equations. In particular, it leads to the well-known covariant conservation law (4.13):

$$(R^{\mu\nu} - {}^1/_2 h^{\mu\nu} R)_{;\nu} \equiv 0. \tag{8.15}$$

In general relativity, the identity (8.15) is usually derived as a consequence of the Bianchi identities for the Riemann curvature tensor. But here it is an expression of the Noether identities and is valid for any symmetric tensor field $h_{\mu\nu}$, for example, the tensor of elastic deformations of a medium.

Suppose now that $\delta\hat{L}/\delta h_{\mu\nu} = \hat{T}^{\mu\nu}$, where $\hat{T}^{\mu\nu}$ is a symmetric tensor describing the sources of a field. Then (8.14) leads to a covariant conservation law for $T^{\mu\nu}$, namely, $T^{\mu\nu}{}_{;\nu} = 0$.

Thus, the covariant, identically satisfied law of conservation of energy in general relativity is a strong conservation law which follows from the Noether identities for the field equations and is a consequence of the invariance of the theory with respect to the infinite group $G_{\infty 4}$.

To obtain an ordinary conservation law (in the sense of the ordinary divergence) in accordance with the general rule for all gauge fields, it is necessary to introduce a pseudocurrent (in this case, a pseudotensor $t^{\mu\nu}$) of the field itself. The total tensor will be conserved in the ordinary sense: $\partial_\nu (T^{\mu\nu} + t^{\mu\nu}) = 0$. However, this procedure does not lead to correct integral conservation laws.

The identities (8.14) make it possible to determine the structure of the supplementary conditions on the field equations. For example, if the field equations have the form $\delta\hat{L}/\delta h_{\mu\nu} = m^2 (h_{\mu\nu} + q\delta_{\mu\nu} h^\rho_\rho)$, then from (8.14) we obtain[4,5]

$$m^2 \nabla_\nu (h_{\mu\nu} + q\delta_{\mu\nu} h^\rho_\rho) = 0. \tag{8.16}$$

If $h_{\mu\nu}$ is identified with the metric tensor $g_{\mu\nu}$, all covariant derivatives acquire the usual meaning. For the metric tensor $g_{\mu\nu}$, the conditions (8.16) can be simplified by choosing a locally geodesic coordinate system, which leads to the Gilbert—Lorentz conditions

$$m^2 \partial_\nu (g_{\mu\nu} + q\delta_{\mu\nu} g^\rho_\rho) = 0. \tag{8.17}$$

We could have proceeded in the opposite direction, i.e., postulated the supplementary conditions (8.17) and the identities relating the supplementary conditions to the field equations in the form

$$\nabla_\nu \left(\delta\hat{L}/\delta h_{\mu\nu} - m^2 (h_{\mu\nu} + q\delta_{\mu\nu} h^\rho_\rho)\right) \equiv 0.$$

Then we would have found that the transformations of $h_{\mu\nu}$ which generate these identities are (8.3) and that the field equations are Einstein's equations.[9]

Integral Conservation Laws in General Relativity. Discussions of the problem of constructing the integral law of conservation of energy in general relativity often begin with the remark that the differential law of conservation of energy in Einstein's space V_4 has an unusual structure: it is not the ordinary, but the covariant divergence of the energy—momentum tensor that is equal to zero. To "rectify" the differential conservation law, one introduces various energy—momentum pseudotensors of the gravitational field, which when added to the energy—momentum tensor of matter lead to the ordinary differential law of conservation of energy. But if one considers not flat, but Riemannian space, a problem of nonuniqueness arises in the integration of this differential conservation law. The nonuniqueness occurs for two reasons: 1) because of the noncovariance of the pseudotensors and the ordinary conservation law $\partial_\mu (T^{\mu\nu} + t^{\mu\nu}) = 0$; 2) because of the nonuniqueness of the integral of a symmetric tensor in Riemannian space. In the case of vector conserved currents, there is no ambiguity, since the covariant derivative of a vector (more precisely, of a vector density) is the same as the ordinary derivative, and an integral of a vector over a 3-dimensional hypersurface Σ is an integral of a scalar and is therefore well-defined. Integral conserved quantities (energy, mass, or momentum) in general relativity depend on the choice of coordinate system. The process of integration itself also depends on this choice and on the method of comparing quantities at different points of V_4, i.e., on the choice of connection coefficients. This ambiguity can be eliminated, but it is then necessary to reconsider the set of integral characteristics that can be obtained from the differential conservation law for the energy—momentum tensor in general relativity.

The point is that integration in a Riemannian space is uniquely defined only for scalars. Since the element of volume is a completely antisymmetric product of differentials of the coordinates, a scalar under an integral sign can be obtained only by contracting the element of volume with antisymmetric tensors. Thus, integration of symmetric tensors of rank $n \geqslant 2$ remains ambiguous. One of the possible methods of unambiguous integration of symmetric tensors is integration with a vector.

Consider Green's formula in Riemannian space:

$$\int_V J^m_{;\,m}\, d^4 v = \oint_\Sigma J^m \varepsilon(N) N_m\, d^3 v, \tag{8.18}$$

where N_m is the normal vector, $\varepsilon(N)$ is the indicator of the vector N_m, and d^4v and d^3v are the invariant elements of volume. For a symmetric second-rank tensor T^{ik}, a formula analogous to (8.18) can be obtained by means of an auxiliary vector:

$$\int (T^{ik}\lambda_k)_{;i}\, d^4v = \int T^{ik}_{;i}\lambda_k d^4v + \int T^{ik}\lambda_{i;k}\, d^4v. \qquad (8.19)$$

It can be seen from Eq. (8.19) that an integral conservation law for a symmetric tensor is meaningful only along certain directions, namely, along the trajectories of the vector field λ_k. If a Riemannian space does not admit the existence of any vector field (which is possible in the general case), then it is impossible to find integral conservation laws in this space. It is true that this limitation is unimportant for a pseudo-Riemannian space having a metric with the Minkowski signature, since the condition of the existence of such a metric is identical with the condition of the existence of a continuous field of directions on the manifold. Thus, in the spaces of general relativity some vector field always exists, and it is therefore always possible to construct integral conserved quantities (at least one). The properties of an integral conserved quantity are determined by the character of the vector field λ_k. Therefore integral conserved quantities, even when obtained by integration of the differential conservation law for the energy—momentum tensor, may not form the energy—momentum vector, but may have a completely different meaning. The concept of 4-momentum (the energy—momentum vector) can be introduced only in the special case when there exist four mutually commuting vector fields whose components reduce to the form $\lambda_k = \delta^k_n$.

As an example, consider scalar particles in a space with the de Sitter metric. For a scalar field, all conserved currents have the form [see Eq. (4.8)] $J^\mu_a = T^\mu_\nu \xi^\nu_a$, where ξ^ν_a is determined by the coordinate transformations $\delta x^\mu = \xi^\mu_a \varepsilon^a$. Suppose that λ_k determines the group of motions of space-time, i.e., is one of the Killing vectors: $\lambda_k = \xi_{ak}$, with $\xi_{a(k;i)} = 0$. Then the relation (8.19) takes a form which is reminiscent of Eq. (8.18):

$$\int T^{ik}_{;i} \underset{a}{\xi_k}\, d^4v = \oint T^{ik} \underset{a}{\xi_k}\, \varepsilon(N) N_i\, d^3v. \qquad (8.20)$$

In the particular case of flat space, ξ^k_a has the form: 1) for displacements, $\xi^k_{(l)} = \delta^k_l$; 2) for rotations, $\xi^k_{in} = \underset{(in)}{L^k_l} x^l$, where $\underset{(in)}{L^k_l} = \delta^k_n g_{il} - \delta^k_i g_{nl}$ is the matrix of Lorentz rotations. From Eq. (8.20), we obtain in the first case

$$\int T^{i(k)}_{;i} d^4v = \oint T^{i(k)}\varepsilon(N) N_i d^3v. \qquad (8.21)$$

If the covariant conservation law $T^{i(k)}_{;i} = 0$ is satisfied, then $\oint T^{i(k)}\varepsilon(N)N_i d^3v = 0$. By integrating over an infinitely remote surface, where T^{ik} vanishes, this result is usually used to obtain an integral conservation law for the 4-momentum P^i:

$$P^{(i)} = \int T^{(i)0}\varepsilon(N)N_0 d^3v. \tag{8.22}$$

We note that the vector index of $P^{(i)}$ is in fact the group index. It has been used to label the Killing vectors corresponding to displacements. Therefore it is placed in parentheses.

In the case of rotations of flat space, Eq. (8.20) gives

$$\int T^{ik}_{;i} \underset{(pq)}{L_{kn}} x^n d^4 v = \int (T^{ik} \underset{(pq)}{L_{kn}} x^n)_{;i} d^4 v - \int T^{ik} \underset{(pq)}{L_{ki}} d^4 v =$$

$$\oint T^{ik} \underset{(pq)}{L_{kn}} x^n \varepsilon(N) N_i d^3 v. \tag{8.23}$$

If the tensor T^{ik} is symmetric, this leads to the conservation law for the angular momentum of the system, $M_{(pq)} = \int M^i_{(pq)}\varepsilon(N)N_i d^3v$:

$$\int \underset{(pq)}{M^i_{;i}} d^4 v = \oint \underset{(pq)}{M^i} \varepsilon(N) N_i d^3 v, \tag{8.24}$$

where

$$M^i_{(pq)} = T^i_p x_q - T^i_q x_p. \tag{8.25}$$

In a space with constant curvature (de Sitter space), the element of length has the form $-ds^2 = \varphi^2 dx^{i^2}$ in stereographic coordinates. Here $\varphi = [1 + (r^2 - x_0^2)/4R^2]^{-1}$, where $r^2 = (x^1)^2 + (x^2)^2 + (x^3)^2$. The de Sitter group is isomorphic to the group of rotations O(5). Its generators can be represented in the form

$$\Pi_l = \varphi^{-1}P_l + (x_k/2R^2)L_{kl}; \quad L_{ik} = x_i P_k - x_k P_i, \tag{8.26}$$

where $P_k = -i\partial/\partial x^k$. Thus, it is only in the limit of flat space ($R \to \infty$) that the operator of 4-momentum Π_l acquires the meaning of an operator of displacement with respect to the coordinates. In curved space, "momentum" is always mixed with "angular momentum." To avoid this, it would be necessary to choose as the group of motions four mutually commuting operators and call them the operators of 4-momentum. However, this is impossible, since the de Sitter group is simple and has rank 2. As a result, the 4-momentum and total angular momentum of the system become components of a 5-dimensional total angular momentum of the system.

Let us write the integral conserved quantities, assuming, as usual, that $T^{ik} = 0$ at spatial infinity. For the components of the 4-dimensional angular momentum, we obtain the usual expression

$$M_{lm} = \int M^0_{(lm)}\, \varepsilon(N) N_0 d^3v = \int \varphi^4 M^0_{(lm)} d^3x = \int (M_{0(lm)}/\sqrt{g_{00}})\varphi^3 d^3x, \quad (8.27)$$

where $M^0_{(lm)}$ is defined by Eq. (8.25), but x now denotes not Cartesian, but stereographic coordinates.

The generalized displacements Π_l give four Killing vectors, which determine the conserved (under the same conditions at spatial infinity) generalized 4-momentum with components

$$\begin{aligned} P^{(0)} = \int \sqrt{-g}\, T^0_k \xi^k_0\, d^3x = \int \varphi^3 T^0_0\, d^3x - \int (r^2/2R^2)\, \varphi^4 T^0_0\, d^3x + \\ \int \varphi^4 (x_0 x_\alpha/2R^2)\, T^0_\alpha\, d^3x = \int \varphi^3 T^0_0\, d^3x - \\ {}^1/_2 \int (x_\alpha M_{0\,(0\alpha)}/\sqrt{g_{00}}\, R^2)\, \varphi^3\, d^3x, \quad \alpha = 1, 2, 3; \end{aligned} \quad (8.28)$$

$$\begin{aligned} P^{(\alpha)} = \int \sqrt{-g}\, T^0_k \xi^k_\alpha\, d^3x = \int \varphi^3 T^0_\alpha\, d^3x - {}^1/_2 \int \varphi^4 T^0_\alpha R^{-2} \sum_{k \neq \alpha} x_k^2\, d^3x + \\ {}^1/_2 \int \varphi^4 \sum_{k\neq\alpha} T^0_\alpha (x_k x_\alpha/R^2)\, d^3x = \int \varphi^3 (T_{0\alpha}/\sqrt{g_{00}})\, d^3x - \\ {}^1/_2 \int (x^k M_{0(\alpha k)}/\sqrt{g_{00}}\, R^2)\, \varphi^3\, d^3x. \end{aligned} \quad (8.29)$$

These expressions are meaningful in the case when the metric of space-time is assumed to be rigidly fixed and not related to the distribution of matter. But if we consider a geometrized theory (the general theory of relativity), T^{ik} is necessarily proportional to the metric tensor g^{ik}, and instead of (8.27)—(8.29) we obtain

$$M_{(0\alpha)} = \int \sqrt{-g}\, \xi^0_{(0\alpha)}\, d^3x = \int \varphi^4 x_\alpha\, d^3x \quad (8.30)$$

(the other components of "angular momentum" are equal to zero) and

$$\begin{aligned} P^{(0)} = \int \varphi^3 d^3 x - \int \frac{r^2}{2R^2} \varphi^4\, d^3x = \int \frac{(1-(r^2+x_0^2))/4R^2}{1+(r^2-x_0^2)/4R^2} \varphi^3\, d^3x; \\ P^{(\alpha)} = \frac{1}{2} \int \frac{x_0 x_\alpha}{R^2 [1+(r^2-x_0^2)/4R^2]} \varphi^3\, d^3x. \end{aligned} \quad (8.31)$$

All the indices on the integral quantities $P^{(i)}$ and $M_{(ik)}$ are group indices, so that $P^{(i)}$ and $M_{(ik)}$ do not change under transformations of the coordinate system. But they depend on the choice of the surface of integration and the basis ξ^μ_a in the Lie algebra of the group of motions. Thus, the integral conserved quantities corresponding to space-time symmetry properties form a multiplet which gives a realization of the regular representation of the Lie algebra of the group of motions of V_4. In Minkowski space, the existence of the energy—momentum 4-vector as an independent integral characteristic of the system is related to two "fortuitous" circumstances: first, the group of motions of Minkowski space (the Poincaré group) has a 4-dimensional invariant subgroup; secondly, Minkowski space itself can be identified with the group space of this invariant subgroup. Therefore the vectors P^i of the Lie algebra of the Poincaré group become space-time vectors. But in the general case, the integral conserved quantities are vectors of a fiber space over V_4, where as the fiber we find the group space of the group of motions. If the structure of the group of motions of a curved space is such that V_4 itself can be identified with this group or with its invariant subgroup, the integral conserved quantities **P** and **M** are space-time vectors, as in flat space. However, they have fewer components than in flat space, since the group of motions of a Riemannian space is smaller than that of a pseudo-Euclidean space.

REFERENCES

[1]S. L. Adler and R. F. Dashen, Current Algebras, Benjamin, New York (1968). V. de Alfaro, S. Fubini, G. Furlan, and C. Rossetti, Currents in Hadron Physics, North-Holland, Amsterdam (1973).

[2]E. Noether, Nachr. Ges. Wiss. Goettingen, Math.-Phys. Kl. 2, 235 (1918).

[3]R. Utiyama, Phys. Rev. 101, 1597 (1956).

[4]N. P. Konopleva and G. A. Sokolik, in: Problemy teorii gravitatsii i élementarnykh chastits (Problems of the Theory of Gravitation and Elementary Particles, Atomizdat, Moscow (1966), p. 22.

[5]N. P. Konopleva, in: Gravitatsiya i teoriya otnositel'nosti (Gravitation and the Theory of Relativity), Nos. 4—5, Kazan State University (1968), p. 67.

[6]V. I. Ogiyevetsky and I. V. Polubarinov, Ann. Phys. (N.Y.) 25, 358 (1963); Nucl. Phys. 76, 677 (1966).

[7]S. L. Glashow and M. Gell-Mann, Ann. Phys. (N.Y.) 15, 437 (1961).

[8]A. G. Iosif'yan and N. P. Konopleva, Dokl. Akad. Nauk SSSR 198, 1036 (1971) [Sov. Phys. Dokl. 16, 432 (1971)].

[9]V. I. Ogievetskiĭ and I. V. Polubarinov, in: Sovremennye problemy gravitatsii (Contemporary Problems of Gravitation), Tbilisi State University (1967), p. 430.

[10]N. P. Konopleva, in: Tezisy dokladov III Mezhvuzovskoĭ nauchnoĭ konferentsii po problemam geometrii (Abstracts of Contributions to the 3rd Inter-University Scientific Conf. on Problems of Geometry), Kazan State University (1967), p. 80.

[11]K. Yano, The Theory of Lie Derivatives and its

Applications, North-Holland, Amsterdam (1957).

[12]N. P. Konopleva, in: Problemy teorii gravitatsii i élementarnykh chastits (Problems of the Theory of Gravitation and Elementary Particles), No. 3, Atomizdat, Moscow (1970), p. 103.

[13]N. P. Konopleva, Dokl. Akad. Nauk SSSR 190, 1070 (1970) [Sov. Phys. Dokl. 15, 134 (1970)].

[14]J. L. Synge, Relativity: The General Theory, North-Holland, Amsterdam (1960).

[15]I. M. Gel'fand and S. V. Fomin, Variatsionnoe ischislenie, Fizmatgiz, Moscow (1961) [English translation: Calculus of Variations, Prentice-Hall, Englewood Cliffs, N. J. (1963)].

[16]R. F. O'Connell and D. R. Tompkins, Nuovo Cimento 38, 1088 (1965); J. Math. Phys. 6, 1952 (1965).

[17]T. W. B. Kibble, J. Math. Phys. 2, 212 (1961).

[18]N. P. Konopleva and G. A. Sokolik, Dokl. Akad. Nauk SSSR 154, 310 (1964) [Sov. Phys. Dokl. 9, 60 (1964)].

[19]N. P. Konopleva, Vest. Mosk. Univ., Ser. Fiz., No. 3, 73 (1965).

[20]V. B. Berestetskiĭ, in: Trudy Pervoĭ shkoly fiziki ITÉF (Proc. of the 1st School of Physics at the Institute of Theoretical and Experimental Physics, Moscow), Atomizdat, Moscow (1973), p. 3.

[21]V. I. Ogievetskiĭ and I. V. Polubarinov, Yad. Fiz. 4, 216 (1966) [Sov. J. Nucl. Phys. 4, 156 (1967)].

CHAPTER III. GEOMETRICAL THEORY OF GAUGE FIELDS

§9. GAUGE FIELDS AND A UNIFIED GEOMETRICAL THEORY OF INTERACTIONS

The usual goal of every geometrical approach to interactions is to find a space in which the studied fields become standard geometrical objects. This makes it possible to apply the powerful and well-developed methods of geometry, algebra, and topology to study the properties of the solutions of the classical equations of motion and field equations, which are frequently nonlinear. The space which is found is often regarded as a formal mathematical configuration space.

A more profound approach to the geometrization of interactions is related to the work of Poincaré[1] and Einstein,[2] who defined the properties of space-time in terms of the properties of measuring instruments and procedures (see Chapter I). By developing these ideas, it can be shown that the geometrization of any interaction implies a transition to frames of reference (or measuring instruments) which move in a force-free manner in the corresponding field. Such frames of reference are said to be inertial with respect to the interaction in question. If, in such a frame of reference, a way of measuring the coordinates of the studied objects without altering the state of inertial motion of the frame of reference can be found, then it is possible to construct a reference space associated with this frame of reference.

The set of reference spaces constructed in this way does not in general form any single space-time. This shows up in the fact that in one of the inertial frames of reference other inertial frames of reference may appear to move noninertially. To take into account and describe this possibility, it is assumed in geometrical theories that there exists some unique space-time in which the trajectories of all inertial frames of reference are geodesics. Then the change in the distance between inertial frames of reference can be described by the equation of geodesic deviation. A nonzero value of the second derivative of the geodesic deviation indicates the existence of a real physical field in which the inertial frames of reference move in a force-free manner. This physical field is described in terms of the curvature or torsion of space-time, i.e., it manifests itself in its nonholonomic character. As is well known, this is the situation in Einstein's theory of gravitation.

Einstein's conception admits a generalization to arbitrary types of interactions. Space-time is then generalized in such a way that the properties of locally inertial frames of reference are described not only by space-time symmetries, but also by internal symmetries. The system of geodesics of 4-dimensional Riemannian space is replaced by more complex force-free configurations — geodesics in a fiber space over V_4. The general principle of constructing spaces in which a given interaction or several interactions vanish makes it possible to express the content of the classical theory of any gauge field in purely geometrical form and to reduce it to the theory of connections of the principal fiber space over V_4. This permits a unified treatment of a number of theories which at first sight are unrelated to one another. We have in mind here the search for a unified treatment of strong, electromagnetic, weak, and gravitational interactions.[3,4]

It is simplest to imagine a fiber space[5] as a combination of two spaces connected by a special mapping. Namely, one of the two spaces is chosen as the base of the fiber space, and a copy of the second space (the fiber) is projected to each point of the base. Thus, a fiber space can be regarded as an ordinary n-dimensional space (the base), whose points are replaced by some m-dimensional spaces F_x, which in general have their own geometry. It is assumed that in each fiber F_x there acts a group G_r which does not affect the base. The base is a manifold which is invariant with respect to the action of the group G_r. The simplest example of a fiber space is the manifold of trajectories of some intransitive group. In this case, the points of the base are ordered according to the invariants of the intransitive group. A fiber is a trajectory of the group on which the group acts transitively.* The Riemannian space of the general theory of relativity can also be regarded as a fiber space if the fiber is taken to be the tangent space to V_4, i.e., use is made of a tangent bundle. As a result, gravitation, not only as a physical field in flat space, but also in the geometrical treatment, comes into line with other fields.

---

*Transitivity means that any point of a space can be carried into any other point by means of the transformations belonging to a given group. Intransitivity means that this is not possible, for example, if the group leaves certain points fixed (a center of rotation).

The geometrical approach makes it possible to construct a classical theory of all gauge fields according to a single principle, changing only G_r in going from one type of interaction to another. If each type of interaction is associated with its own symmetry group (a local gauge group), it is possible to construct a hierarchy of interactions, in which the transition to a stronger interaction involves an enlargement of the symmetry group, while a weakening of the interaction involves a violation of the symmetry and a reduction of the gauge group. However, the feasibility of such a classification of interactions has its limitations, since, as we shall show in §12, there exists a dependence between the space-time and internal symmetry properties of the solutions of the field equations through the structure of the holonomy group, which is a subgroup of the gauge group. The relation between the type of interaction and the internal symmetry is nonunique, since, for example, different fields can have spherically symmetric potentials, while all spherically symmetric solutions correspond to an Abelian holonomy group.[6] In the geometrical approach, a decisive role is played by the tensor dimensions of the fields. These determine the transformation properties of the fields and the form of the equations.

A particular case of a fiber space is a manifold of surfaces embedded in a space of higher dimension. Embedding makes it possible to give a geometrical interpretation of the internal symmetries and mass of a gauge field.[7-10] If a vector potential A_μ^a is identified with the nondiagonal components of the metric $g_{a\mu}$ of a $(r+4)$-dimensional embedding space, we obtain a generalization to gauge fields of the unified theory of gravitation and electromagnetism due to Kaluza and Klein.[11] This approach was first proposed in 1965 by DeWitt[12] and was subsequently developed by Kerner,[13] Trautman,[14] and Cho.[15] Embedding provides a convenient method of studying the topological properties of a fiber space.

A more natural geometrical interpretation of gauge fields is to identify the vector potentials A_μ^a with the connection coefficients of the principal fiber space whose base is Riemannian space-time, the fiber being a finite gauge Lie group G_r.[16,17] In this case, the stress tensor $F_{\mu\nu}^a$ of the gauge field becomes the curvature tensor of the fiber space. The equations of motion of particles interacting with the gauge field acquire the meaning of free equations in the fiber space. This approach reproduces all the results and equations of the Lagrangian variational theory of gauge fields, and it makes it possible to classify the solutions of these equations according to their algebraic and topological properties. A unified theory of all types of interactions using the geometrical interpretation of gauge fields as connection coefficients of a fiber space over V_4 was proposed for the first time in 1965 by Konopleva and Sokolik.[3] A classification of the solutions of the Yang—Mills equations by means of the concept of the holonomy group of a fiber space was first given by Loos[6] in 1965. The algebraic classification of topologically nontrivial, asymptotically flat solutions of the Yang—Mills equations in Euclidean space V_4 was first given by Atiyah and Ward[18] in 1977 and was subsequently generalized by Atiyah, Hitchin, Drinfeld, and Manin[19] in 1978.

As a consequence of the universality of the approach, for all types of gauge fields we find equations of one and the same type, which in the case of isotopic symmetry are well known as the Yang—Mills equations. These equations are reminiscent of Maxwell's equations, but they are highly nonlinear, since they take into account "self-action" of the fields. In other words, each gauge field can act not only on particles and on other objects which are external in relation to the field, but also on itself. The self-action of a field is determined by the structure of its corresponding gauge group. In the case of the electromagnetic field, the gauge group is a one-parameter group, and the corresponding equations are therefore identical with Maxwell's equations (there is no self-action). As a consequence of the nonlinearity of the equations of gauge fields, not all their classical solutions can be interpreted quantum mechanically as solutions describing elementary particles. This transition, the possibility of which is assumed in the correspondence principle, can actually be made only for special configurations of fields which are asymptotically flat. Special interest attaches to particle-like solutions of the classical equations of gauge fields (instantons, having finite action; solitons, having finite energy; and monopoles, vortices, and other defects of the topology of manifolds). Instantons play a major role in quantum field theory, solitons are important in theories with spontaneous symmetry breaking, and the properties of defects in ordered media determine the parameters of phase transitions.

The existing geometrical methods of describing interactions can be classified in a natural way according to the method of constructing the geometry of space-time and the space of internal variables. If the corresponding spaces are homogeneous, the geometrical approach reduces to the algebraic classification of their symmetry groups and the representations of these groups. When using algebraic methods, it is frequently necessary to go over to spaces of higher dimensions, in which curved space-time or the internal space appear as a surface. The Riemannian approach gives, for the characteristics of the manifold, and hence of the interactions, the following objects: the metric tensor (not necessarily symmetric), the connection coefficients (in general, nonsymmetric), and the curvature tensor and its contractions. In Einstein's theory, a symmetric metric tensor is used to describe the properties of the gravitational field. The connection coefficients are not independent, but are determined by the condition that the covariant derivative of the metric tensor is equal to zero. The antisymmetric part of the connection coefficients — the torsion tensor — is taken to be equal to zero. The fact that the covariant derivative of the metric tensor is equal to zero ensures equal-volume transport in a holonomic coordinate system.

Soon after the creation of Einstein's theory, the possibilities of Riemannian geometry which were not used in general relativity attracted the attention of theoreticians, who were striving to find a unified picture of the world. First, there appeared geometrical theories of the electromagnetic field, using the torsion of space-time.[2,20] The geometrical approach also seemed most natural for the creation of a unified theory of classical fields describing not only gravitation and the electromagnetic field individually, but also the interaction between them.[21] The search for a geometrical description of the electromagnetic field was made in four main directions: 1) generalization of parallel transport of vectors in 4-dimensional space-time; 2) generalization of the space-time metric; 3) simultaneous solution of the Maxwell and Einstein equations in the Riemannian space of general relativity as an already unified system of equations of the gravitational and electromagnetic fields (geometrodynamics); 4) generalization of space-time itself, i.e., introduction of an additional, fifth dimension.

In the first case, the connection coefficients of Riemannian space-time V_4 were generalized. For this purpose, in addition to the Christoffel symbols $\Gamma^{\lambda}_{\mu\nu}$ corresponding to the gravitational field (according to Einstein), additional connection coefficients Γ_{μ}, which were identified with the vector potential of the electromagnetic field A_{μ}, were introduced. Thus, space-time V_4 was endowed with a Weyl geometry. In Weyl space, parallel transport changed the length of vectors. This additional transformation (scale gauge) was identified with the gauge transformations in electrodynamics.

Geometrical unified theories with nonsymmetric connection coefficients (torsion) were once again proposed in the 1950s and 1960s. One of the gauge theories of gravitation (Kibble, 1961) also led to nonsymmetric connection coefficients in V_4, but in contrast with earlier work, Kibble identified the torsion with the tensor of the spin density of matter rather than with electromagnetism. An approach which seems very interesting was proposed by Schouten,[22] who used torsion to describe the electromagnetic properties of a continuous medium. Although this approach is macroscopic, it permits a description of not only the properties of a continuous medium, but also the variation of these properties with temperature. This approach was further developed, for example, in Refs. 23 and 24, where it was shown that the torsion can be related to the density of dislocations in the medium. In Ref. 25 it was shown that the torsion can be related to the density of sources of the electromagnetic field. In Ref. 26 invariant quadratic relations were obtained between energy and momentum and between momentum and the stress tensor of the electromagnetic field in a medium, and it was shown that the corresponding well-known linear relations, which are also used in field theory, may not hold in a nonhomogeneous, anisotropic, dispersive medium.

In 1945 Einstein[2] proposed a unified theory of gravitation and electromagnetism, using a nonsymmetric metric tensor $g_{\mu\nu} = g_{(\mu\nu)} + g_{[\mu\nu]}$. In this theory, the electromagnetic field is described by the antisymmetric part of the metric, which is identified with the stress tensor $F_{\mu\nu}$ of the electromagnetic field. The connection coefficients in this theory are also nonsymmetric and include the electromagnetic field. The formalism becomes very cumbersome and complicated here, owing to the ambiguity of the operations of covariant differentiation and the raising and lowering of indices. However, the nonsymmetric theory gives a very interesting dependence between the gravitational and electromagnetic fields. As a result of this dependence, for example, a neutral massive rotating star may, in the course of time, acquire a charge, i.e., an electromagnetic field. Moreover, there exists an upper limit on the electromagnetic field which a gravitating system can have.[21] No other theory gives such a limit.

The geometrodynamics of Rainich, Wheeler, and Misner[27,28] makes use of the ordinary Riemannian geometry of the general theory of relativity, but it generalizes its equations. The system of Einstein equations is of second order in the derivatives of $g_{\mu\nu}$. But if in the combined system of Einstein—Maxwell equations the electromagnetic variables are eliminated, a system of fourth-order equations is obtained for $g_{\mu\nu}$. This system contains sufficient information for a purely geometrical description of both the gravitational and electromagnetic fields. The electromagnetic field in this case is related to the rate of change of the Riemannian curvature of space-time. The traces left by the electromagnetic field on the metric of V_4 are so characteristic that it is possible to reconstruct from them the properties of the field which produced them. In geometrodynamics, there can exist stable, purely field configurations — geons. For an external observer, geons behave like gravitating masses, but there are actually no singularities inside them. They correspond to regular solutions of the Einstein—Maxwell equations and "consist" of closed lines of force of the field. Electric charge also acquires a purely geometrical interpretation in geometrodynamics if allowance is made for the topological properties of V_4. If space-time is given a non-Euclidean topology (with "handles" and "holes"), the flux of the lines of force through each topological handle behaves like a classical electric charge from the point of view of an external observer situated near one neck of the handle. However, the size of this charge has no direct relation to the charges of the elementary particles, such as the electron.

The geometrodynamics of Rainich, Wheeler, and Misner shows that classical physics, once it includes Einstein's theory of gravitation and Maxwell's electrodynamics, is a purely geometrical theory which is unified in a natural way. In other words, classical physics is an aspect of geometry. At the present time, the solutions of the equations of geometrodynamics are being actively investigated, in particular, by the method of Newman and Penrose.[29] The system of Yang—Mills and Einstein equations, considered jointly, also possesses interesting particle-like solutions.[30] However, the Yang—Mills equations in flat V_4 do not give stable configurations like geons.[31] The generalization of geometrodynamics to the case of gauge fields was also considered in Refs. 4 and 32.

The gravitational field, regarded as a gauge field, can correspond to several symmetry groups: 1) the general covariant group; 2) the local Lorentz group; and 3) the group of scale transformations of the interval. In the first case, its properties are determined by the properties of the metric tensor, and this gives the usual Einstein theory; in the second case, they are determined by the properties of the Ricci connection coefficients, and this leads to field equations of fourth order. In the third case, it is assumed that the source of the field is the trace of the energy—momentum tensor and that the carriers are scalar particles. Consequently, the approach based on gauge symmetries can lead to more general theories of gravitation than that of Einstein. The generalized theory of gravitation in which the role of the field variables is played by connection coefficients is similar in its structure to Maxwell's electrodynamics.

Having established the relation between different gauge fields and gravitation, or the structure of V_4, it is possible, first, to take into account the effect of nongravitational interactions on the geometry of space-time, and second, to interpret the internal symmetries and masses of gauge fields as an effective manifestation of the curvature of space-time at small distances. The first aspect reduces geometrically to the properties of the projection of the space in which the gauge field acts onto the tangent space to V_4. The second aspect is related to the properties of the embedding of Riemannian space-time V_4 in a flat space of higher dimension. The additional dimensions which appear here become the arena for the action of the internal symmetries.

Geometrical methods are flourishing in field theory and in the theory of continuous media, and what was once an exotic occupation is now becoming a working tool of theoreticians. In this chapter, we shall show that all the existing methods for the geometrical description of interactions can be formulated in terms of the geometry of fiber spaces.

§10. EXTERIOR FORMS ON A MANIFOLD AND THE STRUCTURE EQUATIONS OF SPACE

Maxwell's Electrodynamics and Exterior Forms. The form of the physical laws is usually established by means of tensor analysis. However, tensor analysis requires the introduction of nonsingular coordinate systems in which the components of vectors and tensors can be specified. At the same time, according to the definition of a differentiable manifold, one nonsingular coordinate system is insufficient to cover a manifold which is topologically not equivalent to an open set in Euclidean space. Therefore, for example, in an arbitrary differentiable manifold it is impossible to describe a field such as the electromagnetic field by specifying its components $F_{\mu\nu}$ in any one concrete coordinate system. If the components are used, they must be specified with respect to the coordinates of several coordinate systems (or local maps), which together cover the entire manifold.

At each point of a neighborhood, there exists

considerable freedom of choice of the coordinate systems. Thus, the components of the electromagnetic field intensity in different coordinate systems are in themselves not so important as the concept obtained by abstraction from them — the intrinsic value F of the field intensity.* To understand the meaning of this term, consider vectors in a manifold. The statement that a vector** v is known means that we can if necessary specify its components in any nonsingular coordinate system. The relation between the vector v and its components is expressed by the equation

$$\mathbf{v} = v_m \mathbf{e}^m. \tag{10.1}$$

The components v_m depend on the choice of the vectors $\mathbf{e}^m$ of the basis of the coordinate system. The simplest basis is formed from the gradient vectors dx^m of the coordinate functions:

$$\mathbf{v} = v_m dx^m. \tag{10.2}$$

Covariant vectors are linear combinations of differentials of the coordinates and are known as differential forms of rank 1, 1-forms, or *Pfaffian forms*.[33]

Instead of using differentials of the coordinates as the basis vectors, we can take an arbitrary set of n linearly independent vectors ω^α (here the index labels the vector, and not its component). With respect to this set of basis vectors, the components of a vector v are defined by the expansion

$$\mathbf{v} = v_\alpha \omega^\alpha. \tag{10.3}$$

The components of the vector gradient in this basis are known as the *Pfaffian derivatives* with respect to ω^α: $\operatorname{grad} f = f_\alpha \omega^\alpha$. A set of n vectors ω^α is sometimes called a nonholonomic basis, in contrast to the holonomic basis formed from the first differentials of the coordinates.

To describe the electromagnetic field, we must use forms of higher rank, or 2-forms. The simplest 2-form α can be represented as an exterior product of two 1-forms ($\mathbf{u} = u_\alpha \omega^\alpha$ and $\mathbf{v} = v_\alpha \omega^\alpha$):

$$\alpha = \mathbf{u} \wedge \mathbf{v} = {}^1\!/_2 (u_\alpha v_\beta - u_\beta v_\alpha) \omega^\alpha \wedge \omega^\beta = {}^1\!/_2 \alpha_{\alpha\beta} dx^\alpha \wedge dx^\beta. \tag{10.4}$$

The operation of exterior product $\wedge$ generalizes the vector product to the case of multiplication of vectors and antisymmetric tensors of arbitrary rank in such a way that after the multiplication we again obtain antisymmetric

*The intrinsic value arises in going over to a coordinate-free expression of the fields on a manifold in terms of p-forms. It is independent of the choice of coordinate system.
**More precisely, a covariant vector.

tensors. The exterior product $\wedge$ is determined by the following requirements: 1) associativity; 2) distributivity, and also, for the product of two vectors, anticommutativity, i.e., $u \wedge v = -v \wedge u$. In particular, $u \wedge u = 0$.

An exterior product of p vectors or a linear combination of these vectors forms a p-form. If each vector is expressed in terms of ω^α, the p-form can be written as

$$a = \sum_{\alpha_1 < \alpha_2 < \ldots < \alpha_p} a_{\alpha_1 \alpha_2 \ldots \alpha_p} \omega^{\alpha_1} \wedge \omega^{\alpha_2} \wedge \ldots \wedge \omega^{\alpha_p} =$$

$$(p!)^{-1} a_{\alpha_1 \alpha_2 \ldots \alpha_p} \omega^{\alpha_1} \wedge \omega^{\alpha_2} \wedge \ldots \wedge \omega^{\alpha_p}. \qquad (10.5)$$

The tensor components are the coefficients of the p-form [in Eq. (10.5), $a_{\alpha_1 \ldots \alpha_p}$]. On a manifold, p-forms, unlike the tensor coefficients, can be chosen in an invariant manner. In addition, it is possible to define differentiation of p-forms in an invariant manner (exterior differentiation).

The operator D of exterior differentiation, when applied to a scalar function (a 0-form), gives a vector (a 1-form). For p-forms (10.5) of higher rank, this operation is generalized by means of the definition[33]

$$(Da) = (p!)^{-1} da_{\alpha_1 \ldots \alpha_p} \wedge dx^{\alpha_1} \wedge \ldots \wedge dx^{\alpha_p}.$$

In terms of tensor components,

$$(Da)_{\alpha_1 \ldots \alpha_{p+1}} = \Sigma (-1)^P \partial a_{\beta_2 \ldots \beta_{p+1}} / \partial x^{\beta_1},$$

where P has the value 0 or +1, depending on whether the permutation of the indices $\alpha_1, \ldots, \alpha_{p+1}$ which gives the indices $\beta_1, \ldots, \beta_{p+1}$ is even or odd.

The operation D, or the *exterior derivative*, is an antisymmetric differentiation which generalizes the ordinary operation of taking the curl of a vector. It is linear: $D(a_1 + a_2) = Da_1 + Da_2$. When applied to a product in which the first factor is a p-form, it gives

$$D(a \wedge b) = (Da) \wedge b + (-1)^p a \wedge Db.$$

Repeated application of exterior differentiation gives zero:

$$D(Da) = 0.$$

Maxwell's equations can be regarded as equations for the coefficients $F_{\mu\nu}$ and $^*F_{\mu\nu}$ (the dual tensor) of two 2-forms[28]

$$f = {}^1\!/_2 F_{\mu\nu} dx^\mu \wedge dx^\nu; \qquad (10.6)$$

$$^*f = {}^1\!/_2 {}^*F_{\mu\nu} dx^\mu \wedge dx^\nu. \qquad (10.7)$$

The exterior derivative of the form (10.6) is given by

$$D\mathfrak{f} = {}^1/_2\,(\partial F_{\alpha\beta}/\partial x^\nu)dx^\nu \wedge dx^\alpha \wedge dx^\beta = {}^1/_6\,(\partial F_{\alpha\beta}/\partial x^\nu + \partial F_{\beta\gamma}/\partial x^\alpha + \partial F_{\gamma\alpha}/\partial x^\beta)dx^\nu \wedge dx^\alpha \wedge dx^\beta. \tag{10.8}$$

According to Maxwell's equations, the coefficients of the 3-form (10.8) reduce to zero: $(\partial F_{\alpha\beta}/\partial x^\nu + \partial F_{\beta\gamma}/\partial x^\alpha + \partial F_{\gamma\alpha}/\partial x^\beta) = 0$. Hence

$$D\mathfrak{f} = 0. \tag{10.9}$$

Equation (10.9) is satisfied identically if $\mathfrak{f} = DA$, i.e.,

$$F_{\alpha\beta} = (\partial A_\beta/\partial x^\alpha) - (\partial A_\alpha/\partial x^\beta).$$

The remaining Maxwell equations in the absence of sources imply that the exterior derivative of the 2-form (10.7) is equal to zero:

$$D{}^*\mathfrak{f} = 0. \tag{10.10}$$

Equations (10.9) and (10.10) constitute a coordinate-free expression of Maxwell's equations in an arbitrary manifold with arbitrarily chosen coordinate systems.

Integration on an arbitrary manifold is not a completely well-defined and unique operation if it is applied to arbitrary tensors. An integration is unique only when a p-form is integrated over a p-dimensional space. In this case, the integration can be performed in such a way that the result is a scalar quantity, which is independent of the coordinate system. Therefore the integral conservation laws in electrodynamics which have invariant meaning are the following:

$$\oint_{c^2} \mathfrak{f} = 4\pi p \text{ (p is the magnetic charge);}$$

$$\oint_{c^2} {}^*\mathfrak{f} = 4\pi q \text{ (q is the electric charge);}$$

c^2 is a closed 2-dimensional surface.

The fact that a vector potential exists means that there is no magnetic charge: $\int_{c^2} \mathfrak{f} = \int_{c^2} D\mathbf{A} = \int_{\partial c^2} \mathbf{A} = 0$, since $\partial c^2 = 0$ (∂c^2 is the boundary of c^2).

If the manifold is metrized, we can introduce an operation of symmetric differentiation of forms δ, which generalizes the operation of taking the divergence of a vector:

$$(\delta\omega)_{k_1 \ldots k_{p-1}} = -\nabla^l \omega_{lk_1 \ldots k_{p-1}}, \; p \text{ is the rank of } \omega.$$

In terms of the invariant operators D and δ, we can define the invariant wave operator $\Delta = D\delta + \delta D$.[33] The operator Δ is a topologically and metrically self-adjoint operator, i.e., the following scalar products are equal:

$(\Delta\alpha, \beta) = (\alpha, \Delta\beta) = (\alpha, *\Delta*^{-1}\beta)$, where the asterisk denotes the operation defined by

$$(*\,\omega)_{j_1 \ldots j_{n-p}} = \delta^{1 \ldots n}_{i_1 \ldots i_p\, j_1 \ldots j_{n-p}}\, \omega^{i_1 \ldots i_p}.$$

The indices of the coefficients $\omega^{i_1 \cdots i_p}$ are raised by means of the metric g^{ik}, and $\delta^{1 \cdots n}_{i_1 \ldots i_n}$ is the n-dimensional Kronecker symbol.

In Euclidean space, solutions of equations of the form $\Delta\omega + \mu = 0$, where μ is a form, are always harmonic forms.

Using the invariant operators D, δ, and Δ, it is easy to show that the free Maxwell equations are not wave equations. They become wave equations under the supplementary conditions $D\delta A = 0$, a particular case of which is the Lorentz conditions $\delta A = 0$. The Lorentz conditions are not the only conditions which distinguish wave solutions. For example, if the vector potential A_μ coincides with the Killing vector, i.e., $\partial_\mu A_\nu = -\partial_\nu A_\mu$, then the condition $\delta A = 0$ is satisfied and leads to wave equations for A_μ. In the general case, wave equations for a vector field have the form

$$\Delta A_\mu = (D\delta + \delta D)\, A_\mu = F_{\mu\nu}{}^{;\,\nu} - (A_\nu{}^{;\,\nu})_{;\,\mu} = -\nabla_\nu \nabla^\nu A_\mu - R^\nu_\mu A_\nu = 0.$$

They can be obtained from the Lagrangian if the Lorentz conditions are taken into account explicitly in the Lagrangian. Here R^ν_μ is the Ricci tensor, and ∇_ν denotes the covariant derivative.

Green's functions can always be constructed for a topological Laplacian. Therefore it is useful to generalize the topological Laplacian to forms with values in some Lie algebra. Such a generalization, by means of an "extension" of the covariant derivative, has been proposed and used for an invariant formulation of gravitational waves.[4,34] The operators D and δ defined in terms of extended covariant derivatives have the same properties as the de Rham operators, i.e., $D^2 = \delta^2 = 0$, $\Delta = D\delta + \delta D$, $D\Delta = \Delta D$, and $\delta\Delta = \Delta\delta$. If the topological de Rham Laplacian in the coordinate representation in Riemannian space has the form

$$(\Delta\omega)_{k_1 \ldots k_p} = -\nabla^i \nabla_i\, \omega_{k_1 \ldots k_p} + \sum_{\nu=1}^{p} (-1)^\nu (\nabla_{k_\nu} \nabla^i - \nabla^i \nabla_{k_\nu})\, \omega_{i k_1 \ldots \hat{k}_\nu \ldots k_p},$$

the topological Laplacian for forms with values in a Lie algebra is defined as

$$(\Delta\omega^a)_{k_1 \ldots k_p} = -\tilde{\nabla}^i \tilde{\nabla}_i\, \omega^a_{k_1 \ldots k_p} + \sum_{\nu=1}^{p} (-1)^\nu (\tilde{\nabla}_{k_\nu} \tilde{\nabla}^i - \tilde{\nabla}^i \tilde{\nabla}_{k_\nu})\, \omega^a_{i k_1 \ldots \hat{k}_\nu \ldots k_p},$$

where $\hat{k}_\nu$ is the omitted index, and $\tilde{\nabla}_i$ denotes the extended covariant derivative. Thus, in curved space the form of the wave operator depends on the tensor dimensions of the fields. The form of the D'Alembertian is preserved in going over to non-Euclidean spaces only for scalar functions in the absence of conformal transformations of the interval. Conformal transformations also change the invariant wave operator for scalar fields.

Method of Exterior Forms. Structure Equations and Mapping of Manifolds. Consider a smooth manifold covered by coordinate neighborhoods. Let us specify in the neighborhood of each point on this manifold a system of Pfaffian forms[35]:

$$\underset{x}{\omega}^{i} = X^{i}_{j}\, dx^{j}; \quad \underset{y}{\omega}^{i} = Y^{i}_{j}\, dy^{j}, \tag{10.11}$$

where dx^i and dy^j are the differentials of the coordinates in the neighborhood of the points x and y, respectively. We call them the principal forms. Suppose that in the intersection of the neighborhoods of the points x and y the corresponding coordinates are related by differentiable transformations $y^i = y^i(x)$. We require that in the intersection of the neighborhoods of the points x and y the forms $\underset{x}{\omega}^{i}$ and $\underset{y}{\omega}^{i}$ are identical (they are invariant forms):

$$\underset{y}{\omega}^{i} = Y^{i}_{j}\frac{\partial y^{j}}{\partial x^{k}}dx^{k} = \underset{x}{\omega}^{i} = X^{i}_{k}\, dx^{k}.$$

From this requirement, we obtain the transformation law for the coefficients of 1-forms:

$$X^{i}_{k} = (\partial y^{j}/\partial x^{k})Y^{i}_{j}. \tag{10.12}$$

The transformation law (10.12) ensures invariance of the forms ω^i in the intersection of different coordinate neighborhoods. By going from one point to another, we can specify ω^i on the entire manifold.

The system of forms ω^i [i.e., the system of equations (10.11)] is completely integrable, since the n variables depend on n differentials. We require here that the matrix of coefficients X^i_j is nondegenerate.

As is well known, the conditions for integrability of systems of equations are determined by their exterior differentiation. For the system of equations (10.11), exterior differentiation leads to the conditions

$$D\omega^{i} = dX^{i}_{j} \wedge dx^{j} = dX^{i}_{j} \wedge \tilde{X}^{j}_{k}\, \omega^{k} = \omega^{k} \wedge \left(-dX^{i}_{j}\tilde{X}^{j}_{k}\right), \tag{10.13}$$

where $D\omega^i$ is the exterior differential of ω^i, and $\tilde{X}^j_k$ is the inverse matrix of X^j_k and satisfies the condition $dx^j = \tilde{X}^j_k\omega^k$. We can add to the expression in parentheses in Eq. (10.13), without altering it, the term $X^i_{kl}\omega^l$, where the matrix X^i_{kl} is symmetric in the lower indices (since we are then adding zero: $X^i_{kl}\omega^l \wedge \omega^k = 0$). Then (10.13) takes the form

$$D\omega^i = \omega^k \wedge \omega_k^i, \tag{10.14}$$

where $\omega_k^i = dX_j^i \tilde{X}_k^j + X_{kl}^i \omega^l$ are two-index forms. The equations (10.14) are called the structure equations.

We can require that ω_k^i, like ω^i, are invariant forms and are uniquely defined in the intersection of the neighborhoods. The transformation law for the coefficients X_{kl}^i is then determined.

The equations (10.14) constitute necessary and sufficient conditions for the complete integrability of the system of forms ω^i, since the exterior differentials of these forms are expanded in terms of the forms themselves. Let us take the exterior differential of (10.14). Since the exterior differential of an exterior differential is identically equal to zero, we obtain

$$0 = \omega^k \wedge (\omega_k^l \wedge \omega_l^i - D\omega_k^i).$$

By the generalized Cartan lemma, the coefficients in parentheses can be expanded in terms of the same forms ω^i:

$$D\omega_k^i - \omega_k^l \wedge \omega_l^i = \omega^l \wedge \omega_{kl}^i, \tag{10.15}$$

where ω_{kl}^i are new 3-index forms. Again, we can require that ω_{kl}^i are invariant in the intersection, i.e., have a global character, and take the exterior differential of (10.15). We then obtain

$$D\omega_{kl}^i = \omega_{kl}^m \wedge \omega_m^i + \omega_k^m \wedge \omega_{ml}^i + \omega_l^m \wedge \omega_{km}^i + \omega^m \wedge \omega_{klm}^i . \tag{10.16}$$

Here we find the 4-index forms ω_{klm}^i.

Thus, we obtain an infinite chain of coupled structure equations for an infinite sequence of forms ω^i, ω_k^i, ω_{kl}^i, ω_{klm}^i, ... specified globally on the entire manifold. This chain of equations determines the structure of the manifold. The chain is broken if, starting with a certain number of indices, the forms of higher order can be expressed in terms of forms of lower orders.

Ordinary differential geometry corresponds in the language of exterior forms and structure equations to the simplest case in which the 2-index forms ω_k^i become linear combinations of the 1-index principal forms ω^i:

$$\omega_k^i = \Gamma_{kl}^i \omega^l . \tag{10.17}$$

The coefficients of the expansion (10.17) are then called connection coefficients, and the forms ω_k^i are called connection forms; in general, Γ_{kl}^i are nonsymmetric and are not related to the metric. The chain of structure equations is replaced in this case by two equations, which were derived by Cartan and called by him the structure equations of space[36]:

$$D\omega^i = \omega^k \wedge \omega_k^i + S_{kl}^i \omega^l \wedge \omega^k; \tag{10.18}$$

$$D\omega_k^i = \omega_k^l \wedge \omega_l^i + R_{klm}^i \omega^m \wedge \omega^l, \tag{10.19}$$

where S_{kl}^i is the torsion tensor, and R_{klm}^i is the curvature tensor. According to Cartan, the forms ω^i determine the displacement of the coordinate origin, $dM = \omega^i \mathbf{e}_i$, and the forms ω_k^i determine the change in the basis vectors (tetrad) in going from one point of the manifold to another: $d\mathbf{e}_i = \omega_i^k \mathbf{e}_k$. For an orthogonal basis in V_4, the forms ω_l^k determine Lorentz rotations of the tetrads. In the general case, the forms ω^i correspond to the space of first differentials, ω^i and ω_j^i correspond to the space of second differentials, and so forth.

The meaning of the structure equations becomes clear if we take the manifold to be some classical Lie group. We introduce on it n linearly independent Pfaffian forms $\Omega^\alpha = \Omega^\alpha(u, du) = a_\beta^\alpha(u) du^\beta$. Let us suppose that the transformations from the group $v^\alpha = \varphi^\alpha(a, u)$ leave Ω^α invariant, i.e., $\Omega^\alpha[\varphi(a, u), d_u\varphi(a, u)] = \Omega^\alpha(u, du)$. Then instead of ω_k^i we have the 2-index forms $\Omega_\beta^\alpha = f_{\beta\gamma}^\alpha \Omega^\gamma$, where $f_{\beta\gamma}^\alpha$ are the structure constants of the group, which simultaneously play the role of connection coefficients on the group. The structure equations (10.14) take the form of the structure equations of the group:

$$D\Omega^\alpha = {}^1/_2 f_{\beta\gamma}^\alpha \Omega^\beta \wedge \Omega^\gamma. \tag{10.20}$$

By taking the exterior differential of (10.20), we obtain a relation for the structure constants of the group — the Bianchi identity.

If we consider the r-dimensional space of a representation of a finite Lie group, the structure equations lead to equations for the generators of this representation. We specify the structure forms of the representation: $\Delta X^J = dx^J - \xi_\alpha^J(x)\Omega^\alpha$. We take the first integrals of these equations, $\tilde{x}^J = f^J(a, x)$, so that for $a = 1$ we have $\tilde{x}^J = x^J$. Then for $d\tilde{x}^J = 0$ we obtain $dx^J = \xi_\alpha^J(x)\Omega^\alpha$. Exterior differentiation of this completely integrable system yields Lie's differential equations:

$$(\partial \xi_{[\alpha}^J / \partial x^K)\, \xi_{\beta]}^K = \xi_\nu^J f_{\alpha\beta}^\nu, \quad J, K = 1, 2, \ldots, r.$$

In the general case, at each fixed point of the manifold (a point is fixed by the condition $\omega^i = 0$) the structure equations for every system of forms are analogous to (10.20). For example, (10.15) becomes $D\omega_k^i = \omega_k^l \wedge \omega_l^i$, where $\omega_k^i = dX_k^l \tilde{X}_l^i$. In other words, at each point of the manifold we have a system of forms such that their exterior differentials are expressed in terms of the forms themselves, the expansion coefficients being constants. This is the same picture as in the theory of Lie groups. Thus, the structure equations lead to the appearance at each point of the manifold of the space of some linear Lie group for which ω^i, ω^i_j, ω^i_{jk},... are structure forms. For example, ω^i and ω^i_j correspond to the second differential Lie group. The dimension of the space of structure forms ω^i, ω^i_j, ω^i_{jk} is $n + n^2 + n^2(n+1)/2$. In other words, over each point of the manifold we have a new space (fiber) with a structure group that acts in it. The manifold is stratified and becomes a fiber manifold. Fibers are defined over each point of the neighborhood of x and over each point of the neighborhood of y. Over each point of the intersection of the neighborhoods, we have two fibers, which are "displaced" with respect to one another in a definite manner (by an element of the group). For a unique definition of the fiber over the intersection of the neighborhoods, we can take, instead of the invariant forms of the group Ω^α satisfying (10.20), forms ω^α specified over the entire manifold, in particular, over the intersection of the neighborhoods. Then the structure equations take the form

$$D\omega^i = \omega^k \wedge \omega^i_k; \tag{10.21}$$

$$D\omega^\alpha = {}^1/_2 f^\alpha_{\beta\gamma}\, \omega^\beta \wedge \omega^\gamma + \omega^k \wedge \omega^\alpha_k, \tag{10.22}$$

where

$$\underset{x}{\omega^\alpha} = \underset{y}{\omega^\alpha}; \quad \underset{x}{\omega^\alpha_k} = \underset{y}{\omega^\alpha_k}.$$

The invariant forms Ω^α are related to arbitrary forms ω^α by the equation $\omega^\alpha = \Omega^\alpha + \Gamma^\alpha_k \omega^k$, where Γ^α_k are the connection coefficients of the fiber space, which specify a mapping of the fibers at two infinitesimally spaced points onto one another. If a point is fixed, i.e., $x^i = \text{const}$ and $\omega^i = 0$, the first system of equations is satisfied identically, and the second becomes the structure equations of the group.

The structure equations (10.21) and (10.22), if they are simply postulated, enable us to consider fiber spaces with an arbitrary Lie group as the structure group. They are called the *fiber-space equations*. It is in arbitrary fiber spaces that gauge fields become connection coefficients.

A mapping of manifolds can be defined in terms of a mapping of the corresponding systems of exterior forms in the following way.[35] Suppose that on the first manifold we are given a system of forms $\Phi^I = U^I_K du^K$, and on the second manifold $\theta^a = t^a_b dt^b$, obeying the structure equations

$$D\theta^a = \theta^b \wedge \theta^a_b; \tag{10.23}$$

$$D\Phi^I = \Phi^K \wedge \Phi^I_K. \tag{10.24}$$

We specify a mapping $u^I = u^I(t)$. Then there is a relation between the differentials, $du^I = (\partial u^I/\partial t^a)dt^a$, which leads to the relations $\Phi^I = U^I_K(\partial u^k/\partial t^a)dt^a$ and $dt^a = \tilde{t}^a_b\theta^b$, i.e., $\Phi^I = U^I_K(\partial u^K/\partial t^a)t^a_b\theta^b$. Thus, the forms on the first manifold are expressed linearly in terms of the forms on the second manifold:

$$\Phi^I = \Lambda^I_a\theta^a. \tag{10.25}$$

This system of forms has the property of correct extension, i.e., exterior differentiation gives an expansion only in the principal forms. There are no other coefficients. Therefore we can apply the generalized Cartan lemma, according to which the coefficients of the principal forms must be expanded in terms of the forms themselves. Indeed, let us differentiate (10.25):

$$D\Phi^J = d\Lambda^J_a \wedge \theta^a + \Lambda^J_a D\theta^a.$$

We replace $D\Phi^J$ and $D\theta^a$ here by their expressions from (10.23) and (10.24). Then

$$\Phi^K \wedge \Phi^J_K = d\Lambda^J_a \wedge \theta^a + \Lambda^J_a\,\theta^b \wedge \theta^a_b. \tag{10.26}$$

All terms in (10.26) are proportional to $\Lambda^K_a\theta^a$. Applying Cartan's lemma, we obtain an equation which must hold for the coefficients Λ^J_a in the mapping of the systems of forms:

$$d\Lambda^J_a + \Lambda^K_a\Phi^J_K - \Lambda^J_b\,\theta^b_a = \Lambda^J_{ab}\theta^b, \tag{10.27}$$

where Λ^J_{ab} are new coefficients.

If the correct extension holds once, then it always holds. By taking the exterior differentials of (10.27), we obtain an infinite chain of coupled equations for the coefficients of the mapping, Λ^J_a, Λ^J_{ab}, Λ^J_{abc}, etc. By truncating this chain of equations, we fix the mapping of the manifolds with accuracy up to infinitesimal quantities of the corresponding order: $\Lambda^J_a \sim \partial u^K/\partial t^a$, $\Lambda^J_{ab} \sim \partial^2 u^K/\partial t^a \partial t^b$, etc.

§11. GAUGE FIELDS AS THE CONNECTION COEFFICIENTS OF THE PRINCIPAL FIBER SPACE OVER V_4

The Concepts of Connection and Fiber Space. The transition to the contemporary global point of view and the replacement of linear groups by a general Lie group have led to major changes in the theory of connections. The concept of a connection arose in the work of Levi-Civita as parallel transport of tangent vectors of a manifold. In his work, the connection was defined by the metric. Weyl generalized the concept of connection and showed that the metric in the definition of connection is unimportant. Thus, there arose the concept of spaces of affine connection. A further generalization occurred with the appearance of König spaces and connection. König introduced a connection in vector bundles and showed that the dimensions of the fiber and the base need not be identical. Subsequently, Cartan introduced the concepts of projective, conformal, and other connections. As was shown by Schouten, these connections can be modeled locally in vector bundles. At the present time, the concept of connection is formulated for the most general fiber bundles, when there is not necessarily a Lie group acting in the fiber and the connection is not necessarily linear. We shall require the concept of a linear connection in a homogeneous fiber bundle, when an arbitrary Lie group G_r acts in the fiber.

A connection in a homogeneous fiber bundle (in general, nonlinear) is introduced as a mapping of the set of paths in the base into the set of diffeomorphisms of a fiber onto a fiber, satisfying certain conditions.[37,38] In other words, the connection determines a mapping of fibers onto one another when they are transported along different paths in the base.

A fiber space[5] is a differentiable c^v-manifold E on which there is specified an equivalence relation R such that: a) the quotient space $B=E/R$, or the basis space, is a differentiable manifold of n dimensions; b) the projection p, i.e., the canonical mapping of the manifold E onto the base B, corresponding to the definition of B as the quotient space, is a c^v-differentiable mapping that is everywhere of rank n. Under these conditions, the structure of a differentiable fiber space on E is determined by the following collection of elements: B — the base; F — the standard fiber; a c^v-differentiable manifold; G_r — a Lie group, which acts in a c^v-differentiable manner on F (the group of automorphisms of F); p — the projection of E onto the base B; and Φ — the family of homeomorphisms of the topological product $U \times F$ onto the inverse image $p^{-1}(U)$, where U is an open set of the space B, a number of conditions on Φ being satisfied. If F is the group space G_r, we call E the *principal fiber space*; if F is the tangent space to the base, we speak of the *tangent bundle*; when F is the representation space of G_r, we say that E is the *associated fiber space*.

In going from one point of the base to another, we have an isomorphic mapping of the fibers "attached" at these points onto one another. If this mapping is the identity mapping, i.e., motion in the base does not produce a transformation of the fiber, this means geometrically that the fiber space is a direct sum of two spaces. This is a trivial case. Gauge groups and their associated gauge fields lead to a nontrivial example of fiber spaces. Indeed, let us choose the base to be space-time V_4 and the fiber to be the group space of some semisimple Lie group.* With the motion of a point in V_4, the fiber, as we have said, is transformed isomorphically. In our case, this transformation belongs to G_r and depends on the point of V_4. These two properties characterize local gauge groups. The connection coefficients specify a mapping of the fibers belonging to infinitesimally spaced points of the base. The same role is played by gauge fields, in relating the representation spaces of the gauge group. If the fiber is the space in which the group of isotopic transformations acts (isospace), the connection between the isospaces at different points of V_4 is given by a Yang—Mills field. If the fiber coincides with the tangent space, the mapping of fibers is given by the Ricci connection coefficients, which are identified with the gravitational field. Analogous statements can be made for other gauge fields.

Definition of Connection in Terms of Vector Fields in a Fiber Space. Consider a space tangent to a fiber space. Locally, it decomposes into two spaces, forming a direct sum: the space tangent to the base, and the space tangent to the fiber. Let us choose in the space tangent to the base n linearly independent vectors $\mathbf{e}_\mu$, and in the space tangent to the fiber r vectors $\mathbf{e}_a$, which are simultaneously tangent to one-dimensional subgroups of G_r. Together these $(n+r)$ vectors form a basis of the space tangent to the fiber space (tetrad): $\{\mathbf{e}_1, \ldots, \mathbf{e}_{n+r}\}$. The manifold $L_0(p)$ — the space of tetrads — has dimension $n(n+r)$. Vectors tangent to the fiber are called *vertical vectors*. If a mapping of fibers is given along a path in the base, this means that these fibers are "pierced" by a family of curves invariant with respect to the action of the group. The components of the vector field $\mathbf{v}$ tangent to these curves satisfy the equations

$$v^\mu \mathbf{e}_\mu + v^a \mathbf{e}_a = \mathbf{v}, \quad \mu = 1, \ldots, n; \; a = 1, \ldots, r; \tag{11.1}$$

$$dv^\mu + v^\nu \omega^\mu_\nu = \underset{1}{v}^\mu dt;$$

$$dv^a + v^\varkappa \omega^a_\varkappa + v^b f^a_{bc} \omega^c = \underset{1}{v}^a dt, \tag{11.2}$$

where f^a_{bc} are the structure constants of G_r, and ω^c, ω^μ_ν, $\omega^a_\varkappa$ are exterior differential forms.

Along a path in the base, we put $\omega^\mu = v^\mu dt$, where t is a

*If the group G_r is not semisimple, there may not exist a Riemannian space associated with it, since the metric on the group is then degenerate.

parameter. The form of the coefficients of v^b in (11.2) is a consequence of the invariance of the field v with respect to the action of the group and means that all the vectors "protruding" from the fiber have one and the same projection in the base. It can be shown that the converse also holds: if a vector field is given by Eqs. (11.1) and (11.2), their integral curves give an isomorphism of the fibers onto each other. These integral curves are called *horizontal paths*. To determine transport of fibers along an arbitrary path in the base or, equivalently, to define a connection in the entire fiber space, and not only along individual curves, it is necessary to specify n vector fields satisfying equations of the type (11.1) and (11.2):

$$\underset{\varkappa}{\mathbf{v}} = \underset{\varkappa}{v}^{\mu}\, \mathbf{e}_{\mu} + \underset{\varkappa}{v}^{a}\, \mathbf{e}_{a}, \quad \varkappa = 1, \ldots, n; \tag{11.3}$$

$$d\underset{\lambda}{v}^{\mu} + \underset{\lambda}{v}^{\varkappa}\, \omega^{\mu}_{\varkappa} = \underset{\lambda}{v}^{\mu}_{,\varkappa}\, \omega^{\varkappa}; \tag{11.4}$$

$$d\underset{\lambda}{v}^{a} + \underset{\lambda}{v}^{\varkappa}\, \omega^{a}_{\varkappa} + \underset{\lambda}{v}^{b} f^{a}_{bc}\, \omega^{c} = \underset{\lambda}{v}^{a}_{,\varkappa}\, \omega^{\varkappa}. \tag{11.5}$$

Contracting the relation (11.3) with quantities $\overset{\varkappa}{v}_{\lambda}$ satisfying the condition $\overset{\varkappa}{v}_{\lambda}\, \underset{\varkappa}{v}^{\mu} = \delta^{\mu}_{\lambda}$, we obtain

$$\mathbf{E}_{\lambda} = \mathbf{e}_{\lambda} + \overset{\varkappa}{v}_{\lambda}\, \underset{\varkappa}{v}^{a}\, \mathbf{e}_{a} = \mathbf{e}_{\lambda} + \Gamma^{a}_{\lambda}\, \mathbf{e}_{a}.$$

The vectors $\mathbf{E}_{\lambda}$ are projected into n vectors in the base. The geometrical object

$$\Gamma^{a}_{\mu} = \overset{\varkappa}{v}_{\mu}\, \underset{\varkappa}{v}^{a} \tag{11.6}$$

plays the role of connection coefficients of the fiber space.

It follows from Eq. (11.6) that the horizontal and vertical components of the vector field, $\underset{\varkappa}{v}^{\mu}$ and $\underset{\varkappa}{v}^{a}$, are interrelated:

$$\underset{\varkappa}{v}^{a} = \Gamma^{a}_{\mu}\, \underset{\varkappa}{v}^{\mu}. \tag{11.7}$$

Consequently, knowing the connection coefficients Γ^{a}_{μ}, from the projection of the vector field into the base (v^{μ}) we can determine its projection into the fiber (v^{a}) and thereby reconstruct the vector field completely, i.e., "elevate" it to a fiber space. Therefore we say that specification of Eqs. (11.4) and (11.5) implies specification of n elevations of the vector fields, this being equivalent to specification of the connection coefficients (11.6).

Specification of n elevations of the vector fields makes it possible to determine a connection along any path in the base. If the mapping of fibers is made by means of some pseudogroup, the components of $\underset{\varkappa}{v}$ are specialized in such a way that they determine one-parameter transformations of this pseudogroup.

If the vector fields $\underset{\varkappa}{v}^a$ and $\overset{\varkappa}{v}_\mu$ can be identified with the tetrads, from the definition (11.7) we obtain the relation $\overset{\varkappa}{v}^b \underset{\varkappa}{v}^a = \Gamma^a_\lambda \Gamma^b_\mu \underset{\varkappa}{v}^\lambda \overset{\varkappa}{v}^\mu$, which is equivalent to $g^{ab} = g^{\lambda\mu} \Gamma^a_\lambda \Gamma^b_\mu$ or $g_{ab} g^{\lambda\mu} \Gamma^a_\lambda \Gamma^b_\mu = r$.

Vector Potentials of Gauge Fields as Connection Coefficients of a Fiber Space. Differentiating the expression (11.6) for Γ^a_μ and using the structure equations of a fiber space together with Eqs. (11.4) and (11.5), it is easy to obtain the transformation law for this quantity ($\mu = 1, \ldots, 4$; $a = 1, \ldots, r$):

$$d\Gamma^a_\mu - \Gamma^a_\nu \omega^\nu_\mu + f^a_{bc} \Gamma^b_\mu \omega^c + \omega^a_\mu = \Gamma^a_{\mu\nu} \omega^\nu. \tag{11.8}$$

Taking the particular values of the forms $\omega^\nu = 0$, $\omega^a_\mu = -\partial_\mu \varepsilon^a$, $\omega^c = -\varepsilon^c$, and $\omega^\nu_\mu = \partial \delta x^\nu / \partial x^\mu$, we obtain the well-known transformation law for the vector potentials of gauge fields A^a_μ:

$$\delta A^a_\mu - A^a_\nu (\partial \delta x^\nu / \partial x^\mu) - f^a_{bc} A^b_\mu \varepsilon^c - \partial_\mu \varepsilon^a = 0.$$

Note that allowance is made here for transformations of the vector potential with respect to both the group index and the space-time index.

Since from the point of view of differential geometry the character of a quantity is determined by its transformation law, we can identify the vector potentials of a gauge field A^a_μ with the connection coefficients of the principal fiber space Γ^a_μ whose base is V_4 and whose structure group is a gauge group G_r.[4,39] If this identification is made, the equations of motion of particles interacting with the gauge field become free equations. We shall demonstrate that this is so.

Covariant derivatives in a fiber space are constructed for arbitrary geometrical objects on the basis of their transformation law with respect to G_r and arbitrary continuous coordinate transformations in the base. Use is made here of the concept of a Lie derivative. We shall give an algorithm for finding the covariant derivative, from which it becomes clear that the covariant derivative in a fiber space is identical with the Yang—Mills covariant derivative, which includes the interaction with the gauge field A^a_μ.

The field of an arbitrary geometrical object in a fiber space is given by equations of the form

$$dY^J + \Phi^J_a (Y) \omega^a = Y^J_\mu \omega^\mu, \tag{11.9}$$

where J are the components of the object, $a = 1, \ldots, r$, $\mu = 1, \ldots, n$ (for V_4, $n = 4$), and $\Phi_a^J(Y)$ are arbitrary functions of Y. If the functions Φ_a^J are linear in Y, we say that a linear object is given. The covariant differential is obtained by replacing the arbitrary forms ω^a in (11.9) by the invariant forms $\tilde{\omega}^a = \omega^a - \Gamma_\mu^a \omega^\mu$. This differential has the form

$$\nabla Y^J = dY^J + \Phi_a^J(Y)\,\tilde{\omega}^a = (Y_\mu^J + \Gamma_\mu^a \Phi_a^J)\,\omega^\mu.$$

The covariant derivative of an arbitrary geometrical object is the coefficient of ω^μ and therefore has the form

$$Y_{;\mu}^J = Y_\mu^J + \Gamma_\mu^a \Phi_a^J(Y). \tag{11.10}$$

In particular, for the linear representations of G_r given by the relations

$$\bar{\delta}\psi = \underset{a}{I}\,\varepsilon^a\,\psi - (\partial\psi/\partial x^\mu)\,dx^\mu$$

or

$$\bar{\delta}\psi - \underset{a}{I}\varepsilon^a\,\psi = -(\partial\psi/\partial x^\mu)\,dx^\mu,$$

from (11.9) and (11.10) we obtain

$$\psi_{;\mu} = -(\partial\psi/\partial x^\mu) + \underset{a}{I} A_\mu^a\,\psi. \tag{11.11}$$

Here $\underset{a}{I}$ is the generator of the representation of G_r. It is easy to see that the expression (11.11) is identical with the covariant Yang—Mills derivative. Thus, if the interaction with a gauge field is introduced by replacing the ordinary derivatives by the covariant derivatives in the Yang—Mills sense (minimal interaction), the corresponding equations (or Lagrangians) can be regarded as free, but defined in a fiber space with the connection object $\Gamma_\mu^a = A_\mu^a = \underset{\varkappa}{v^a}\,\underset{\varkappa}{v_\mu}$.

The curvature tensor of the fiber space is determined from the structure equations (10.22), but with the arbitrary forms ω^a replaced by the invariant forms $\tilde{\omega}^a$. In this case, instead of (10.22) we obtain the structure equations

$$D\tilde{\omega}^a = {}^1\!/_2 f_{bc}^a\,\tilde{\omega}^b \wedge \tilde{\omega}^c + R_{\mu\nu}^a\,\omega^\mu \wedge \omega^\nu, \tag{11.12}$$

where $R_{\mu\nu}^a$ is the curvature tensor of the fiber space, which can be expressed in terms of connection coefficients as follows:

$$R_{\mu\nu}^a = \partial_{[\mu}\,\Gamma_{\nu]}^a - {}^1\!/_2 f_{bc}^a\,\Gamma_{[\mu}^b\,\Gamma_{\nu]}^c. \tag{11.13}$$

If the tangent space is chosen as the fiber, the quantities Γ_μ^a become Ricci coefficients (or Christoffel symbols), and $R_{\mu\nu}^a$ reduces to the ordinary curvature tensor.

It is easy to see that $R^a_{\mu\nu}$ is expressed in terms of Γ^a_μ in exactly the same way that the stress tensor $F^a_{\mu\nu}$ of the gauge field is expressed in terms of the vector potential A^a_μ:

$$F^a_{\mu\nu} = A^a_{[\nu,\mu]} - {}^1/_2 f^a_{bc} A^b_{[\mu} A^c_{\nu]}. \tag{11.14}$$

If $\Gamma^a_\mu = A^a_\mu$, then $F^a_{\mu\nu} = R^a_{\mu\nu}$. Thus, identification of the vector potentials of gauge fields with connection coefficients of a fiber space leads to identification of the stress tensor of a gauge field with the curvature tensor of this space.

The second covariant derivative of a linear geometrical object (like the second covariant derivative in the Yang—Mills sense), being antisymmetrized, can be expressed in terms of the curvature tensor of the fiber space:

$$\psi^J_{;[\mu\nu]} = \underset{a}{I^J_K} R^a_{\mu\nu} \psi^K. \tag{11.15}$$

Thus, localization of gauge groups, i.e., the introduction over each point of space-time of a Lie group G_r (or the corresponding algebra) in such a way that, referred to different points, these groups are isomorphic to one another, implies the transition to the principal fiber space. The base of this space is space-time V_4, and the fiber is the group G_r, regarded as a manifold. Locally, i.e., in the neighborhood of each point, the fiber space is a direct sum of two spaces $G_r \times V_4$, and the tangent space is a direct sum of two tangent spaces $T_{V_4} \times A_r$ (T_{V_4} is the tangent space to V_4, and A_r is the Lie algebra of the group G_r). Introduction of the gauge group described by the vector potential A^a_μ implies the introduction of a connection of the fiber space. If the stress tensor of the gauge field is $F^a_{\mu\nu} = 0$, then the connection which is introduced has zero curvature. Minimal interactions with the gauge field, corresponding to replacement of the ordinary derivatives by covariant derivatives in the Yang—Mills sense, imply a transition to covariant differentiation in the fiber space. In this case, the equations which include the interaction with A^a_μ remain free.

From the geometrical point of view, the equations of gauge fields imply that we are considering connections on the principal fiber space such that the divergences of the corresponding curvature tensors are equal to zero, $R^{\mu\nu}_{a;\nu} = 0$, or to a specified quantity, $R^{\mu\nu}_{a;\nu} = J^\mu_a$. In the case of the tangent bundle over V_n with $n \leqslant 3$, these equations reduce identically to the condition of constant scalar curvature, $R = \text{const}$. For $n > 3$, the condition $R = \text{const}$ is a consequence of the fact that the divergence of the curvature tensor is equal to zero.

Gauge fields can be classified according to the

algebraic properties of the field tensor. There are two types of electromagnetic fields: wave and nonwave fields.[40] There are three types of gravitational fields, and wave, nonwave, and mixed fields can be present in each of them.[41] There are four types of Yang—Mills fields. Each of the types II—IV can include both wave and nonwave fields. Type I is a nonwave field.[42] Allowance for the differential properties of the field tensor permits a more detailed algebraic classification (taking into account the properties of the holonomy groups of the fiber space).

§12. CLASSIFICATION OF SOLUTIONS OF THE CLASSICAL EQUATIONS OF GAUGE FIELDS

Introduction. As a result of the geometrical approach, three types of classifications of the solutions of the classical equations of gauge fields have become possible. The first type uses the mathematical fact that the components of the curvature tensor of an arbitrary manifold (in our case, a fiber space), together with their covariant derivatives, form the algebra of the holonomy group of this manifold. From this point of view, the equations of a gauge field lead to restrictions on the holonomy group and on the structure of the manifold. If additional gauge conditions are present, there is a dependence between the space-time and internal symmetry properties of gauge fields. This makes it possible to obtain certain physically justified criteria for selecting solutions. This classification was first proposed by Loos[6] in 1965 and was developed in the work of his school.

Another classification is the classification of gauge fields according to the algebraic properties of the field tensor and is constructed by analogy with the Petrov classification of gravitational fields. This classification is a local one. It makes it possible to relate the invariants constructed from the components of the field tensor and the eigenvectors of the field tensor to the symmetry properties of the solutions and to the possible types of asymptotic behavior of gauge fields (Eguchi,[42] 1976).

The third type of classification of solutions uses the concept of homotopy groups of a manifold, i.e., groups of deformations of closed paths. This approach is useful in the description of processes in macroscopic ordered media, in particular, in the theory of phase transitions. The possibility of studying the properties of a continuous medium by means of homotopy groups and other topological characteristics is explained by the fact that a nonhomogeneous space can be regarded as a model of a continuous medium, and the defects of the structure of a continuous medium can be described as a breakdown of the connection of a manifold (topological defects). These manifest themselves physically in exactly the same way as the external sources of gauge fields. Therefore it becomes possible to establish an analogy between the properties of elementary particles and the properties of defects in ordered media, and also a topological classification of particle-like solutions of the equations of gauge fields. Topological classifications are asymptotic, in the sense that they explore the behavior of gauge fields in regions where the field tensor is equal to zero and the vector potential becomes the connection of absolute parallelism.[18]

In what follows, we consider the first and third classifications.

Classification of Gauge Fields According to Holonomy Groups. Definitions. The holonomy group of a fiber space is defined as follows.[5] For a given point z in the principal fiber space E, the holonomy group $\mathcal{H}_z$ of a given connection at the point z is the set of elements $g \in G$ such that the points z and zg^{-1} can be joined by a horizontal path.

More perspicuously, the holonomy group can be obtained as follows. We choose a vector in the space tangent to the fiber (i.e., in the Lie algebra) with origin at the point z of a fixed fiber and carry the chosen fiber around a closed path in the base. With the transport of the fiber, the origin of the vector tangent to the fiber (vertical vector) is carried around a horizontal path. As a result of the transport, the fiber is transformed, and the point z of the initial fiber goes into $z' = zg^{-1}$, where $g \in G$.

The holonomy group is the group of transformations of the vectors of the space tangent to the fiber (i.e., the vectors of the Lie algebra) obtained as a result of their parallel transport around all possible closed paths in the base starting from a given point x. If we consider only paths contained in a given neighborhood of the point, the holonomy group is called local. If the transport is made only along paths which can be contracted into a point, the holonomy group is called restricted. If the holonomy group is defined over the entire space and the choice of paths is not restricted, it is called simply the holonomy group. For transport of a vector space, the holonomy group consists of linear transformations of the vectors and is said to be homogeneous. If an affine space rather than a vector space is transported, the holonomy group is nonhomogeneous, since it can contain displacements of the tangent space.

The structure of the holonomy group reflects the properties of the space in some region, although it is itself defined at each point of this region individually.

Every element of a restricted nonhomogeneous holonomy group $\rho_x(V_n)$, where V_n is a differentiable manifold, is a product of a finite number of elements obtained from the elements of the local nonhomogeneous holonomy groups ρ_y^* ($y \in V_n$) by transport along paths joining the point y to the point x. The same applies to a restricted homogeneous holonomy group $\sigma_x(V_n)$.

Holonomy groups at different points z are conjugate subgroups of G_r, and if z and z' belong to the base, then $\mathcal{H}_z = \mathcal{H}_{z'}$, while if z and z' belong to a single fiber and $z' = z\gamma$, then $\mathcal{H}_{z'} = \gamma^{-1}\mathcal{H}_z\gamma$, where $\gamma \in G_r$. The properties of the holonomy group are intimately related to the topological properties of the base, namely, to the topological invariant of the base — the Poincaré group. The Poincaré group is defined by the number of classes of closed paths on the manifold which cannot be transformed into one another by a continuous deformation (i.e., are not homotopic to one another). A restricted homogeneous holonomy group σ_z is a normal divisor of $\mathcal{H}_z$. Then there exists a mapping (homomorphism) $f : \pi_x$ onto $\mathcal{H}_z/\sigma_z$, where π_x is the Poincaré group of the base at the point $x = pz$. Thus, by varying the topology of V_4, i.e., π_x, allowance can also be made for the discrete symmetries of elementary particles.

<u>Holonomy Group, Gauge Group, and Structure of a Manifold.</u> It was shown by Cartan[36] that if an affine connection without torsion is given on the group space of a finite Lie group G_r, then the restricted holonomy group is identical with the derived subgroup of its adjoint group. In other words, if G_r is known, the restricted holonomy group is constructed by first finding the group of all automorphisms G_r (the adjoint group), consisting of transformations of the form $\gamma^{-1}G\gamma$, where $\gamma \in G$, and then finding the derived subgroup of this group, containing all its commutators, i.e., elements of the form $a \cdot b \cdot c \ldots a^{-1} \cdot b^{-1} \cdot c^{-1}$, where a, b, c belong to the adjoint group and are arbitrary in number.

The derived subgroup is a normal divisor of the corresponding group. Therefore, for a simple group the derived subgroup is the same as the group itself. But in this case the adjoint group is also the same as the group itself. Consequently, a simple gauge group can be regarded not only as the fundamental group of its associated Riemannian space, but also as the holonomy group of this space. The reconstruction of a gauge group from the restricted holonomy group $\mathcal{H}$ in general involves the search for the normalizers of this group.

Any Lie group can be realized in the form of the holonomy group of a space of affine connection but for Riemannian spaces (i.e., spaces provided with a metric) this is not so. The choice of the holonomy group strongly restricts the possible structure of the Riemannian space, but in certain cases (when the holonomy group is small or exceptional, i.e., does not coincide with the orthogonal group in the tangent space) determines it completely.[43] The following groups can be exceptional holonomy groups of a Riemannian space: $U(n/2)$; $SU(n/2)$; $\mathrm{Sp}(n/4)$, $\mathrm{Sp}(1) \cdot \mathrm{Sp}(n/4)$; G_2 $(n = 7)$; $\mathrm{Spin7}$ $(n = 8)$; $\mathrm{Spin9}$ $(n = 16)$. An attempt to associate gauge groups and internal symmetries with such holonomy groups would lead to spaces of higher dimension than 4 (except for a solvable group and its subgroups, for which $n = 4$ is possible).

In an arbitrary manifold V_n, the holonomy group characterizes the structure of parallel transport and the degree of deviation of the geometry of V_n from Euclidean geometry. For an analytic connection of zero curvature, the restricted holonomy group reduces to the identity transformation.

If the holonomy group is smaller than the group G_r acting in the fiber, we can readily reduce the group acting in the fiber to the holonomy group, since all the transformations of vectors taken over all of V_4 belong to the holonomy group and do not extend outside this group. Geometrically, the restricted nature of the holonomy group in comparison with the largest possible group $GL(n)$ means that it is possible to fix a part of the tetrads in the fiber, which do not transform under parallel transport. This happens, for example, as a result of a certain degree of homogeneity or symmetry of the base.

The structure of a manifold is not determined uniquely by the holonomy group. It is possible to introduce different connections on one and the same manifold, and in general they give different holonomy groups. Therefore, knowing only the connection (or its holonomy group), it is not possible to say what the manifold is. There can be different manifolds with the same holonomy group. As a simple example, we can introduce a Euclidean connection on one section of a hyperboloid. The curvature tensor for this connection is equal to zero. At the same time, for the connection induced on a 2-dimensional hyperboloid embedded in a 3-dimensional Euclidean space, the hyperboloid is a space of constant curvature, i.e., the curvature tensor of the induced connection is nonzero.

The holonomy groups of manifolds are also related to their groups of motions. A stationary subgroup of a group of motions is simultaneously a subgroup of the holonomy group. It is easy to see this from the following considerations.

If in a manifold we have the field of a covariantly constant tensor (for example, the field of the metric tensor in Riemannian space), this field determines at a point x of the manifold a tensor which is invariant with respect to the holonomy group $\mathcal{H}_x(V_n)$, and, conversely, every tensor at the point x which is invariant with respect to the holonomy group $\mathcal{H}_x(V_n)$ generates under parallel transport in the manifold a field whose covariant derivative is equal to zero. Thus, as a consequence of the existence of the covariantly constant field of the metric tensor on a manifold, the holonomy group at each point x_0 of this manifold coincides with the group of motions of a flat space having the same metric $g_{\mu\nu}(x_0)$ or with one of its subgroups. It follows from this that for a pseudo-Riemannian space V_4 and a Riemannian connection (i.e., Christoffel symbols) the holonomy group can be only the Lorentz group or its subgroup. The exceptional holonomy groups mentioned at the beginning correspond to exceptional connections.

If the groups describing the internal symmetries of elementary particles are regarded as holonomy groups of a pseudo-Riemannian space, and the elementary particles as their representations, we obtain the concept of non-point-like particles. The dependence of the wave functions of the particles on a finite region enters through the structure of the holonomy groups in terms of whose representations they are defined.

Some Non-Spherically-Symmetric Solutions of Yang—Mills Equations with Point Sources. Positive Definiteness of the Field Energy and Fixed Space-Time Symmetry of Solutions as Restrictions on the Structure of the Holonomy Group. It is convenient to make use of the concept of the holonomy group of a fiber space for the analysis of the classes of solutions of the classical equations of gauge fields, since the solution of these equations means, in geometrical language, the determination of the connection of the principal fiber space over V_4. This analysis shows that physical requirements such as positive definiteness of the field energy, Lorentz conditions (i.e., a restriction on the spin carried by a field), and the choice of a definite space-time symmetry of the solutions constitute restrictions on the possible structure of the corresponding holonomy group. Thus, we have a dependence between the space-time and internal symmetry properties of the solutions.

In a Riemannian space, the curvature tensor and its covariant derivatives or, more precisely, their contractions with arbitrary vectors v^λ, w^μ, $u_1^{v_1}$, $u_2^{v_2}$,... at a point x of the form

$$R^{\alpha}_{\beta\lambda\mu} v^{\lambda} w^{\mu};\ \ldots\ u^{v_k}_{k} \ldots u^{v_1}_{1} \nabla_{v_k} \ldots \nabla_{v_1} R^{\alpha}_{\beta\lambda\mu} v^{\lambda} w^{\mu} \tag{12.1}$$

determine the elements of the Lie algebra $d\sigma^{*}_{x}$ of the local homogeneous holonomy group.

In a similar way, in a fiber space the Lie algebra of the local homogeneous holonomy group is determined by the curvature tensor and its covariant derivatives (i.e., by the stress tensor $F_{\mu\nu}$ of the gauge field and its covariant derivatives). From this point of view, the equations of a gauge field constitute restrictions on the covariant derivatives of the curvature tensor. It is possible to determine from them the Lie algebra of the holonomy group if the space-time symmetry of the solutions is known. In other words, if we seek solutions of Yang—Mills equations possessing a given symmetry in V_4, we are restricting the gauge group.

For example, the following theorems hold for gauge fields with point sources[6,44]: 1) spherically symmetric analytic solutions for a gauge field of a point charge have Abelian holonomy groups; 2) for a gauge field with an Abelian holonomy group, the Yang—Mills equations and the Bianchi identities reduce to Maxwell's equations. Therefore all spherically symmetric solutions of Yang—Mills equations with a point source are Coulomb-like, or, more precisely, can be reduced to Coulomb form by choosing the gauge. In the particular cases of gauge symmetry O(3) and SU(3), this was demonstrated by Ikeda and Miyachi and by Loos [see (2.26) and (2.30)]. These Coulomb-like solutions correspond to long-range forces like the electromagnetic and gravitational fields. To obtain a gauge field of non-electromagnetic type, we must find a solution with a non-Abelian holonomy group.[6]

We shall consider an example of such a solution with a non-semi-simple holonomy group. We shall write both the vector potentials and the components of the stress tensor in operator form, $\Gamma, F_{\mu\nu}$, without particularizing the form of the matrices in which the expansion must be made. The Yang—Mills equations in operator form have the structure

$$\nabla_{\nu} \mathbf{F}^{\mu\nu} = \mathbf{J}^{\mu}. \tag{12.2}$$

Let t, r, θ, φ be spherical coordinates in space-time V_4, which for simplicity is assumed to be flat (Minkowski space). We shall seek non-spherically-symmetric solutions of these equations for a point charge in the form

$$\boldsymbol{\Gamma}_t = [f(\theta)/r]\mathbf{A};\ \ \boldsymbol{\Gamma}_r = 0;\ \ \boldsymbol{\Gamma}_\theta = 0;\ \ \boldsymbol{\Gamma}_\varphi = h(\theta)\mathbf{B}, \tag{12.3}$$

where f and h are real functions depending only on θ, and

A and **B** are constant matrices.

The current density $\mathbf{J}^\mu$ is nonzero only on the world line of the charge, $x_1 = x_2 = x_3 = 0$. Then

$$\left.\begin{aligned} &\mathbf{F}_{tr} = (f/r^2)\,\mathbf{A};\ \mathbf{F}_{t\theta} = -(f'/r)\,\mathbf{A};\\ &\mathbf{F}_{t\varphi} = (fh/r)\,[\mathbf{AB}];\ \mathbf{F}_{\varphi r} = 0;\\ &\mathbf{F}_{\varphi\theta} = -h'\,\mathbf{B};\ \mathbf{F}_{r\theta} = 0. \end{aligned}\right\} \qquad (12.4)$$

For the current density, we obtain

$$\left.\begin{aligned} &\mathbf{J}^t = (r^3)^{-1}(f''\,\mathbf{A} + f'\,\mathbf{A}\cos\theta - (fh^2/\sin^2\theta)\,[\mathbf{B}\,[\mathbf{AB}]]);\\ &\mathbf{J}^r = 0;\ \mathbf{J}^\theta = 0;\\ &\mathbf{J}^\varphi = -(r^4\sin^2\theta)^{-1}(h''\,\mathbf{B} - h'\,\mathbf{B}\cos\theta) - f^2 h\,[\mathbf{A}\,[\mathbf{AB}]]. \end{aligned}\right\} \qquad (12.5)$$

The equations (12.5) constitute a system of nonlinear partial differential equations for the functions $f(\theta)$ and $h(\theta)$. We require that the functions $f, f', fh/\sin\theta$, and $h'/\sin\theta$ are bounded in the interval

$$0 < \theta < \pi. \qquad (12.6)$$

Such solutions will be called *regular*. We discard the trivial solutions $f(\theta) = 0$ and $h(\theta) = 0$.

For a point charge, the equations (12.5) lead to the relations

$$[\mathbf{B}\,[\mathbf{AB}]] = \alpha\mathbf{A};\ [[\mathbf{AB}]\mathbf{A}] = \beta\mathbf{B}, \qquad (12.7)$$

where α and β are real numbers. The equations (12.7) show that the holonomy group is generated by the operators **A**, **B**, and [**AB**] (the derived subgroup).

We introduce the new variable $z = \cos\theta$. Then the equations (12.5) take the form

$$f'' - 2zf'/(1 - z^2) - \alpha fh^2/(1 - z^2)^2 = 0; \qquad (12.8)$$

$$h'' + \beta f^2 h/(1 - z^2) = 0, \qquad (12.9)$$

where the primes denote differentiation with respect to z, and the conditions of regularity of the solutions reduce to the following conditions:

$$f,\ f'(1 - z^2)^{1/2},\ fh(1 - z^2)^{-1/2} \text{ and } h \text{ are bounded for } -1 < z < 1. \qquad (12.10)$$

The operators **A** and **B** can be normalized in such a way that $|\alpha|$ and $|\beta|$ are equal to unity (if they do not vanish). Then there are three possibilities: 1) $\alpha = 0,\ |\beta| = 1$; 2) $|\alpha| = |\beta| = 1$; 3) $|\alpha| = 1,\ |\beta| = 0$.

Suppose that $\alpha = 0$ and $|\beta| = 1$. The only regular solution of (12.8) is $f(z) = \text{const}$. For $\beta = -1$, (12.9) has a nonzero regular solution only when $f = 0$. This solution is $h = h_0 + h_1 z$, where h_0 and h_1 are constants. For $\beta = 1$,

(12.9) has a regular solution only when

$$f^2 = m(m+1);\ m \geqslant 0 \text{ and is integral.} \qquad (12.11)$$

This solution has the form

$$h(z) = (1 - z^2)(d/dz)P_m'(z), \qquad (12.12)$$

where $P_m(z)$ is the Legendre polynomial of degree m.

For $\alpha = 0$ and $\beta = 1$, it follows from (12.7) that **B** and [**AB**] generate an Abelian invariant subgroup $\mathcal{H}$. Now it can be shown that Tr $\mathbf{B}^k = 0$ for any positive integer k, from which it follows that **B** is a nilpotent matrix. Consequently, $\exp(\omega\mathbf{B})$ is a polynomial, so that $\mathcal{H}$ is noncompact. In accordance with (12.7), we have

$$\mathbf{A} = \mathbf{L}_1 + a_2\mathbf{L}_2 + a_3\mathbf{L}_3;\ \mathbf{B} = b_2\mathbf{L}_2 + b_3\mathbf{L}_3, \qquad (12.13)$$

where $\mathbf{L}_1$, $\mathbf{L}_2$, and $\mathbf{L}_3$ are the generators of $\mathcal{H}$, satisfying the commutation relations

$$[\mathbf{L}_1, \mathbf{L}_2] = \mathbf{L}_3;\ [\mathbf{L}_2, \mathbf{L}_3] = 0;\ [\mathbf{L}_3, \mathbf{L}_1] = \mathbf{L}_2. \qquad (12.14)$$

Here the Lie algebra is considered over the field of real numbers. Therefore a_2, a_3, b_2, and b_3 are real.

If the fiber is a 2-dimensional complex linear vector space, it is always possible to choose a basis in it such that

$$\mathbf{L}_1 = \pm\begin{pmatrix}\gamma + i & 0\\ 0 & \gamma\end{pmatrix};\ \mathbf{L}_2 = \begin{pmatrix}0 & 1\\ 0 & 0\end{pmatrix};\ \mathbf{L}_3 = \pm\begin{pmatrix}0 & i\\ 0 & 0\end{pmatrix}, \qquad (12.15)$$

where γ is a complex number. The representation (12.15) has an invariant subspace, but because of the noncompactness of $\mathcal{H}$ it is not completely reducible. Irreducible 2-dimensional representations of the Lie algebra of $\mathcal{H}$ do not exist.

An analysis of the other cases ($|\alpha| = |\beta| = 1$; $|\alpha| = 1$, $|\beta| = 0$; and $|\alpha| = |\beta| = 0$) shows that the most general solution for a gauge field of a point charge is given by Eqs. (12.11) and (12.12). This solution is not spherically symmetric. The holonomy group corresponding to it is not semisimple and not compact. Examination of the asymptotic behavior of the components of this solution shows that one of them behaves at infinity like a Coulomb potential (although the behavior is quite complicated near $r = 0$), while the remaining components are short-range. The solution (12.11) and (12.12) can be interpreted as describing two "bubbles" at large distance from one another and interacting with each other "almost" electromagnetically. Near the "bubbles" and inside them, the field is short-range, like the nuclear forces. In other words, the "bubbles" (or Yang—Mills "particles") are purely field configurations, which in this sense are reminiscent of Wheeler's geons.

Uzes[44] considered solutions of the Yang—Mills equations having planar symmetry. The presence of a plane of symmetry was defined by the condition of vanishing of the covariant derivatives with respect to y and z of the stress tensor of the field and its covariant derivatives. It was shown that for solutions of the equations (12.2) having this property the holonomy group is Abelian, and the Yang—Mills equations reduce to Maxwell's equations if there is at least one region of space-time in which there are no sources of the field.

An important point in the physical interpretation of the solutions of the equations of gauge fields is the requirement of positive definiteness of the energy density of the field. In the absence of sources, positive definiteness of the energy density ensures that the holonomy groups of the corresponding solutions are semisimple and compact. The solutions (12.11) and (12.12) do not satisfy this requirement. Therefore they contain nonvanishing components which do not contribute to the energy—momentum tensor. Another example of the same kind is the regular solution of the free Yang—Mills equations, with a non-Abelian holonomy group, satisfying at a fixed instant of time $x^0 = t_0$ the requirement

$$\mathbf{F}_{01} = b\mathbf{F}_{23}; \ \mathbf{F}_{02} = b\mathbf{F}_{31}; \ \mathbf{F}_{03} = b\mathbf{F}_{12}, \qquad (12.16)$$

where $b = \text{const}$.

An analysis shows that the conditions (12.16) are satisfied at all instants of time $x^0 > t_0$ if $b = \pm i$ and the vector potentials (connection coefficients) satisfy for $b = \mathrm{i}$ the condition

$$\partial_0 \mathbf{\Gamma}_1 = \mathrm{i}\,(\partial_2 \mathbf{\Gamma}_3 - \partial_3 \mathbf{\Gamma}_2 - [\mathbf{\Gamma}_2 \mathbf{\Gamma}_3]\,) \ \text{cycl.}$$

This solution has as the holonomy group the complexified holonomy group of the 3-dimensional subspace $x^0 = \text{const}$. This holonomy group is noncompact. The energy—momentum tensor of the gauge field in this case is equal to zero.

If we regard as physical only those solutions which give a positive energy density of the field, we can specify whole classes of solutions which are rejected by this requirement. For example, it can be shown that there do not exist particle-like (i.e., regular and, in a certain sense, localizable) solutions of the free Yang—Mills equations whose holonomy group is compact and semisimple, and the field is of one of the following types: 1) a constant field, for which there exists a gauge such that the vector potentials do not depend on the time; 2) a static field, for which the stress tensor $\mathbf{F}_{\varkappa\lambda}$ is covariantly constant in time; and 3) a stationary field, whose stress tensor satisfies the condition $\nabla_t F_{\kappa\lambda} = [\mathbf{T}, F_{\kappa\lambda}]$, where $\mathbf{T}$ is an operator field $\mathbf{T}(x)$ having special properties (for example, $\nabla_\kappa \mathbf{T}$ belongs to the Lie algebra of the holonomy group).

Thus, the simplicity and compactness of the holonomy group strongly restrict the choice of solutions of the classical Yang—Mills equations. These are in essence physical requirements, which guarantee the positive definiteness of the energy density of the field. In its turn, the structure of the holonomy group depends strongly on the space-time symmetry of the solutions. Thus, the requirement of positive definiteness of the energy also imposes restrictions on the possible space-time configuration of the gauge field.

Topological Classification of Defects in Ordered Media. Topological concepts are used in gauge field theories for the classification of point singularities and for the study of the global properties of gauge fields.[45] The geometrical construction of a fiber space over V_4, whose standard fiber is the space of internal states of the system, makes it possible to develop an analogy between the properties of elementary particles and defects in ordered media. This analogy is attractive not only as a route to a unified physical picture of the world, but also as a new source of physical models for testing field-theoretic ideas, which are amenable to experimental investigation.

An important aspect of investigations into the physics of phase transitions and the properties of condensed media is the study of the defects which occur in the ordered phase and lead to the formation of various observable macrostructures. It is then necessary to construct a classification of elementary defects and to examine the mechanisms by which accumulations of them are formed. The elementary defects (points, lines, walls, etc.) can be classified by making use of the continuity of the properties of the medium, i.e., topological characteristics.

The type of ordering in a medium is described by an order parameter. For example, in an ordered alloy the order parameter is a real scalar, in a superfluid it is a complex scalar, and in an isotropic ferromagnet it is a vector. The order parameter is defined at each point of an ordered medium and, by definition, it characterizes the internal state of the medium at this point. In the absence of distortions, the internal state must be the same at all points of a sample. The appearance of defects is accompanied by a variation of the internal state of the medium from point to point.

The internal state of a medium can be represented as a

point in a certain abstract space of internal degrees of freedom. The subspace formed by the points corresponding to all possible values of the order parameter with the same amplitude is also called the manifold of internal states.[46] For example, for a real scalar order parameter, the manifold of internal states consists of two points (±1). For a complex scalar order parameter, it is a circle; for a vector order parameter, it is a sphere, and so forth. The manifold of internal degrees of freedom has important topological characteristics: the dimension and connection of the manifold. In the theory of critical phenomena near the point of a phase transition, a major role is played by the dimension of the internal space. The connection of the internal space is important in the theory of defects.

Consider a linear defect (vortex) in a 3-dimensional volume of a superfluid. To describe this defect, we surround it by a closed loop. The phase change $\Delta\varphi/2\pi$ of the complex order parameter in traversing a closed contour is a topological invariant: to every traversal around the vortex along a closed contour in ordinary space there corresponds a certain closed path in the manifold of internal states. This closed path (more precisely, the class of equivalent closed paths into which the given path can be reduced by a continuous deformation inside the manifold of internal states) topologically characterizes the linear defect. If a closed path can be continuously deformed into a point of the manifold, the linear defect is not topologically stable, since it can be continuously deformed into the absence of defects. If a closed path in the manifold cannot be continuously contracted into a point, the linear defect is said to be topologically stable.

We generalize this construction to spheres and defects of arbitrary dimensions d and d'. We surround the defect by a subspace of dimension r such that $d' + 1 + r = d$. In 3-dimensional space, a defect wall is surrounded by two points (i.e., a 0-dimensional sphere S_0), a linear defect by a closed loop (a 1-dimensional sphere S_1), and a point defect by a sphere (a 2-dimensional sphere S_2).

At each point of the surrounding subspace S_r, there exists some internal state, which is represented by a point in the manifold of internal states F. Thus, a mapping of S_r into F is determined. The possible mappings of S_r into F can be divided into classes of equivalent mappings which admit continuous deformation into one another on F. The set of these classes is called the r-th homotopy group of the manifold F and is denoted by $\pi_r(F)$.

In the case of a 3-dimensional volume of a superfluid,

it is known that the manifold of internal states is a 1-dimensional sphere $F = S_1$ and $\pi_0(S_1) = 0, \pi_1(S_1) = Z$, and $\pi_2(S_1) = 0$, where 0 denotes the trivial group consisting of a single element, and Z is the additive group of integers. It follows from this that in a superfluid there do not exist stable walls or stable point defects, but stable vortex lines are possible, and these can be characterized by an integer (positive or negative) — the strength of the vortex. In 3-dimensional isotropic ferromagnets, for which $F = S_2$, stable point defects are possible, but there are no stable linear defects.

By introducing the concept of an n-component order parameter, it is possible to study a large class of systems, including those analyzed above as particular cases when $n = 1$ (a real scalar order parameter), $n = 2$ (a complex scalar), and an ordinary vector with $n = d$. For an n-vector order parameter, the manifold of internal states is $F = S_{n-1}$, since the amplitude is assumed to be constant. As is well known,[47] $\pi_r(S_m) = 0$ for $r < m$, while $\pi_m(S_m) = Z$. Consequently, topologically stable defects have dimension $d' = d-n$. This means that topologically stable defects do not exist for $n > d$. For $0 < n < d$, there is one type of defect (points for $n = d$, lines for $n = d - 1$, walls for $n = d - 2$, etc; other defects can occur for $d > 4$). For $n < 0$, there are no stable defects.

We note that the boundary of the region in the $(n - d)$ plane in which defects can exist (the "triangle of defects") is the diagonal $n = d$, which plays an important role in critical phenomena, and the line $n = 0$, which describes disordered systems. The case $n = 0$ corresponds to a stable defect whose dimension is the same as that of the space. Therefore the entire system can be regarded as the center of a defect with a disordered structure.

In a uniaxial nematic liquid crystal, the order parameter is a linear element, i.e., a vector without direction. For an n-component order parameter, the manifold of internal states $F = P_{n-1}$ is a real $(n - 1)$-dimensional projective space. For ordinary nematics in 3-dimensional space, we have $F = P_2$, i.e., a projective plane. For 2-dimensional nematics, $F = P_1 = S_1$. Then $\pi_1(P_m) = Z_2$ and $\pi_r(P_m) = \pi_r(S_m)$ for $r > 1$, where Z_2 is the 2-element group of integers (mod 2). Therefore, for example, in ordinary 3-dimensional nematics, in addition to point defects, which are common for them and for the corresponding vector systems, there can occur topologically stable linear defects which have the property of being proper antiparticles. Two nematic linear defects can decay into point defects.

In the superfluid A phase of ^{3}He, if the nuclear spin

degrees of freedom are neglected, the orbital order parameter is a system of three orthogonal vectors. The manifold of internal states is $F = SO(3) = P_3$. Therefore the A phase can be regarded as a variety of nematic with high dimension. It is predicted that in a 3-dimensional volume of ^{3}He there will be no walls and points, and also lines, which are proper antiparticles. If we attempt to construct a point defect for one of the three orthogonal vectors, string singularities attached to the point defect necessarily appear for the other two vectors. This situation is reminiscent of the Dirac monopole.

Topological Classification of Particle-Like Solutions. Particle-like solutions of the Yang—Mills equations in Euclidean space-time (instantons and monopoles) can be classified in a natural way by means of characteristic classes. Characteristic classes (in particular, Chern classes) are the simplest global invariants which measure the deviation of the structure of a fiber space from a direct product. They are intimately related to the concept of curvature. A characteristic class is the total curvature corresponding to some connection. At the same time, characteristic classes determine the behavior of smooth vector fields with isolated zeros on a compact orientable manifold. The relation between the simplest characteristic class — the Euler characteristic $\chi(M)$ of a manifold M — and the number of isolated zeros of a smooth vector field v is given by Hopf's theorem: $\chi(M) = \Sigma$ zeros of v. The properties of magnetic monopoles and instantons can be treated from a unified point of view by appealing to the global properties of gauge transformations.

Consider a d-dimensional Euclidean space, which is conformal to a sphere S^d. It can be covered by several local maps. Suppose that each map covers a region R_i and that in the overlap region $R_i \cap R_j$ of neighboring maps transition functions φ_{ij} are given. Gauge fields in R_i and R_j are defined in such a way that for $x \in R_i \cap R_j$ we have

$$\mathbf{A}_\mu^{(i)}(x) = \varphi_{ij}^{-1}(x)\,\mathbf{A}_\mu^{(j)}(x)\varphi_{ij}(x) + \varphi_{ij}^{-1}(x)\partial_\mu\varphi_{ij}(x).$$

The global gauge transformations consist of deformations of the regions R_i and the ordinary gauge transformations for each $\mathbf{A}_\mu^{(i)}$. We shall say that a set of gauges which can be globally reduced to each other forms a gauge type. In the Abelian case, this approach makes it possible to describe the Dirac monopole without string singularities.[48] The gauge type is characterized in this case by the Dirac quantization condition.

The gauge type on a compactified Euclidean space S_4 is uniquely characterized by the homotopy class $\pi_3(G)$ of transition functions φ_{12} which map a region $R_1 \cap R_2$ homeomorphic to S^3 onto a semisimple group G if G is connected. It can be determined by means of the second Chern class $c_2 = (1/8\pi^2)$ Tr $(\Omega \wedge \Omega)$, where $\Omega = {}^1/_2 F_{\mu\nu} dx^\mu \wedge dx^\nu$ is the curvature form of a fiber space. The gauge type is characterized by the number

$$q = \int c_2 = \int_{R_1} c_2 + \int_{R_2} c_2 - \int_{R_1 \cap R_2} c_2 = (1/8\pi^2) \int_{S^3} (J_\mu^{A(1)} - J_\mu^{A(2)})\, d\sigma^\mu =$$

$$- (1/24\pi^2) \int_{S^3} \varepsilon_{\mu\nu\alpha\beta} \mathrm{Tr}(\varphi_{12}^{-1} \partial_\nu \varphi_{12} \varphi_{12}^{-1} \partial_\alpha \varphi_{12} \varphi_{12}^{-1} \partial_\beta \varphi_{12})\, d\sigma^\mu, \quad (12.17)$$

where $J_\mu^A = \varepsilon_{\mu\nu\alpha\beta}$ Tr $(A_\nu \partial_\alpha A_\beta + {}^2/_3 A_\nu A_\alpha A_\beta)$. In quantum field theory, J_μ^A is called the *anomalous current*. The second Chern class can be expressed in terms of the divergence of the anomalous current through the identity

$$c_2 = (1/32\pi^2) \varepsilon_{\mu\nu\alpha\beta} \mathrm{Tr}(F_{\mu\nu} F_{\alpha\beta}) d^4x = (1/8\pi^2) \partial_\mu J_\mu^A d^4x. \quad (12.18)$$

(Here $F_{\mu\nu} = \partial_{[\mu} A_{\nu]} + [A_\mu, A_\nu]$.)

In Maxwell's electrodynamics, we have an analogous identity:

$$F_{\mu\nu} {}^*F^{\mu\nu} \equiv \partial_\mu (\varepsilon^{\mu\nu\tau\lambda} A_\nu F_{\tau\lambda}) \equiv \partial_\mu J^{\mu A}.$$

For arbitrary gauge fields in a pseudo-Euclidean space V_4, we have

$$F_{\mu\nu}^{a} {}^* F_a^{\mu\nu} \equiv \partial_\mu [\varepsilon^{\mu\nu\tau\lambda} (A_\nu^a F_{a\tau\lambda} + {}^1/_3 f_{abc} A_\nu^a A_\tau^b A_\lambda^c)] = \partial_\mu J^{\mu A}.$$

For $F_{\mu\nu}^{a} {}^* F_a^{\mu\nu} = 0$, i.e., $\mathbf{E}^a \mathbf{H}_a = 0$, the anomalous current is locally conserved ($\partial_\mu J^{\mu A} = 0$). Therefore we have the conserved charge

$$Q_m = \int \varepsilon^{0ikl} (A_i^a F_{akl} + {}^1/_3 f_{abc} A_i^a A_k^b A_l^c) d^3x.$$

The condition $Q_m = \mathrm{const}$ in electrodynamics means that the number of vortex lines in a closed system is conserved.

The solution of the Yang—Mills equations describing a magnetic monopole has the form ('t Hooft[49] and Polyakov[50])

$$\left.\begin{aligned} &\varphi^a(x) = F x^a / r; \ \hat{\varphi}^a(x) = x^a / r; \\ &A_0^a = 0; \ A_i^a = (1/e)\, \varepsilon_{aij} x^j / r^2. \end{aligned}\right\} \quad (12.19)$$

Here $i, j = 1, 2, 3$ are spatial indices, $a, b = 1, 2, 3$ are indices in isospace, and φ^a are scalar Higgs fields. The magnetic charge can be expressed in terms of the Higgs fields by the formula

$$M = (1/4\pi) \int k_0\, d^3x = -(1/8\pi e) \int \varepsilon_{0\nu\rho\sigma}\, \varepsilon_{abc}\, \partial^\nu \hat{\varphi}^a \partial^\rho \hat{\varphi}^b \partial^\sigma \hat{\varphi}^c\, d^3x. \quad (12.20)$$

Here e is the electron charge. Thus, magnetic charge without singular strings can be interpreted as a topological characteristic of the Higgs fields. However, it should be noted that in going over to another coordinate system the expression (12.20) for M changes. It explicitly contains the vector potentials of the gauge field, but the total magnetic charge does not change. The magnetic charge is also conserved in time, regardless of the equations of motion ($\dot{M} \equiv 0$). The Dirac quantization conditions mean that $eM = q$. The current k_0 vanishes everywhere, apart from the zeros of the Higgs field, where it is singular. Recalling that Higgs fields play the role of sources in the Yang—Mills equations and that the spinor wave functions of electrons in Maxwell's equations play the same role, the relation between the origin of magnetic charge in the Yang—Mills theory and the origin of strings of magnetic flux in electrodynamics becomes clear. Magnetic charge, like the flux, occupies the region of space where the wave functions of the particles interacting with the gauge field vanish. The number of quanta of magnetic flux in a string, like the number of quanta of magnetic charge in a monopole, is determined by a certain integral over a closed surface enclosing the region of zeros of the wave functions of the particles. In electrodynamics this closed surface is S^1, and in the Yang—Mills theory it is S^2.

In a d-dimensional conformally flat Euclidean space, the gauge type of regular gauge fields is characterized by the homotopy class $\pi_{d-1}(G)$ of transition functions defined on S^{d-1}. If d is even, the homotopy class can be described by the $(d/2)$-th Chern class, defined as

$$[(-1)^{d/2}/(2\pi \mathrm{i})^{d/2}\,(d/2)!] \sum \delta^{j_1 \dots j_{d/2}}_{i_1 \dots i_{d/2}} \Omega^{i_1}_{j_1} \wedge \dots \wedge \Omega^{i_{d/2}}_{j_{d/2}}, \quad (12.21)$$

where the summation is taken over all ordered subsets $(i_1 \dots i_{d/2})$ of $d/2$ elements belonging to the set $(1 \dots r)$, where r is the dimension of the matrices of the gauge group. The $(d/2)$-dimensional δ symbol implies a summation over all permutations $(j_1 \dots j_{d/2})$ of $(i_1 \dots i_{d/2})$, the odd permutations being taken with a minus sign; Ω is the curvature form of the matrix representation of the gauge group G. For a semisimple group G, all the arguments given above can easily be generalized. But if the dimension of the manifold is odd, the homotopy class $\pi_{d-1}(G)$ cannot be represented in terms of the curvature form. This happens because any characteristic class that can be represented in terms of $F_{\mu\nu}$ must be an even form.

Consider now a $(d-1)$-dimensional sphere S^{d-1} in a d-dimensional Euclidean space and Yang—Mills fields on S^{d-1}.

The gauge type of a bundle over S^{d-1} is determined by the homotopy group $\pi_{d-2}(G)$. If d is even, there is no representation of $\pi_{d-2}(G)$ in terms of the curvature, since $d-1$ is odd. If d is odd, we can define the $[(d-1)/2]$-th Chern class, which is obtained from the general formula (12.21) with the substitution $d \to d-1$. This makes it possible to generalize the concept of magnetic flux in a 3-dimensional space determined by the first Chern class for a U(1) bundle to spaces of higher dimensions. It is easy to see that the integral of the $[(d-1)/2]$-th Chern class over the sphere S^{d-1} is independent of the choice of sphere, as long as the singularities are unaffected. Thus, if the dimension is odd, the $[(d-1)/2]$-th Chern class can be used to describe the gauge type. If the gauge field has a nontrivial $[(d-1)/2]$-th Chern class, its behavior at infinity is not purely of the gauge type. Consequently, it is not possible to compactify the space, and the topology of the base is different from the topology of S^d.

An instanton can be defined as a classical gauge field in a configuration such that its nontrivial topological characteristic is $\pi_{d-1}(G)$ in a d-dimensional Euclidean space. Then a monopole can be defined as a classical gauge field (possibly singular) in a configuration such that its nontrivial topological characteristic is $\pi_{d-2}(G)$ in a d-dimensional Euclidean space. Thus, if we characterize pseudoparticles by some topological charge expressed in terms of $F_{\mu\nu}$, according to these definitions a pseudoparticle will be called an instanton if the dimension of space-time is even, and a monopole if the dimension of space-time is odd. For example, magnetic charge in 3-dimensional space is characterized by the homotopy group $\pi_1(G)$, which corresponds to the class of closed loops. As is well known, $\pi_1(SU(n)) = 0$ and $\pi_1(SO(n)) = Z_2$ for $n \geqslant 3$. Consequently, there are no monopoles in an SU(n) gauge theory in 3-dimensional space. The first Chern class vanishes identically for SU(n). For SO(n), it is in general not possible to write $\pi_1(SO(n))$ in terms of the curvature tensor. This was shown by Wu and Yang[48] for SO(3). If we consider a generalized monopole in 5-dimensional space, it will be characterized by the homotopy group $\pi_3(G)$ and, therefore, by the second Chern class. It is well known that $\pi_3(SU(n)) = \pi_3(\mathrm{Sp}(n)) = Z$ for $n \geqslant 2$, and $\pi_3(SO(n)) = Z$ for $n \leqslant 5$. Then a 5-dimensional monopole can have an infinite number of gauge types. This is also true for 4-dimensional instantons.

Let us consider the topology of the multi-instanton solutions[51]

$$A_\mu = (i/2)\,\sigma_a A_\mu^a;\; A_\mu^a = \eta_{a\mu\nu}\,\partial_\nu \ln\rho;\; \rho = \sum_{i=1}^{n} \lambda_i^2/(x-x_i)^2, \quad (12.22)$$

where σ_a are the Pauli matrices, and $\eta_{a\mu\nu} = \varepsilon_{0a\mu\nu} - {}^1/_2\varepsilon_{abc}\varepsilon_{bc\mu\nu}$; we have $\varepsilon_{0123} = 1$ and $(\partial_\mu\omega)\,\omega^{-1} = (i/2)\,\sigma_a\eta_{a\mu\nu}\,2x_\nu/x^2$, where $\omega = (x_0 + ix_a\sigma^a)/\sqrt{x^2}$.

The function ρ is the general solution of the equation $(1/\rho)\Box\rho = 0$, which is equivalent to the condition of self-duality of $F_{\mu\nu}$. Equation (12.22) can be transformed to the form

$$A_\mu = -(1/\rho) \sum_{i=1}^{n} [\lambda_i^2/(x-x_i)^2]\,(\partial_\mu\,\omega_i)\,\omega_i^{-1}, \quad (12.23)$$

where

$$\omega_i = [(x-x_i)_0 + i\,(x-x_i)_a\sigma_a]/[(x-x_i)^2]^{1/2}.$$

The stress tensor of the gauge field has the form

$$F_{\mu\nu} = i\,\frac{2}{\rho}\left[\sum_{i=1}^{n} \frac{\lambda_i^2}{(x-x_i)^4} - \frac{1}{\rho}\sum_{i,\,j=1}^{n} \frac{\lambda_i^2\,\lambda_j^2\,(x-x_i,\,x-x_j)}{(x-x_i)^4\,(x-x_j)^4}\right]\eta_{a\mu\nu}\,\sigma^a.$$

Clearly, $F_{\mu\nu}$ is not singular at $x = x_i$, although the vector potentials A_μ are singular [see (12.23)], since the singularities in (12.23) can be eliminated by a purely gauge transformation. It is convenient to study the global properties of this solution by the method of homotopy classes outlined above.

We partition the entire space into n regions $R_1, \ldots, R_n$ in such a way that each region R_i contains only one singular point x_i and its intersection with the neighboring regions occurs only at infinity. It is necessary to arrange the regions R_i in such a way that only neighboring regions R_i and R_{i+1} intersect. A regular gauge field is defined in each region R_i as

$$A_\mu^{(i)} = \frac{1}{\rho}\sum_{i\neq l}\left[\frac{\lambda_l^2}{(x-x_l)^2}\,\omega_i^{-1}\,\partial_\mu\,\omega_i - \frac{\lambda_l^2}{(x-x_l)^2}\,\omega_i^{-1}\,(\partial_\mu\,\omega_l)\,\omega_l^{-1}\,\omega_i\right]. \quad (12.24)$$

The functions ω_i are singular at $x = x_i$. In each intersection $R_i \cap R_{i+1}\,(\simeq S^3)$, we define the transition function $\varphi_{i,\,i+1} = \omega_{i+1}^{-1}\,\omega_i$. Then

$$A_\mu^{(i)} = \varphi_{i,\,i+1}^{-1}\,A_\mu^{(i+1)}\,\varphi_{i,\,i+1} + \varphi_{i,\,i+1}^{-1}\,\partial_\mu\,\varphi_{i,\,i+1}. \quad (12.25)$$

We see that for $|x| \to \infty$ we have $\varphi_{i,\,i+1} \to 1$ and $A_\mu^{(i)} \to O\,(1/|x|^2)$. The vector potential (12.25) leads to the same field tensor as (12.23), since they are related to each other by a (singular) gauge transformation

$$A_\mu^{(i)} = \omega_i^{-1}\,\hat{A}_\mu\,\omega_i + \omega_i^{-1}\,\partial_\mu\,\omega_i,$$

where $\hat{A}_\mu$ is the solution (12.22).

The global gauge type of this solution is characterized by the integral of the second Chern class. At the same time, $q = \sum_{i=1}^{n-1}$ (integrals of the homotopy classes of $\varphi_{i,\,i+1}$) $= n - 1$, since the integral of the homotopy class (or the number of turns) of $\varphi_{i,\,i+1}$ is equal to unity for each i.

We choose a coordinate system such that the singular points x_i and x_{i+1} lie on the time axis $x_0 = t$, with $t_i > 0$ and $t_{i+1} < 0$. For simplicity, we assume that the intersection $R_i \cap R_{i+1}$ is a hypersphere between t_{i+1} and t_i. Then

$$\varphi_{i,\,i+1} = \{[r^2 + (t - t_{i+1})^2]\,[r^2 + (t - t_i)^2]\}^{-1}[\,(t - t_{i+1}) \times (t - t_i) + r^2 - \mathrm{i} x_a \sigma_a\,(t_{i+1} - t_i)];\; r^2 = x_1^2 + x_2^2 + x_3^2.$$

For any hypersphere S^3 between t_i and t_{i+1}, we have $\varphi_{i,\,i+1} \to 1$ for $r \to \infty$ and $\varphi_{i,\,i+1} \to -1$ for $r \to 0$. Thus, when x ranges over S^3, the value of the function $\varphi_{i,\,i+1}$ ranges over the SU(2) group manifold, and each point on SU(2) occurs once and only once. Therefore the integral of the homotopy class is equal to unity. This result holds for any intersection, since continuous deformations of this region produce continuous changes of $\varphi_{i,\,i+1}$ if the singularities are not affected.

From the point of view discussed above, tunneling from one classical vacuum to another vacuum related to the first one by a gauge transformation (a process described by an instanton) corresponds to the presence in some space-time region of homotopically nontrivial transition functions. Therefore the n-instanton solution (12.23) can be generalized by writing the vector potential for any combination of instantons and anti-instantons. Such a field configuration is not a solution of the field equations, although the action integral remains finite. The simplest example of an instanton—anti-instanton configuration gives

$$A_\mu = -\frac{1}{\rho}\left[\frac{\lambda_1^2}{(x-x_1)^2}(\partial_\mu \omega_1)\,\omega_1^{-1} + \frac{\lambda_2^2}{(x-x_2)^2}(\partial_\mu \omega_2)\,\omega_2^{-1} + \frac{\lambda_3^2}{(x-x_3)^2}(\partial_\mu \omega_{32})\,\omega_{32}^{-1}\right], \tag{12.26}$$

where $\omega_{32} = \omega_2 \omega_3^{-1} \omega_2^{-1}$. Here we have used two transition functions φ_{12} and φ_{23}^{-1}. The inverse transition function corresponds to an anti-instanton. Solutions of the type (12.26) are used to study the interaction of instantons.

Solitons. Solitons satisfy the condition $H = \mathrm{const}$ (the energy is constant). They were originally obtained in 2-dimensional models of field theory as topologically nontrivial solutions of the sine-Gordon equation

$$\partial^2\psi/\partial t^2 - \partial^2\psi/\partial\xi^2 = -\gamma \sin\psi. \qquad (12.27)$$

The geometrical meaning of this equation is that it describes a 2-dimensional surface of constant curvature in harmonic coordinates.[52] Indeed, the 2-dimensional interval in harmonic coordinates has the form $ds^2 = \sin^2(\psi/2)du^2 + \cos^2(\psi/2)dv^2$, and the only component of the curvature tensor R_{1212} is proportional to the scalar curvature $K = (\partial^2\psi/\partial t^2 - \partial^2\psi/\partial\xi^2)/\sin\psi$. Introducing the notation $u = t$, $v = \xi$, and $K = -\gamma$, this gives (12.27). The static solutions of the 2-dimensional equation (12.27) also describe one-dimensional motion of a mathematical pendulum if $\psi = \alpha$, this being the angle of deviation from a position of unstable equilibrium. The soliton solution $\alpha = 4 \operatorname{arctg} \exp(\sqrt{\gamma} t)$ corresponds to the case in which the motion of the pendulum begins from the upper (unstable) equilibrium position with zero velocity. Clearly, the pendulum returns to this same equilibrium position after an infinite time, but the angle α then changes by 2π. Therefore we can say formally that the pendulum has gone over from an equilibrium position with $\alpha = 0$ to another equilibrium position with $\alpha = 2\pi$. In the language of quantum theory, this process corresponds to a transition from one vacuum to another.

In 2-dimensional gauge field theories with spontaneous symmetry breaking, we have the Hamiltonian of a scalar field,

$$H = \frac{1}{2}\int_{-\infty}^{\infty} dx \left[\pi^2 + \left(\frac{d\varphi}{dx}\right)^2 - \mu^2\varphi^2 + \frac{\lambda\varphi^4}{2}\right], \text{where } \pi = \frac{\partial\varphi}{\partial t}.$$

If the vacuum expectation value of the field φ is not equal to zero ($\bar{\varphi}^2 = \mu^2/\lambda$), there is a topologically nontrivial extremal of the potential energy, which is determined by the equation

$$\varphi_c'' + \mu^2\varphi_c - \lambda\varphi_c^3 = 0. \qquad (12.28)$$

The topological nontriviality consists in the fact that for $x \to \pm\infty$ the solution tends to different constants $\varphi_c(\pm\infty) = \pm\bar{\varphi}$ (like the phase of the mathematical pendulum). The solution $\varphi_c(x) = (\mu/\sqrt{\lambda})\tanh(\mu x/\sqrt{2})$ contains one "twist" of the boundary values (one kink) on the interval $-\infty < x < \infty$. In more complicated cases, there can be solutions containing several twists of the boundary values of the field on an interval of variation of the independent variable. Such solutions are called n-*kink* (or n-*soliton*) solutions. As a rule, they are not static. The number of solitons is called the *topological quantum* number.

Topologically nontrivial soliton solutions arise in field theory when a supplementary condition of constancy of the field amplitude, $\langle\varphi^2\rangle = \text{const}$, is imposed but the direction of the vector φ in the space of fields is not fixed. On the boundaries of the domain of definition, φ may have different directions.

Quantization of soliton solutions leads to the appearance of particles with extremely large mass. In fact, the fluctuations of the field in the neighborhood of a soliton solution behave like "particles" in a potential well, whose dimensions are determined by the field gradients. The spectrum of oscillatory energy levels near φ_c leads to the mass spectrum[50,53]

$$M_n = \frac{2\sqrt{2}}{3}\frac{\mu^3}{\lambda} + \frac{\sqrt{3}+2}{2\sqrt{2}}\mu + \frac{\mu}{\sqrt{2}}\sqrt{n(4-n)}. \qquad (12.29)$$

This mass spectrum contains a dependence on the dimensions of the well, $l \sim \sqrt{2}/\mu$, and on the reciprocal of the coupling constant λ. Consequently, if the coupling constant is small, the masses (energies) of "extended particles" may be very large. Equation (12.29) is associated with the hypothesis of the existence of strongly interacting particles constructed from weakly interacting particles.[54]

In the real 4-dimensional case, there exist analogous solutions concentrated around a finite closed contour (string) undergoing periodic nonlinear oscillations. These solutions describe strings (or vortices in superfluids) interacting through a massless scalar field. Since the energy and action are related by the equation $H_{\pm} - 2U = L_{\mp}$, where U is the potential, the plus and minus signs signify Euclidean and pseudo-Euclidean signatures of V_4, and there exists a relation between solitons and instantons.

§13. GAUGE FIELDS AND THE STRUCTURE OF SPACE-TIME

Introduction. There are several ways of establishing a connection between a gauge field and the structure of space-time. If the fiber is projected onto the tangent space to the base, i.e., the internal and space-time variables are identified, the gauge field manifests itself in deviations of the geometry and topology of V_4 from the Euclidean forms (through curvature, torsion, or nonmetric transport, depending on the properties of the projection, as well as the appearance of nonzero topological charges). The simplest example is the SL(2, C) gauge theory of gravitation, in which the gauge group is taken to be the Lorentz group, which transforms the local tetrad basis of the tangent space at each point of V_4. For the metric connection, this theory generalizes Einstein's theory. In a similar way, we can obtain Weyl's geometrical treatment of electrodynamics, which involves nonmetric connection coefficients.

A second method is to identify the vector potentials of gauge fields with connection coefficients of a configuration space of higher dimension in which V_4 is regarded as a hypersurface. In this case, the tangent space to V_4 is part of the tangent space to V_{r+4}. The connection coefficients describing the gravitational field form part of the components of the total connection coefficients. If it is assumed that the wave functions have a periodic dependence on the internal variables, it is possible with this approach to obtain a generalization of the mass operator.[16]

A third method is to identify the vector potentials of gauge fields with nondiagonal components of the metric of a $(r+4)$-dimensional configuration space in which V_4 is regarded as a surface. With this procedure, it is possible to obtain a geometrical interpretation of the mass of a gauge field and to show that the Lagrangian of the system of gravitational and Yang—Mills fields is the scalar curvature of a $(r+4)$-dimensional fiber space.[9,12-15] This approach generalizes the Kaluza—Klein unified theory of gravitation and electromagnetism to non-Abelian gauge fields.

A generalization of geometrodynamics is the simultaneous solution of the system of Yang—Mills—Einstein equations and the simultaneous solution of the system of equations for scalar (Higgs), vector (gauge), and tensor (gravitational) fields. Here we obtain new types of particle-like solutions of the nonlinear classical equations discussed in the preceding section.

Gauge Fields and Non-Euclidean Space-Time. An interpretation of gauge fields as a manifestation of a non-Euclidean character of space-time V_4 can be obtained by using the projection of the fiber onto the tangent space to the base by means of tetrads. In the general case, it turns out that gauge fields give rise to a geometry in V_4 with torsion and with a nonzero covariant derivative of the metric tensor.[3,4]

In the ordinary approach of differential geometry, the geometry of a manifold is completely determined by specification of three quantities[55]: the metric tensor $g_{\mu\nu}$, its covariant derivative $Q_{\mu\nu\tau} = -g_{\nu\tau;\mu}$, and the torsion tensor $S^{\lambda}_{\mu\nu}$. If $Q_{\mu\nu\tau} = 0$, the transport is said to be *metric*. If $Q_{\mu\nu\tau} = -\Gamma_{\mu} g_{\nu\tau}$, where Γ_{μ} is some vector, the transport is called *semimetric*, and for $S^{\lambda}_{\mu\nu} = 0$ it is said to be of Weyl type. If $S^{\lambda}_{\mu\nu} = 0$, the transport is *symmetric*;

for $S^{\lambda}_{\mu\nu} = p_{[\mu}\delta^{\lambda}_{\nu]}$, it is semisymmetric. All these varieties of geometries in V_4 can be obtained by means of a mapping of the fiber onto the base.

The total connection coefficient of V_4 has the form

$$\Gamma^{\varkappa}_{\mu\lambda} = \begin{Bmatrix} \varkappa \\ \mu\lambda \end{Bmatrix} + S_{\mu\lambda}{}^{\varkappa} - S_{\lambda}{}^{\varkappa}{}_{\mu} - S_{\mu}{}^{\varkappa}{}_{\lambda} + \frac{1}{2}(Q_{\mu\lambda}{}^{\varkappa} + Q_{\lambda}{}^{\varkappa}{}_{\mu} - Q^{\varkappa}{}_{\mu\lambda}), \quad (13.1)$$

where $\begin{Bmatrix} \varkappa \\ \mu\lambda \end{Bmatrix}$ is the Christoffel symbol of the second kind:

$$\begin{Bmatrix} \varkappa \\ \mu\lambda \end{Bmatrix} = \frac{1}{2} g^{\varkappa\sigma} (\partial_{\mu} g_{\lambda\sigma} + \partial_{\lambda} g_{\sigma\mu} - \partial_{\sigma} g_{\mu\lambda}) - \Omega^{\varkappa}_{\mu\lambda} + g_{\mu\tau} g^{\varkappa\sigma} \Omega^{\tau}_{\lambda\sigma} +$$

$$g_{\lambda\tau} g^{\varkappa\sigma} \Omega^{\tau}_{\mu\sigma} \overset{h}{=} \frac{1}{2} g^{\varkappa\sigma} (\partial_{\mu} g_{\lambda\sigma} + \partial_{\lambda} g_{\sigma\mu} - \partial_{\sigma} g_{\mu\lambda}) . \quad (13.2)$$

Here $\Omega^{\tau}_{\mu\sigma}$ are the coefficients of nonholonomic character of the coordinate system, the symbol $\overset{h}{=}$ indicates equality in a holonomic coordinate system, and

$$S^{\varkappa}_{\mu\lambda} = \Gamma^{\varkappa}_{[\mu\lambda]} + \Omega^{\varkappa}_{[\mu\lambda]} \overset{h}{=} \Gamma^{\varkappa}_{[\mu\lambda]}. \quad (13.3)$$

We note that if $g_{\lambda\sigma}$, $Q^{\varkappa\lambda}_{\mu}$, and $S^{\varkappa}_{\mu\lambda}$ are assumed to be given then not only the part of $\Gamma^{\varkappa}_{\mu\lambda}$ which changes sign but also $\Gamma^{\varkappa}_{(\mu\lambda)}$ will depend on the choice of the torsion $S^{\varkappa}_{\mu\lambda}$. This dependence is absent if and only if $S_{\mu\lambda\varkappa}$ is a completely antisymmetric tensor. We note also that the Christoffel symbols are symmetric only in a holonomic coordinate system.

Suppose that the vectors of the tangent space satisfy the relations

$$\Phi^{i} = h^{i}_{\mu}\Phi^{\mu}; \; \Phi_{i} = h^{\mu}_{i}\Phi_{\mu}; \quad (13.4)$$

$$\nabla_{\mu}\Phi^{i} = h^{i}_{\nu}\nabla_{\mu}\Phi^{\nu}. \quad (13.5)$$

Then we can express the gauge field $A_{\mu}(kl)$ in terms of the tetrads and the connection coefficients $\Gamma^{\nu}_{\mu\tau}$ in the base:

$$\Gamma^{\tau}_{\mu\nu} = h^{\tau}_{i}\partial_{\mu}h^{i}_{\nu} - h^{\tau}_{i}A_{\mu}(pq)\underset{(pq)}{I^{i}_{k}}h^{k}_{\nu}. \quad (13.6)$$

Using (13.6), it is easy to obtain the covariant derivative of the metric tensor:

$$g_{\lambda\nu;\mu} = 2A_{\mu}(pq)\, h_{i\,(\lambda}\, \underset{(pq)}{I^{i}_{k}}\, h^{k}_{\nu)} . \quad (13.7)$$

Analogous formulas can be obtained for the mapping onto the base of the space of a 4-dimensional representation of an arbitrary group G_r. For this, it is sufficient to replace the matrices of the representation of the Lorentz group by matrices of the representation of G_r: $\underset{a}{I^{i}_{k}}$ and to replace the gauge field $A_\mu(pq)$ by the vector potential $\overset{a}{A}_\mu$. Formulas for the group SU(2) can be obtained by identifying the components of $A_\mu(pq)$ with two vector potentials: the "electric" potential $A_\mu^a = A_\mu(0, a)$, and the "magnetic" potential $^*A_\mu^a = \varepsilon^{abc}A_\mu(b, c)$. If the representation of G_r is such that all the matrices $\underset{a}{I}{}^i_k$ are real and antisymmetric, (13.7) reduces to

$$g_{\lambda\nu;\mu} = 0. \tag{13.8}$$

For symmetric connection coefficients $\Gamma^\lambda_{\mu\nu}$, it follows from Eq. (13.6) that

$$\Gamma^\tau_{[\mu\nu]} = h^\tau_i \partial_{[\mu} h^i_{\nu]} - h^\tau_i A^a_{[\mu} \underset{a}{I}{}^i_k h^k_{\nu]} = 0, \tag{13.9}$$

and

$$A^a_\mu h_{i[\tau} \underset{a}{I}{}^i_k h^k_{\nu]} = \left(h_{\tau i}\partial_{[\mu} h^i_{\nu]} + h_{\nu i}\partial_{[\tau} h^i_{\mu]} - h_{\mu i}\partial_{[\nu} h^i_{\tau]}\right) + {}^1/_2\,(Q_{\tau\mu\nu} - Q_{\nu\mu\tau}). \tag{13.10}$$

If the matrices $\underset{a}{I}{}^i_k$ are symmetric, Eq. (13.10) gives

$$Q_{\nu\mu\tau} - Q_{\tau\mu\nu} = 2\,(h_{\tau i}\partial_{[\mu}h^i_{\nu]} + h_{\nu i}\partial_{[\tau}h^i_{\mu]} - h_{\mu i}\partial_{[\nu}h^i_{\tau]}) = 4\Delta_{\mu,\tau\nu} \tag{13.11}$$

If $\underset{a}{I}{}^i_k$ are antisymmetric, we obtain $Q_{\mu\nu\tau} = 0$ and

$$A^a_\mu h_{i\tau} \underset{a}{I}{}^i_k h^k_\nu = \Delta_{\mu,\tau\nu}. \tag{13.12}$$

Here $\Delta_{\tau,\mu\nu}$ is the Ricci connection coefficient. Using the quantities $\underset{a}{\omega}_{\tau\nu} = h_{\tau i}\underset{a}{I}{}^i_k h^k_\nu$ and the inverse quantities $\overset{a}{\omega}{}^{\nu\lambda}$, which are related to the former by the equations $\underset{a}{\omega}_{\nu\lambda}\overset{b}{\omega}{}^{\nu\lambda} = \lambda\delta^b_a$ and $\overset{a}{\omega} = g_{ab}\underset{b}{\omega}$, where $g_{ab} = f^l_{am}f^m_{lb}$ is the metric on the group, we we can rewrite Eq. (13.12) in the form $A^a_\mu = (1/\lambda)\overset{a}{\omega}{}^{\nu\lambda}\Delta_{\mu,\nu\lambda}$. Similarly, Eq. (13.6) can be represented in the form

$$A^a_\mu = \lambda^{-1}\overset{a}{\omega}{}^{\nu\lambda}\,(h_{\lambda i}\partial_\mu h^i_\nu - \Gamma_{\lambda,\mu\nu}). \tag{13.13}$$

The Ricci rotation coefficients are nonzero only in a nonholonomic coordinate system ($\Omega^\tau_{\mu\nu} = h^\tau_i\partial_{[\mu}h^i_{\nu]} \neq 0$). They describe transport of an orthogonal system of tetrads. In a holonomic coordinate system with $\Gamma^\tau_{[\mu\nu]} = 0$, it follows from (13.11) and (13.12) that $Q_{\nu\mu\tau} = Q_{\tau\mu\nu}$ and $A^a_\mu = 0$. If $\Gamma^\tau_{[\mu\nu]} \overset{h}{=} S^\tau_{\mu\nu} \neq 0$, we obtain for the symmetric and antisymmetric $\underset{a}{I}{}^i_k$, respectively, the equations

$$Q_{\nu\mu\tau} - Q_{\tau\mu\nu} \overset{h}{=} 2\,(S_{\nu\tau\mu} + S_{\nu\mu\tau} - S_{\tau\mu\nu}); \tag{13.14}$$

$$\overset{a}{A}_{\mu} = (2\lambda)^{-1} \overset{h}{\ } \overset{a}{\omega}{}^{\nu\tau} (S_{\nu\tau\mu} + S_{\nu\mu\tau} - S_{\tau\mu\nu}). \qquad (13.15)$$

Thus, if V_4 is a Riemannian space without torsion, the gauge fields can be related only to the nonholonomic character of the coordinate system in the same way as inertial forces.

The equations (13.15) are equivalent to the conditions of covariant constancy of the tetrads:

$$h^i_{\nu;\mu} = \partial_\mu h^i_\nu - A^a_\mu \underset{a}{I}{}^i_k h^k_\nu - \Gamma^\lambda_{\mu\nu} h^i_\lambda = 0; \qquad (13.16)$$

$$h^\lambda_{i;\mu} = \partial_\mu h^\lambda_i + h^\lambda_k A^a_\mu \underset{a}{I}{}^k_i + \Gamma^\lambda_{\mu\nu} h^\nu_i = 0. \qquad (13.17)$$

The conditions (13.16) and (13.17) ensure that a local basis of the representation of G_r is transformed into itself under parallel transport in V_4. At the same time, they imply that the tetrads at each point $x = \text{const}$ satisfy the equations (10.27) for the coefficients of differentiable mappings. Does there exist a system of tetrads satisfying our conditions? The conditions for integrability of (13.16) have the form

$$h^\lambda_{i;[\mu\nu]} = R_{\mu\nu\tau}{}^\lambda h^\tau_i + F^a_{\mu\nu} \underset{a}{I}{}^k_i h^\lambda_k + S_{\mu\nu}{}^\sigma h^\lambda_{i;\sigma} = 0,$$

where $F^a_{\mu\nu}$ is the stress tensor of the gauge field, and

$$R_{\nu\mu\tau}{}^\lambda = \partial_{[\nu} \Gamma^\lambda_{\mu]\tau} + \Gamma^\lambda_{[\nu|\sigma|} \Gamma^\sigma_{\mu]\tau} + \Omega^\sigma_{\nu\mu} \left(\Gamma^\lambda_{\sigma\tau} + A^a_\sigma \underset{a}{\omega}{}^\lambda_\tau \right)$$

(the vertical lines designate indices which do not participate in the antisymmetrization). For $S^\sigma_{\mu\nu} = 0$, as a consequence of (13.16) and (13.17), this gives a relation between the curvature tensor of V_4 and the stress tensor of the gauge field:

$$R_{\nu\mu\lambda}{}^\tau h^i_\tau = F^a_{\mu\nu} \underset{a}{I}{}^i_k h^k_\lambda \qquad (13.18)$$

or

$$R_{\nu\mu\lambda}{}^\tau = \underset{a}{\omega}{}^\tau_\lambda F^a_{\mu\nu}. \qquad (13.19)$$

Fulfillment of (13.19) also ensures the integrability of (13.17). It follows from Eq. (13.19) that if $\underset{a}{I}{}^i_k$ is an antisymmetric matrix, then $R_{\nu\mu\lambda}{}^\lambda = 0$, which is a necessary and sufficient condition for equal-volume transport.

The Ricci curvature tensor is obtained by contracting (13.19) with $g^{\lambda\nu}$:

$$R^\tau_\mu = F^a_{\mu\nu} \underset{a}{\omega}{}^{\tau\nu}. \qquad (13.20)$$

The expression for the scalar curvature has the form

$$R = F^a_{\mu\nu} \underset{a}{\omega}{}^{\mu\nu} \qquad (13.21)$$

This quantity is invariant with respect to local gauge transformations and can be used as a linear Lagrangian for an arbitrary gauge field. By varying (13.21) with respect to the metric, we can obtain an analog of Einstein's equations for an arbitrary gauge field:

$$R^{\tau}_{\mu} - {}^1/_2\, \delta^{\tau}_{\mu} R = F^{a}_{\mu\nu} \underset{a}{\omega}^{\tau\nu} - {}^1/_2\, \delta^{\tau}_{\mu} F^{a}_{\lambda\nu} \underset{a}{\omega}^{\lambda\nu} = {}^{*}F^{a\tau\nu}\, {}^{*}\underset{a}{\omega}_{\mu\nu} = 0\,.$$

Thus, if a 4-dimensional representation of G_r is given by antisymmetric real matrices, we obtain in V_4 metric transport with torsion. If the matrices $\underset{a}{I}{}^{i}_{k}$ are symmetric, the covariant derivative of the metric tensor is nonzero, and we do not have equal-volume transport. We can obtain equal-volume transport only by introducing a corresponding symmetric transport $\overset{s}{\Gamma}{}^{\lambda}_{\mu\nu}$, for which we must choose a special nonholonomic coordinate system.

E x a m p l e. W e y l ' s T h e o r y.[20] In the interpretation of the electromagnetic field in terms of the 4-dimensional geometry of space-time, Weyl, Eddington, and Fock made use of additional connection coefficients Γ_{μ} corresponding to the group of invariance with respect to extensions of the interval: $ds'^2 = \sigma(x) ds^2$.

Projection of the 4-dimensional representation of this Abelian gauge group onto the tangent space gives in V_4 the expressions[4]

$$\left.\begin{aligned} &\Gamma^{\lambda}_{\mu\nu} = h^{\lambda}_{i}\, \partial_{\mu} h^{i}_{\nu} - \mathrm{i}e\delta^{\lambda}_{\nu} A_{\mu}\,; \\ &\Gamma^{\varkappa}_{\mu\varkappa} = {}^1/_2\, \partial_{\mu} \ln|-g| + 2Q_{\mu} = {}^1/_2\, \partial_{\mu} \ln|-g| - 4\mathrm{i}eA_{\mu}; \\ &S^{\lambda}_{\mu\nu} = -\mathrm{i}eA_{[\mu}\, \delta^{\lambda}_{\nu]};\ Q_{\mu\lambda\nu} = -2\mathrm{i}eg_{\lambda\nu} A_{\mu}. \end{aligned}\right\} \qquad (13.22)$$

Thus, the electromagnetic field corresponds to semimetric transport in V_4 ($Q_{\mu} = -2\mathrm{i}eA_{\mu}$) with torsion ($S_{\mu} = -3\mathrm{i}eA_{\mu}$).

Since Q_{μ} was found to be nonzero, covariant differentiation cannot be permuted with raising and lowering of indices. Transport does not conserve volume, since $F_{\mu\nu} = \frac{1}{4} R_{\mu\nu\tau}{}^{\tau} \neq 0$. In a geodesic coordinate system, $\Gamma^{\lambda}_{\mu\nu} = 0$ and $A_{\mu} = -(\mathrm{i}/8e)\, \partial_{\mu} \ln|-g|$. It is easy to see that in this case $F_{\mu\nu} = 0$.

The relations (13.22) are invariant with respect to noncoordinate transformations of the metric of the form

$$g'_{\mu\nu} = \sigma(x) g_{\mu\nu} \qquad (13.23)$$

and their associated transformations of A_{μ}:

$$A'_{\mu} = A_{\mu} - (\mathrm{i}/2e)\partial_{\mu} \ln \sigma(x) = A_{\mu} + e^{-1} \partial_{\mu} \alpha(x), \qquad (13.24)$$

where $\alpha(x) = -(\mathrm{i}/2) \ln \sigma(x)$.

Thus, gradient invariance in electrodynamics, corresponding to gauge transformations of the wave functions

$$\psi' = \exp\,[i\alpha\,(x)\,]\psi = \exp\left[\frac{1}{2}\ln\sigma\,(x)\right]\psi, \qquad (13.25)$$

is a consequence of invariance of the connection coefficients of V_4 with respect to conformal transformations of the metric, $g'_{\mu\nu} = \exp\,[2i\alpha\,(x)\,]g_{\mu\nu}$. If the transformations (13.23) are regarded as coordinate transformations, the long-established conformal invariance of Maxwell's equations becomes a consequence of the incorporation of Lorentz invariance and gradient invariance in a single symmetry group. Conformal transformations of the coordinates are also associated with the interpretation of the electromagnetic field as a gauge field corresponding to the group of motions of a straight line, $x' = \alpha\,(x) + \beta$.[56] In this case, the gauge group is solvable. Its holonomy group is Abelian (a normal divisor).

Theory of Gravitation as an SO(3, 1) Gauge Field Theory. If the theory of gravitation is constructed as the theory of a gauge field associated with the local Lorentz group transforming the orthogonal basis of the tangent space to V_4, the basic field variables are the Ricci connection coefficients $\Delta_\mu\,(ik)$. These transform orthogonal bases belonging to different points of V_4 into one another and play the role of vector potentials of the gravitational field. The physically measurable quantities characterizing a real gravitational field are the components of the curvature tensor,

$$R_{\mu\nu}\,(ik) = \partial_{[\nu}\,\Delta_{\mu]}\,(ik) - {}^1\!/_2 f^{ik}_{lm\,pq}\Delta_{[\mu}\,(lm)\,\Delta_{\nu]}\,(pq). \qquad (13.26)$$

The Lagrangian analogous to the Yang—Mills Lagrangian has the form[57,58]

$$L = (\lambda^2/16\pi)R_{\mu\nu}\,(ik)R^{\mu\nu}\,(ik). \qquad (13.27)$$

The gravitational field equations are obtained by variation of (13.27) with respect to the connection $\Delta_\mu\,(ik)$. They have the form

$$R^{\mu\nu}\,(ik)_{;\,\mu} = 0. \qquad (13.28)$$

We shall show how the gauge theory of gravitation is related to Einstein's theory.

The Ricci connection coefficients can be expressed in terms of the Christoffel symbols as follows:

$$\Delta_\mu\,(ik) = h^k_\tau\,\partial_\mu h^\tau_i + \Gamma^\tau_{\mu\nu}\,h^\nu_i\,h^k_\tau, \qquad (13.29)$$

where h^τ_i are the tetrad vectors, and $\Gamma^\tau_{\mu\nu}$ are the Christoffel symbols. If (13.29) is satisfied, the tetrad components of the curvature tensor are related to the components of the Riemann tensor by the equation

$$h^i_\lambda h^\mu_k R_{\sigma\tau}(ik) = R_{\sigma\tau\lambda}{}^{\mu}. \qquad (13.30)$$

Then instead of (13.28) we have

$$R^{\mu\nu\tau\lambda}{}_{;\mu} = 0. \qquad (13.31)$$

However, using the symmetry properties of the curvature tensor, it can be shown that in an arbitrary pseudo-Riemannian manifold V_4 the following identity holds:

$$C^{\mu}{}_{\nu\tau\lambda;\,\mu} \equiv R^{\mu}{}_{\nu\tau\lambda;\,\mu} - {}^1/_6\, g_{\nu\,[\lambda}\, R_{;\,\tau]}, \qquad (13.32)$$

where $C^{\mu}{}_{\nu\tau\lambda}$ is the conformal curvature tensor. Since for $n \leqslant 3$ all V_n are conformally flat, i.e., $C^{\mu}{}_{\nu\tau\lambda} = 0$, (13.31) reduces to $R = \text{const}$. In a harmonic coordinate system, this equation has the form of the sine-Gordon equation and admits particle-like solutions — solitons. For $n > 3$, the condition $R = \text{const}$ is obtained from (13.31) by contraction with respect to ν and τ with allowance for the contracted Bianchi identities. Therefore we can assume that in the absence of sources the equations of the gauge theory of gravitation have the form

$$C^{\mu}{}_{\nu\tau\lambda;\mu} = 0. \qquad (13.33)$$

For gravitational fields of general form, of Petrov type N, the equations (13.33) are conformally invariant.

If $g_{\mu\nu}$ is regarded as another independent variable describing the gravitational field, the total Lagrangian is a sum of Einstein and Yang—Mills terms[4,32,57]:

$$\mathcal{L} = R + (\lambda^2/16\pi) R_{\mu\nu\tau\lambda} R^{\mu\nu\tau\lambda}.$$

By varying $\mathcal{L}$ with respect to $g^{\mu\nu}$, we obtain a generalization of Einstein's equations:

$$R_{\mu\nu} = (\lambda^2/2\pi) C_{\mu\tau\lambda\nu} R^{\tau\lambda}. \qquad (13.34)$$

Clearly, all Einstein spaces in a vacuum satisfy (13.31) and (13.34). Conformally Einstein spaces satisfying (13.31) are of wave type (type N). Using the doubly dual curvature tensor $\widetilde{R}^{\mu\nu\tau\lambda} = {}^1/_4 \eta^{\mu\nu\alpha\beta}\eta^{\tau\lambda\gamma\delta} R_{\alpha\beta\gamma\delta}$ and the fact that (13.34) implies $R = 0$, it can be shown that (13.31) and (13.34) imply that

$$\widetilde{R}^{\mu\nu\tau\lambda}_{;\,\mu} \equiv 0; \; \widetilde{R}_{\mu\nu} = (\lambda^2/2\pi)\, C_{\mu\tau\lambda\nu}\, \widetilde{R}^{\tau\lambda}.$$

Thus, the system of equations (13.31)-(13.34) is symmetric with respect to the substitution $\widetilde{R}^{\mu\nu\tau\lambda} \to \widetilde{R}^{\mu\nu\tau\lambda}$ and has a class of doubly self-dual and anti-self-dual solutions $R^{\mu\nu\tau\lambda} = \pm\widetilde{R}^{\mu\nu\tau\lambda}$. Doubly dual solutions in a pseudo-Riemannian space have a number of properties which are characteristic of instantons in a Riemannian space. For example, they give zero energy—momentum tensor of the gravitational field, and the Lagrangian (13.27) reduces to the divergence of the anomalous current. Therefore, as in the case of instantons, the action integral is determined by the boundary values of the field and is finite for a compact orientable manifold. At the same time, the self-duality condition $R_{\alpha\beta\gamma\delta} = \pm R^{*}_{\alpha\beta\gamma\delta}$ in a pseudo-Riemannian space is satisfied only by Minkowski space. Doubly dual gravitational fields, like instantons, minimize the action integral with the Lagrangian (13.27). These results can be obtained by making use of the identities[32]

$$R_{\mu\nu\tau\lambda}\widetilde{R}^{\alpha\nu\tau\lambda} \equiv {}^1/_4\,\delta^{\alpha}_{\mu}R_{\gamma\nu\tau\lambda}\widetilde{R}^{\gamma\nu\tau\lambda} \equiv {}^1/_4\,\delta^{\alpha}_{\mu}\left(-C_{\gamma\nu\tau\lambda}C^{\gamma\nu\tau\lambda} + 2R^{\nu}_{\gamma}R^{\gamma}_{\nu} - {}^2/_3\,R^2\right);$$

$$R_{\mu\nu\tau\lambda}R^{\alpha\nu\tau\lambda} + \widetilde{R}^{\alpha\nu}{}_{\tau\lambda}\widetilde{R}_{\mu\nu}{}^{\tau\lambda} \equiv {}^1/_2\,\delta^{\alpha}_{\mu}R_{\gamma\nu\tau\lambda}R^{\gamma\nu\tau\lambda};$$

$$\left(R_{\alpha\beta\gamma\delta} + \widetilde{R}_{\alpha\beta\gamma\delta}\right)\left(R^{\alpha\beta\gamma\delta} + \widetilde{R}^{\alpha\beta\gamma\delta}\right) \equiv 8\left(R^{\beta}_{\alpha}R^{\alpha}_{\beta} - {}^1/_4\,R^2\right).$$

The conditions

$$R^{\mu\tau\lambda\delta} = \widetilde{R}^{\mu\tau\lambda\delta} \tag{13.35}$$

and

$$R^{\mu\tau\lambda\delta} = -\widetilde{R}^{\mu\tau\lambda\delta} \tag{13.36}$$

determine the curvature tensor as simple and isotropic bitensors, respectively,[41] and are equivalent to the conditions

$$C^{\mu\tau\lambda\sigma} = {}^1/_{12}\,(g^{\mu\sigma}g^{\tau\lambda} - g^{\mu\lambda}g^{\tau\sigma})R \tag{13.37}$$

and

$$\delta^{\mu}_{[\lambda}R^{\tau}_{\sigma]} + \delta^{\tau}_{[\sigma}R^{\mu}_{\lambda]} = {}^1/_2\,\delta^{\mu}_{[\lambda}\delta^{\tau}_{\sigma]}R. \tag{13.38}$$

It follows from (13.37) that $R = 0$ and $C^{\mu\tau\lambda\sigma} = 0$. Thus, (13.35) leads to conformally flat spaces with zero scalar curvature. Equation (13.34) reduces to $R_{\mu\nu} = 0$. Einstein's theory admits a unique empty space of this type — flat space-time. But if V_4 satisfying (13.35) is not empty, its metric can be obtained from the vacuum metric $\overset{0}{g}_{\mu\nu}$ by a conformal transformation $g_{\mu\nu} = e^{2\sigma}\overset{0}{g}_{\mu\nu}$, where the scalar function $\sigma(x)$ obeys the equation $\Box\sigma + \overset{0}{g}{}^{\mu\nu}\sigma_{\mu}\sigma_{\nu} = 0$. The energy—momentum tensor of matter in this case is related to $\sigma(x)$ by the equation $T_{\mu\nu} = 2\sigma_{\nu;\mu} - 2\sigma_{\nu}\sigma_{\mu} + \overset{0}{g}_{\mu\nu}\sigma_{\alpha}\sigma^{\alpha}$. Here $\sigma_{\nu} = \partial_{\nu}\sigma$ and $\sigma_{\nu;\mu} = \overset{0}{\nabla}_{\mu}\sigma_{\nu}$. If σ_{ν} is the Killing vector, then $\Box\sigma = 0$, $\sigma_{\alpha}\sigma^{\alpha} = 0$, and $T_{\mu\nu} = -2\sigma_{\mu}\sigma_{\nu}$. These conditions are satisfied by the energy—momentum tensor of electromagnetic radiation, written in the Newman—Penrose formalism.

Equation (13.38) leads to $R^{\mu}_{\nu} = {}^1/_4\,\delta^{\mu}_{\nu}R$, which is a necessary and sufficient condition for V_4 with a local Minkowski metric to be an Einstein space.[41] A physical model of such a space is a homogeneous space which is completely filled with a uniform density of matter.

For the fields (13.36), the equation of geodesic deviation takes the form

$$d^2\eta^\mu/ds^2 + \omega^\mu_\tau \eta^\tau = 0, \tag{13.39}$$

where $\omega^\mu_\tau = C^\mu_{\nu\tau\lambda}\,(dx^\nu/ds)\,(dx^\lambda/ds) + \delta^\mu_\tau R/12$. The quantity $C^\mu_{\nu\tau\lambda}$ is conformally invariant. The quantity R is transformed under conformal transformations of the interval $ds^{2\prime} = e^{2\sigma}ds^2$ to $R' = e^{-2\sigma}(R + 6\Box\sigma + 6\sigma_\mu\sigma^\mu)$. The requirement $R = \mathrm{const}$ under conformal transformations leads to a restriction on $\sigma(x)$: $\Box\sigma + \sigma_\nu\sigma^\nu + R\,(1 - e^{2\sigma})/6 = 0$.

In the Euclidean theory of gravitation, a regular localized solution (instanton) was constructed for the first time by Hawking[59] by means of analytic continuation of the Schwarzschild metric into the Euclidean region. Hawking's solution has finite action. The Euler characteristic for this solution is $\chi = 2$.

Great interest also attaches to the study of the properties of singularities of gravitational fields in terms of their asymptotic properties in pseudo-Riemannian V_4. In the general theory of relativity, by integrating the components of the curvature tensor we obtain the mass of the source of the field. Similarly, in any locally gauge-invariant theory, by integrating the field components we obtain the characteristics of the sources of the gauge field, in complete accordance with the program of geometrization of interactions.

Motion of Particles in Gauge and Gravitational Fields. It is well known that the equations of motion of particles in general relativity follow from the field equations. We shall show that this is true for all gauge fields if not only Einstein's equations, but also the equations of the gauge field are given.

The energy—momentum tensor of a gauge field is defined as the variational derivative of the Lagrangian (5.11) with respect to the metric. By (8.14), it is covariantly conserved and, by definition, is locally gauge-invariant. Einstein's equations with the energy—momentum tensor of the gauge field on the right-hand side have the form

$$R^\mu_\nu = \varkappa\left(F^{\mu\tau}_a F^a_{\nu\tau} - {}^1/_4\,\delta^\mu_\nu F^{\lambda\tau}_a F^a_{\lambda\tau}\right). \tag{13.40}$$

Taking the covariant divergence of (13.40) and using the fact that $R = 0$, we obtain

$$ {}^1/_2 F^{\mu\tau}_a F^a_{[\nu\tau;\,\mu]} + F^{\mu\tau}_a{}_{;\,\mu} F^a_{\nu\tau} = 0. \tag{13.41}$$

Einstein's equations are field equations for the metric; the Yang—Mills equations determine the connection, in general independently of the metric. If the Yang—Mills equations are free and there exists a nonsingular vector potential, the relations (13.41) are satisfied identically. But if there are particles in addition to the fields, the right-hand sides of the equations of the gauge field contain the current, and Einstein's equations (13.40) contain the energy—momentum tensor of the particles. Then (13.41) becomes

$$T^{\mu}_{\nu;\mu} = J^{\mu}_{a} F^{a}_{\nu\mu}, \tag{13.42}$$

from which we obtain a differential analog of the Lorentz equations describing the motion of particles in an electromagnetic field:

$$\nabla_0 T^0_i = \dot{p}_i = J^{\mu}_{a} F^{a}_{i\mu} + \nabla_k T^k_i. \tag{13.43}$$

In the case of gravitation, instead of (13.40)—(13.42) we obtain

$$R^{\mu}_{\nu} = (\lambda^2/4\pi)\left(R^{\mu}_{\tau\lambda\sigma} R_{\nu}{}^{\tau\lambda\sigma} - {}^1/_4\, \delta^{\mu}_{\nu} R^{\alpha}{}_{\tau\lambda\sigma} R_{\alpha}{}^{\tau\lambda\sigma}\right); \; R^{\mu}{}_{\tau\lambda\sigma;\,\mu} R_{\nu}{}^{\tau\lambda\sigma} = 0$$

and for $T^{\mu}_{\nu} \neq 0$ we have $T^{\mu}_{\nu;\,\mu} = (\lambda^2/4\pi\varkappa) R_{\tau[\sigma;\,\lambda]} R_{\nu}{}^{\tau\lambda\sigma}$.

It is interesting to note that the equations of motion of the particles (13.42) are free not only in the absence of a gauge field, but also when $J^{\mu}_{a} F^{a}_{i\mu} = 0$. In the Abelian case (electrodynamics), the equations of the gauge field follow from Einstein's equations (13.40) and in this sense are not independent.[32]

Geometrical Origin of the Internal Symmetries and Mass of a Gauge Field. An arbitrary Riemannian space can be treated locally as a surface embedded in a 10-dimensional Euclidean space. There are then six additional dimensions, which can be interpreted as internal degrees of freedom of elementary particles. It is convenient to choose in the containing space a coordinate system such that the coordinate vectors added to the space-time vectors are orthogonal to them. In that case, the transformations of the "internal" degrees of freedom do not affect space-time. In other words, the transformations of the gauge group commute with the transformations of the Lorentz group. In the 6-dimensional internal space resulting from the embedding, there acts the orthogonal group O(6), which has as a subgroup the group SU(3), which classifies the states of elementary particles. Different methods of embedding give different internal symmetry groups. The dynamical symmetries described by chiral groups of the type SU(2) × SU(2) or SU(3) × SU(3) can also be obtained by embedding.

If V_4 has some degree of symmetry, its embedding class

can be smaller than six. Thus, the static spherically symmetric Schwarzschild field can be embedded in a flat space of six dimensions. Its embedding class is two. The set of 15 generators of the orthogonal group O(6) contains one transformation in a 2-dimensional plane which commutes with the Lorentz group. This transformation can be regarded as an internal symmetry.[10]

For certain gravitational fields, it is possible to give lower or upper limits on the embedding class. For example, there does not exist a Riemannian or pseudo-Riemannian manifold V_4 with $R_{\mu\nu} = 0$ which can be embedded in a 5-dimensional flat space. The only V_4 with $R_{\mu\nu} = \lambda g_{\mu\nu}$ that can be embedded in a 5-dimensional space is a space of constant curvature. The only types of solutions of Einstein's equations with noncoherent matter ($p = 0$) that can be embedded in a 5-dimensional space are the Friedman cosmological models. If the matter rotates, V_4 cannot be embedded in a 5-dimensional space. Thus, the properties of the internal symmetries resulting from embedding are determined by the properties of matter in Riemannian space-time.

The number of additional dimensions increases sharply if all space-time, and not only the neighborhood of some point, is treated as a surface in a flat space of higher dimension. In general, such a global embedding of V_4 with a hyperbolic signature* requires a space having more than 230 dimensions. However, the number of dimensions required for a global embedding of contemporary cosmological models is often much smaller. The dimensions of the region ΔV_4 having internal symmetry properties (or orthogonality of V_4 and the additional subspace) can be related to the range of the gauge field associated with the internal symmetry or with the mass of its quanta.[7-9]

Thus, embedding makes it possible to interpret internal symmetries of elementary particles as a consequence of the curvature of space-time at small distances. This relation should manifest itself more strongly in regions with large curvature. In this approach, there may be a relationship between the cosmological properties of the Universe and the properties of elementary particles.

Kaluza—Klein Unified Theory of Gravitation and Electromagnetism. The 5-dimensional theories developed by Einstein, Kaluza, O. Klein, Mandel, Fock, and Rumer,[2,11,60,61] which provide a geometrical unification of electrodynamics and gravitation, are based on the observation of F. Klein that every mechanical problem of the motion of a material point involving a space with a larger number of dimensions can be

*I.e., when the metric reduces to the form $g_{\mu\nu} = (-1, -1, -1, 1)$.

reduced to the determination of the path of a light ray passing through a corresponding medium. Instead of a 4-dimensional Riemannian space, in 5-dimensional optics (or 5-optics) one considers a 5-dimensional manifold in which the fifth coordinate is proportional to the action S: $x^5 = S/mc$. The trajectories of charged massive particles interacting with the electromagnetic field in V_4 then correspond to trajectories of a light ray propagating in a 5-dimensional Riemannian manifold with the metric tensor

$$g_{\mu\nu} = \begin{pmatrix} g_{ik} + g_i g_k & g_i \\ g_k & 1 \end{pmatrix}. \qquad (13.44)$$

The contravariant components $g^{\mu\nu}$ have the form

$$g^{\mu\nu} = \begin{pmatrix} g^{ik} & -g^i \\ -g^k & 1 + g^{ik} g_i g_k \end{pmatrix}. \qquad (13.45)$$

Here $g_i = (e/mc^2)A_i$, where A_i is the electromagnetic vector potential, and e is the electric charge.

The existence of a quantum of action S is reflected in the topological closure of the fifth coordinate. In other words, the coordinate line of the fifth coordinate is a circle S^1. It is easy to see that the components of the 5-dimensional metric tensor do not depend on x^5. This condition reflects the independence of observed phenomena on the fifth coordinate and is called the condition of cylindricity. It implies the orthogonality of V_4 and x^5 in the chosen metric.

O. Klein and Fock showed that the quantum-mechanical problem of the motion of a spin-0 particle can be formulated as the problem of propagation of scalar waves in a 5-dimensional space if the dependence of the scalar wave function Ψ on the fifth coordinate is subjected to the condition of cyclicity:

$$\Psi(x^1, x^2, x^3, x^4, x^5) = \psi(x^1, x^2, x^3, x^4) \exp(imcx^5/\hbar).$$

In fact, the wave equation for 5-dimensional space, $\partial^2\Psi/\partial x^{\alpha 2} = 0$, corresponds in this case to the Proca equation in 4-dimensional space: $[\Box - (mc/\hbar)^2]\psi = 0$.

Similar considerations apply to vector fields. Owing to the condition of cylindricity, the Proca equations for massive vector mesons in the 4-dimensional formulation become 5-dimensional Maxwell equations. Thus, 5-optics brings together electrodynamics and the dynamics of massive vector mesons into a unified 5-dimensional Maxwell theory. Physically, this means that in the presence of short waves $\lambda \ll \hbar/mc$ the electromagnetic theory of light must take into account not only ordinary photons, but also "heavy photons" — the vector mesons. Rumer[60] noted that a similar situation occurs in the description of sound waves in an unbounded plane-parallel layer. If the sound field contains only long waves $\lambda > l$, where l is the thickness of the layer and λ is the wavelength of sound, then there are only ordinary 2-dimensional phonons, which propagate without dispersion. But if the sound field also contains short waves $\lambda < l/n$, it is necessary to allow for additional 2-dimensional "heavy phonons," which propagate with dispersion. There is then an alternative: 1) to dispense with the 2-dimensional description of the sound field and go over to a 3-dimensional equation; or 2) to preserve the 2-dimensional description and introduce "heavy phonons" in addition to the ordinary 2-dimensional phonons.

The theory of 5-optics leads to the appearance of an additional scalar gravitational field χ, which in the classical theory cannot be separated from the ordinary gravitational field related to the 4-dimensional metric tensor, but which can be separated when the 5-dimensional theory is quantized. This field is related to the 5-dimensional metric tensor by the equation $g_{55} = 1 + \chi$. The presence of the χ field in 5-optics leads to the disappearance of the singularity in the Coulomb potential at $r = 0$ in the problem of the field of a charged point mass.

We note that the vector potentials of the electromagnetic field in the Kaluza—Klein theory are treated as components of the metric of a (4 + 1)-dimensional configuration space. Therefore gradient transformations of the second kind are tensor transformations of the nondiagonal components of the 5-dimensional metric tensor. The quantities A_μ are nonzero in this approach only in special (noninertial) frames of reference, and in the presence of a 5-dimensional principle of equivalence they might be regarded as nonphysical quantities, together with $F_{\mu\nu}$, which play the role of connection coefficients. However, the presence of the tensor field $g_{\mu\nu}$, the vector field A_μ, and the scalar field χ makes the Kaluza—Klein theory attractive for generalizations to the case of gauge theories with spontaneous symmetry breaking in Riemannian V_4 and also supersymmetric theories.

Generalization of the Kaluza—Klein Theory to Arbitrary Gauge Fields. The structure of the containing space which arises from the embedding of Riemannian V_4 is locally identical to the structure of a fiber space. The introduction of a metric in the containing space makes it possible to describe the properties of any gauge field as the properties of the metric of the containing space.[12-15]

In going over to a $(r + 4)$-dimensional space generalizing

the 5-dimensional Kaluza—Klein space, the metric tensor must be represented in the form

$$\left.\begin{array}{l} g_{\alpha\beta}=\left(\begin{array}{c|c} g_{\mu\nu}+g_{ab}A^a_\mu A^b_\nu & g_{ab}A^a_\mu \\ \hline g_{ab}A^b_\nu & g_{ab} \end{array}\right), \quad \begin{array}{l} \mu,\nu=1\ldots 4;\ a,b=1,\ldots,r; \\ \alpha,\beta=1,\ldots,r+4. \end{array} \\ g^{\alpha\beta}=\left(\begin{array}{c|c} g^{\mu\nu} & -g^{\mu\nu}A^b_\nu \\ \hline -g^{\mu\nu}A^a_\mu & g^{ab}+g^{\mu\nu}A^a_\mu A^b_\nu \end{array}\right). \end{array}\right\} \tag{13.46}$$

Coordinate transformations preserve the local orthogonality of V_4 and V_{n-4} if $x^{\mu'} = f(x^\mu)$ and $x^{a'} = \varphi(x^\mu, x^a)$.

It was shown by Kerner[13] that the scalar curvature of a fiber space with the metric (13.46) has the form

$$R = K + g^{\mu\nu}g^{\lambda\sigma}g_{ab}F^a_{\nu\lambda}F^b_{\mu\sigma} + g^{\mu\nu}g_{ab}A^a_\mu A^b_\nu.$$

By a special choice of the coordinate system, it is possible to eliminate the last term and write the scalar curvature in the invariant form

$$R = K + g^{\mu\nu}g^{\lambda\sigma}g_{ab}F^a_{\nu\lambda}F^b_{\mu\sigma} = K + L_A,$$

where K is the scalar curvature of the base, and L_A is the Lagrangian of the gauge field. It is easy to see that the Lagrangian $L = \sqrt{-g}\,R$, where g is the determinant of the metric tensor of V_4, provides a unified description of the system of gravitational and gauge fields. Variation of L with respect to A^a_μ and $g^{\mu\nu}$ leads to a generalization of the Yang—Mills equations to Riemannian space-time, $g^{\nu\lambda}F^a_{\mu\lambda;\,\nu} = 2g^{\nu\lambda}f^a_{bc}A^b_\lambda F^c_{\mu\nu}$, and to Einstein equations for the case when the source of the gravitational field is the energy—momentum tensor of the gauge field, Eq. (13.40).

§14. ELECTRODYNAMICS OF A CONTINUOUS MEDIUM IN A GEOMETRICAL SETTING

Dislocations of a Medium as Sources of the Electromagnetic Field. We have already mentioned that the geometry of a manifold is completely determined by specification of three quantities: a symmetric second-rank tensor field, its covariant derivative, and the torsion tensor. In the case of a 3-dimensional continuous medium, these geometrical concepts correspond to the deformation tensor (a symmetric second-rank tensor), its covariant derivative, and the tensor of the density of dislocations.

Specification of the torsion can be replaced by specification of the covariant divergence of a certain antisymmetric tensor density. Therefore Maxwell's equations in a medium can be treated purely geometrically as equations which determine the torsion in terms of the sources of the field. In this case, the components of the torsion tensor play the role of coefficients of proportionality between the intensity of the electromagnetic field in the medium and the current.

As is well known, the classical equations for the electromagnetic field in a medium (or in a vacuum in the presence of material sources) relate two physically different types of quantities: the characteristics of the field and the characteristics of the medium. To determine the field by means of these equations, it is necessary to specify either the sources and the characteristics of the medium themselves or expressions for them in terms of the characteristics of the field. In the second case, in addition to the field equations, one either postulates relations giving a linear connection between the current and the field (such as Ohm's law) or represents the current as the motion of charged particles, for which an equation of motion is written. To make the electrodynamics of a continuous medium a purely geometrical theory, it is necessary to express the sources of the field in terms of geometrical quantities in such a way that the field equations become free. Since a nonuniform and anisotropic continuous medium can be treated as a model of a non-Euclidean space, this approach makes it possible to assign a geometrical interpretation to the characteristics of the medium. There is then a natural connection between the sources and the fields.

Suppose that the equations for the electromagnetic field in a medium have the form

$$\hat{G}^{\mu\nu}{}_{;\,\nu} = 0; \tag{14.1}$$

$$F_{[\mu\nu;\,\tau]} = 0, \tag{14.2}$$

where the semicolons denote covariant differentiation with respect to the total connection with torsion, the square brackets denote alternation with respect to all indices inside the brackets, and the cap denotes a tensor density. The expression for the field tensor has the form (here $x^4 = \mathrm{i}ct$)

$$\left.\begin{aligned} F_{\mu\nu} &= \begin{pmatrix} 0 & B^z & -B^y & -\mathrm{i}E_x \\ -B^z & 0 & B^x & -\mathrm{i}E_y \\ B^y & -B^x & 0 & -\mathrm{i}E_z \\ \mathrm{i}E_x & \mathrm{i}E_y & \mathrm{i}E_z & 0 \end{pmatrix}; \\ G^{\mu\nu} &= \begin{pmatrix} 0 & H_z & -H_y & -\mathrm{i}D^x \\ -H_z & 0 & H_x & -\mathrm{i}D^y \\ H_y & -H_x & 0 & -\mathrm{i}D^z \\ \mathrm{i}D^x & \mathrm{i}D^y & \mathrm{i}D^z & 0 \end{pmatrix}. \end{aligned}\right\} \tag{14.3}$$

The covariant differentiation in Eqs. (14.1) and (14.2) are taken with respect to an unknown connection, which in the general case has the form (13.1). Expanding the left-hand side of (14.1), we obtain

$$\hat{G}^{\nu\lambda}{}_{;\nu} = \underline{\partial_\nu (\hat{G}^{\nu\lambda}) + \Omega^{\lambda}_{\nu\tau} \hat{G}^{\nu\tau} - 2\Omega_{\nu\tau}{}^{\tau} \hat{G}^{\nu\lambda}} + S^{\lambda}_{\nu\tau} \hat{G}^{\nu\tau} - 2S_{\nu\tau}{}^{\tau} \hat{G}^{\nu\lambda}. \quad (14.4)$$

In (14.4), the sum of the underlined terms depends only on the choice of the coordinate system (curvilinear, rotating, etc.) and is independent of the choice of parallel transport. Therefore in a holonomic coordinate system (14.1) is equivalent to

$$\partial_\nu (\hat{G}^{\nu\lambda}) = 2S_{\nu\tau}{}^{\tau} \hat{G}^{\nu\lambda} - S_{\nu\tau}{}^{\lambda} \hat{G}^{\nu\tau}. \quad (14.5)$$

Identifying the right-hand side of (14.5) with the current of the sources, we find a relation between the current and the field components, which is a generalization of Ohm's law:

$$\hat{J}^{\lambda} = 2S_{\nu\tau}{}^{\tau} \hat{G}^{\nu\lambda} - S_{\nu\tau}{}^{\lambda} \hat{G}^{\nu\tau} = 2{}^{*} S^{\nu\lambda\tau} {}^{*}G_{\nu\tau}. \quad (14.6)$$

Similarly, the second pair of Maxwell's equations, $\partial_{[\mu} F_{\tau\nu]} = 0$, leads to the relation

$$2S_{\nu\tau}{}^{\tau} {}^{*}F^{\nu\lambda} = S_{\nu\tau}{}^{\lambda} {}^{*}F^{\nu\tau}, \quad (14.7)$$

where ${}^{*}F^{\nu\lambda} = (1/2\sqrt{-g})\varepsilon^{\nu\lambda\mu\tau} F_{\mu\tau}$, in which $\varepsilon^{\nu\lambda\mu\tau}$ is the discriminant tensor. In the presence of magnetic sources, we obtain, instead of (14.7),

$$2S_{\nu\tau}{}^{\tau} {}^{*}F^{\nu\lambda} - S_{\nu\tau}{}^{\lambda} {}^{*}F^{\nu\tau} = {}^{*}J^{\lambda}_{m}. \quad (14.8)$$

The system of equations (14.1) and (14.2) is dually invariant if the conditions (14.6) and (14.7) hold in the presence of sources of only electric type, or if (14.6)–(14.8) hold in the presence of sources of both types. Equation (14.6) leads to the relations[25]

$$J^{i} = 2\mathrm{i} S_{k0}{}^{[i} D^{k]} - \varepsilon^{ikl} S_{kl}{}^{n} H_{n}; \quad (14.9)$$

$$J^{0} = 2\mathrm{i} S_{ki}{}^{[k} D^{i]} - \varepsilon^{ikl} S_{[ik}{}^{0} H_{l]}. \quad (14.10)$$

Thus, an electric current can be excited by either electric or magnetic fields, and the current density is linearly related to the intensities of these fields. Equations (14.9) and (14.10) establish a relation between the electrodynamics of a continuous medium and the theory of dislocations in it, since the torsion is expressed in terms of the density of dislocations and is a measurable characteristic.

In the particular case of a semisymmetric geometry, i.e., when the torsion tensor can be represented in the form $S_{\nu\sigma}{}^{\mu} = \frac{1}{4} S_{[\nu} \delta^{\mu}_{\sigma]}$, we obtain

$$J^{\mu} = S_{\nu} G^{\nu\mu}; \tag{14.11}$$

$$S_{\nu} {}^{*}F^{\nu\mu} = 0 \tag{14.12}$$

or

$$J^{i} = \mathrm{i} S_{0} D^{i} - \varepsilon^{ikl} S_{k} H_{l}; \quad J^{0} = -\mathrm{i} S_{k} D^{k};$$

$$S_{k} B^{k} = 0; \; S_{0} B^{i} = \mathrm{i} \varepsilon^{ikl} S_{k} E_{l}.$$

We denote the invariants of the electromagnetic field in the medium by

$$I_{1} = G^{\mu\nu} {}^{*}G_{\mu\nu} = \mathrm{i}\,(HD); \quad I_{2} = {}^{*}F^{\mu\nu} F_{\mu\nu} = \mathrm{i}(BE); \quad L = G^{\mu\nu} F_{\mu\nu} = 2(BH - DE).$$

It follows from (14.11) and (14.12) that

$$(L/2) S_{\lambda} = F_{\lambda\mu} J^{\mu}; \tag{14.13}$$

$$(I_{1}/4)\, S_{\tau} = {}^{*}G_{\tau\mu} J^{\mu}. \tag{14.14}$$

Contracting (14.11) with S_{μ}, we obtain $S_{\mu} J^{\mu} = 0$, i.e., the current and torsion 4-vectors are mutually orthogonal. It follows from (14.14) that

$$(I_{1}/4)\, S_{i} = -J^{0} H_{i} - \mathrm{i} \varepsilon_{ikl}\, J^{k} D^{l},$$

i.e., for $I_{1} \neq 0$ the torsion 3-vector is reminiscent of the Lorentz force acting on a magnetic charge.

Contracting (14.14) with ${}^{*}F^{\tau\nu}$, we find a relation between the magnetic energy density and the current-density vector:

$$(HB) = (\mathrm{i}/J^{0}) \varepsilon_{ikl} J^{i} D^{k} B^{l} = (\mathrm{i}/S_{0}) \varepsilon^{ikl} S_{i} E_{k} H_{l}.$$

Thus, the geometrical properties of the medium are related to the character of the energy processes in it. If we represent the electrodynamics of an arbitrary continuous medium in the form of a free theory, in the case of a semisymmetric geometry the components of the torsion vector S_{i} are, apart from a factor of the ratio of the Lorentz force to the invariant, Lagrangian of the electromagnetic field, while the time component S_{0} is the ratio of the Joule losses to this invariant. The conductivity is equal to S_{0} [see (14.11)]. From the standpoint of geometrical theory, dislocations behave like external sources.

Invariant Properties of the Energy—Momentum Tensor in the Electrodynamics of a Continuous Medium. As is well known, the concept of a photon in a medium is poorly defined. This is partly due to the ambiguity in the definition of the energy—momentum tensor in an arbitrary medium. Using the results of the preceding sections, we define the energy—momentum tensor of the electromagnetic field in an arbitrary medium as the variational derivative of the Lagrangian[26]

$$\hat{L} = -(1/16\pi)\sqrt{-g}\, G^{\mu\nu} F_{\mu\nu} \qquad (14.15)$$

with respect to the metric $g_{\mu\nu}$. The tensors $G^{\mu\nu}$ and $F_{\mu\nu}$ have the form (14.3). The energy—momentum tensor takes the form

$$\hat{T}^{\mu\nu} = \sqrt{-g}\, T^{\mu\nu} = -(\sqrt{-g}/8\pi)\, g_{\tau\lambda}\,(G^{\mu\tau} F^{\nu\lambda} + {}^*G^{\mu\tau}\, {}^*F^{\nu\lambda}). \qquad (14.16)$$

By (8.14), the energy—momentum tensor (14.16) is covariantly conserved and, by definition, is symmetric and locally gauge-invariant. It makes it possible to obtain invariant quadratic relations between the energy and momentum of a photon in a medium, these being analogous to the equation of the mass shell in quantum mechanics. These relations hold for the flux densities of momentum and energy, i.e., they hold for unit volumes of the medium rather than for individual particles. In simple media (uniform, isotropic, nondispersive, stationary, and transparent), and also in a vacuum, it is possible to "take the square root" of these quadratic relations and obtain an analog of the well-known quantum-mechanical linear relations between the energy and momentum of a photon, $E = cp$, where E is the energy of the photon, p is its momentum, and c is the velocity of light. In the general case, it is not possible to do this.

The law of conservation of energy and momentum in the medium has the form

$$0 \equiv \nabla_\mu T^{\mu\nu} \equiv -(1/8\pi)\,[g_{\tau\lambda} F^{\nu\lambda} \nabla_\mu G^{\mu\tau} + {}^1/_2\, g^{\nu\lambda} G^{\mu\tau} \nabla_{[\mu} F_{\lambda\tau]} +$$
$$g_{\tau\lambda} G^{\nu\lambda} \nabla_\mu F^{\mu\tau} + {}^1/_2\, g^{\nu\lambda} F^{\mu\tau} \nabla_{[\mu} G_{\lambda\tau]}]. \qquad (14.17)$$

The identity (14.17) relates the conservation laws for the energy—momentum tensor and the field equations. It is easy to see that to obtain the conservation law $\nabla_\mu T^{\mu\nu} = 0$ in an arbitrary medium, it is necessary to have four groups of equations determining the quantities $\nabla_\mu G^{\mu\tau}$, $\nabla_{[\mu} G_{\lambda\tau]}$, $\nabla_\mu F^{\mu\tau}$, and $\nabla_{[\mu} F_{\lambda\tau]}$. The usually employed material equations $B = \mu H$ and $D = \varepsilon E$ constitute a particular case of conditions of the form

$$G^{\mu\tau} = g^{\mu\lambda} g^{\tau\sigma} F_{\lambda\sigma} \qquad (14.18)$$

and are valid in simple media and in a vacuum. The conditions (14.18) reduce the number of necessary equations by half. The energy—momentum tensor in this case takes the form

$$\tilde{T}^\mu_\nu = -(1/4\pi)\,(G^{\mu\tau} F_{\nu\tau} - {}^1/_4\, \delta^\mu_\nu\, G^{\alpha\tau} F_{\alpha\tau}). \qquad (14.19)$$

The tensor $\tilde{T}^\mu_\nu$ is covariantly conserved and satisfies the identities

$$\widetilde{T}^{\alpha}_{\mu}\widetilde{T}^{\mu}_{\nu} \equiv (1/16\pi)^2\,\delta^{\alpha}_{\nu}(I_1\ \ I_2 + L^2). \qquad (14.20)$$

An electromagnetic wave in a vacuum is defined as a field for which all the invariants are equal to zero, i.e., $I_1 = I_2 = L = 0$. We generalize this condition to a continuous medium and define an electromagnetic wave in the medium as a field for which $I_1 = I_2 = L = 0$. Then it follows from the identity (14.20) that

$$\sum_i \widetilde{T}^0_i \widetilde{T}^i_0 + (\widetilde{T}^0_0)^2 = 0. \qquad (14.21)$$

The components of $\mathbf{p}^A = \widetilde{T}^i_0 = (1/4\pi c)[\mathbf{E}\times\mathbf{H}]$ and $\mathbf{p}^M = \widetilde{T}^0_i = (c/4\pi)[\mathbf{B}\times\mathbf{D}]$ represent the momentum flux-density vectors in the form of Abraham ($\mathbf{p}^A$) and in the form of Minkowski ($\mathbf{p}^M$), respectively. It follows from (14.21) that $\mathbf{p}^A$ and $\mathbf{p}^M$ effectively behave like covariant and contravariant spatial components of a single 4-dimensional vector. Equation (14.21) can be regarded as an analog of the mass-shell equation $\mathbf{p}^2 + m^2 = 0$ for the flux densities of the corresponding quantities. In simple media, $\mathbf{p}^M = -c^2n^2\mathbf{p}^A$, and (14.21) becomes $(\mathbf{p}^M)^2 = c^2n^2(\widetilde{T}^0_0)^2$ or $(\mathbf{p}^A)^2 = (1/c^2n^2)(\widetilde{T}^0_0)^2$, where $n^2 = \varepsilon\mu$. Taking the square root in each of these relations, we obtain

$$|\mathbf{p}^M| = cn\widetilde{T}^0_0; \qquad (14.22)$$

$$|\mathbf{p}^A| = (1/cn)\,\widetilde{T}^0_0. \qquad (14.23)$$

On integration over a 3-dimensional volume, $\mathbf{p}^M$ leads to the momentum vector $\mathbf{P}^M$, which, by virtue of Noether's theorem, is conserved (apart from surface tensions) and in simple media satisfies the relation

$$|\mathbf{P}^M| = (c/4\pi)\left|\int [\mathbf{B}\times\mathbf{D}]\,d^3v\right| = cn\int \widetilde{T}^0_0\,d^3v = \mathscr{E}nc. \qquad (14.24)$$

The relation (14.24), being chronometrically invariant, holds in any coordinate system associated with the given frame of reference and has a quantum analog. Indeed, replacing $\mathbf{P}^M$ by $\hbar p$ and $\mathscr{E}$ by $\hbar\omega$ (where p is the wave vector and ω is the frequency of a photon), we obtain from (14.24) an analog of the dispersion equation $p = \omega n(\omega)$. Using $\mathbf{p}^A$, we can construct the invariant surface momentum

$$\mathbf{P}^A = (4\pi c)^{-1}\int_\sigma \varepsilon_{ikl}[\mathbf{E}\times\mathbf{H}]^i\,d\sigma^{kl} = (cn)^{-1}\int_\sigma \widetilde{T}^0_0\,\mathbf{p}^A\,\mathbf{N}d\sigma/|\mathbf{p}^A| \qquad (14.25)$$

(where $\mathbf{N}$ is the normal to the surface σ), which obeys the relation $P^A = (nc)^{-1}\mathscr{E}$.

Comparing (14.24) and (14.25), we see that the momentum density p^M is the momentum per unit volume of the medium, while the momentum density p^A is the momentum of a unit surface layer. This accounts for the difference between the dispersion laws for these quantities.

We now derive an invariant relation between the flux of momentum density and the stress tensor. In the case of a plane wave, i.e., for $I_1 = I_2 = L = 0$, it follows from (14.20) that

$$\sum_k \widetilde{T}^i_k \widetilde{T}^k_i + \widetilde{T}^i_0 \widetilde{T}^0_i = 0 \tag{14.26}$$

(there is no summation over i here!). In simple media, $\widetilde{T}^i_0$ and $\widetilde{T}^0_i$, and also $\widetilde{T}^i_k$ and $\widetilde{T}^k_i$, are proportional to each other. Therefore it follows from (14.26) that

$$\sum_k (\widetilde{T}^k_i)^2 = c^2 n^2 (\widetilde{T}^i_0)^2 = (c^2 n^2)^{-1} (\widetilde{T}^0_i)^2. \tag{14.27}$$

Choosing the coordinate system such that $\sum_k$ reduces to a single term and designating the corresponding direction by the vector $l^k/|l|$, we find from (14.17) that

$$\widetilde{T}^k_i = (cn)^{-1} (l^k/|l|) \widetilde{T}^0_i = (v^k/c) P^{\mathrm{M}}_i, \tag{14.28}$$

where v^k is the group velocity of light. In an arbitrary medium, there does not exist any proportionality between energy and momentum, but the quadratic relations (14.20) are valid. If the frame of reference is changed, the linear relations (14.22), (14.23), and (14.28) are violated, but the quadratic relations (14.20), (14.21), and (14.26) remain valid in the general case for media with the material equations (14.18). In the case of a plane wave, the relation between the energy density of the field and the stress tensor has the form $(T^0_0)^2 = \sum_{i,k} T^k_i T^i_k$. If the field is of non-wave type, in the corresponding frame of reference, i.e., for $T^i_0 = 0$, we have $3(T^0_0)^2 = \sum_{i,k} T^k_i T^i_k$.

We now consider the properties of a different energy—momentum tensor, which can be obtained by varying the Lagrangian $L' = (1/4\pi) G^{\alpha\beta} {}^*F_{\alpha\beta}$ with respect to the metric. The action integral corresponding to this Lagrangian describes the topological properties of the field. We introduce the notation

$$\delta L'/\delta g^{\alpha}_{\mu} = K^{\mu}_{\alpha} = (1/8\pi)(G^{\mu\sigma} {}^*F_{\alpha\sigma} - {}^*G^{\mu\sigma} F_{\alpha\sigma}). \tag{14.29}$$

The tensors T^{μ}_{α} and K^{μ}_{α} have different symmetry properties. If T^{μ}_{α} is symmetric with respect to the substitution $G^{\mu\sigma} \rightarrow F^{\mu\sigma}$ and is dually invariant, i.e., does not change under the substitution $G^{\mu\sigma} \rightarrow {}^*G_{\mu\sigma}$ and $F_{\mu\sigma} \rightarrow {}^*F^{\mu\sigma}$, then K^{μ}_{α} is symmetric with respect to another substitution $G^{\mu\sigma} \rightarrow {}^*F^{\mu\sigma}$ and changes sign under dual conjugation $G^{\mu\sigma} \rightarrow {}^*G_{\mu\sigma}$ and $F_{\mu\sigma} \rightarrow {}^*F^{\mu\sigma}$. In simple media with the material equations (14.18), we have

$K^{\mu}_{\alpha}=0$, and this does not contribute to the energy and momentum of the field. But if the properties of the medium are such that

$$G^{\mu\sigma}=g^{\mu\lambda}g^{\sigma\rho*}F_{\lambda\rho}, \tag{14.30}$$

then $T^{\mu}_{\alpha}=0$ and $K^{\mu}_{\alpha}=(1/4\pi)\left(F^{\mu\sigma}F_{\nu\sigma}-\frac{1}{4}\delta^{\mu}_{\nu}F^{\alpha\beta}F_{\alpha\beta}\right)=\widetilde{K}^{\mu}_{\alpha}$. Thus, in media obeying the relation (14.30), the energy and momentum of the field are determined by the tensor $\widetilde{K}^{\mu}_{\alpha}$. In such media, all the field invariants are equal in magnitude: $-I_1=I_2=L'=I$. The tensor $\widetilde{K}^{\mu}_{\alpha}$ satisfies the quadratic invariant relations

$$\widetilde{K}^{\alpha}_{\mu}\widetilde{K}^{\mu}_{\nu}\equiv-(1/128\pi^2)\,\delta^{\alpha}_{\nu}\,I^2, \tag{14.31}$$

which for the photon, i.e., for $I=0$, lead to the same relations between the energy and momentum of the field, and also between the momentum and the stress tensor, as those that were obtained above using $\widetilde{T}^{\mu}_{\alpha}$.

In the most general case, when no material equations are presupposed, the following quadratic invariant relations hold:

$$K^{\alpha}_{\mu}T^{\mu}_{\nu}+T^{\alpha}_{\mu}K^{\mu}_{\nu}\equiv-(\beta/8)\delta^{\alpha}_{\nu}[(FF)(G^*G)+(F^*F)(GG)];$$

$$\mathrm{Tr}\,(T^2)\equiv(\beta/8)[(G^*G)(F^*F)+(GF)^2+(FF)(GG)+(G^*F)^2];$$

$$\mathrm{Tr}\,(K^2)\equiv(\beta/8)[-(G^*G)(F^*F)+(GF)^2-(FF)(GG)+(G^*F)^2];$$

$$\mathrm{Tr}\,(T^2)-\mathrm{Tr}\,(K^2)\equiv(\beta/4)[(GG)(FF)+(G^*G)(F^*F)];$$

$$\mathrm{Tr}\,(T^2)+\mathrm{Tr}\,(K^2)\equiv\mathrm{Tr}\,[T\,(G)T(F)]\equiv(\beta/4)[(GF)^2+(G^*F)^2].$$

Here the indices of summation over pairs of indices inside the parentheses are omitted, $T\,(G)=-(1/4\pi)[G^{\mu\tau}G_{\nu\tau}-(1/4)\delta^{\mu}_{\nu}\,(GG)]$, and $T\,(F)=-(1/4\pi)[F^{\mu\tau}F_{\nu\tau}-(1/4)\delta^{\mu}_{\nu}\,(FF)]$.

Thus, in an arbitrary medium the quadratic invariant relations connect not the energy and momentum separately, but the energy, momentum, and stress tensor of the medium. From a geometrical point of view, the choice of material equations implies the introduction of a metric, this being such that its components describe not the gravitational field, but the dielectric and magnetic properties of the medium. In simple media, (14.18) leads to the diagonal metric proposed by Tamm[62]: $g_{\mu\nu}=\sqrt{\mu}\,(1,1,1,(\varepsilon\mu)^{-1})$. The vacuum is described by a plane metric.

REFERENCES

[1]H. Poincaré, La Science et l'Hypothèse, Flammarion, Paris (1902).

[2]A. Einstein, Collected Works [Russian translation], Vol. 2,

Nauka, Moscow (1966).
[3]H. A. Sokolik and N. P. Konopleva, Nucl. Phys. 72, 667 (1965).
[4]N. P. Konopleva, "Geometrical description of gauge fields" [in Russian], in: Proc. of the Intern. Seminar on Vector Mesons and Electromagnetic Interactions, JINR, Dubna (1969).
[5]A. Lichnerowicz, Théorie Globale des Connexions et des Groupes d'Holonomie, Consiglio Nazionale delle Ricerche, Rome (1955).
[6]H. G. Loos, Nucl. Phys. 72, 677 (1965); Nuovo Cimento 58A, 365 (1968); 53A, 201 (1968).
[7]J. Rosen, Rev. Mod. Phys. 37, 204 (1965). Y. Ne'eman, Rev. Mod. Phys. 37, 227 (1965).
[8]J. Rosen, N. Rosen, and Y. Ne'eman, in: Coral Gables Conf. on Symmetry Principles at High Energy, Miami, 1964, Freeman, San Francisco (1964).
[9]N. P. Konopleva, in: Tezisy dokl. Mezhdunar. seminara "Funktsional'nye metody v kvantovoi teorii polya i statistike" (Abstracts of Contributions to the Intern. Seminar on Functional Methods in Quantum Field Theory and Statistics), FIAN, Moscow (1971).
[10]L. de Broglie, D. Bohm, P. Hillion, F. Halbwachs, T. Takabayasi, and J.-P. Vigier, Phys. Rev. 129, 438 (1963).
[11]Th. Kaluza, Zum Unitätsproblem der Physik, Berichte, Berlin (1921), p. 966.
[12]B. DeWitt, Dynamical Theories of Groups and Fields, Gordon and Breach, New York (1965).
[13]R. Kerner, Ann. Inst. Henri Poincaré 9A, 143 (1968).
[14]A. Trautman, Rep. Math. Phys. 1, 29 (1970).
[15]Y. M. Cho, J. Math. Phys. 16, 2029 (1975).
[16]N. P. Konopleva and G. A. Sokolik, Dokl. Akad. Nauk SSSR 154, 310 (1964).
[17]H. Kerbrat-Lunc, C. R. Acad. Sci. 259, 3449 (1964).
[18]M. F. Atiyah and R. S. Ward, Commun. Math. Phys. 55, 117 (1977).
[19]M. F. Atiyah, N. J. Hitchin, V. G. Drinfeld, and Yu. I. Manin, Phys. Lett. 65A, 185 (1978).
[20]H. Weyl, Gravitation und Elektrizität, Sitzungsber. Preuss. Akad. Wiss., Berlin (1918).
[21]M.-A. Tonnelat, Les Principes de la Théorie Electromagnétique et de la Relativité, Masson, Paris (1959).
[22]J. A. Schouten, Tensor Analysis for Physicists, 2nd Ed., Clarendon Press, Oxford (1954).
[23]K. Kondo, Memoirs of the Unifying Study of the Basic Problems in Engineering Sciences by Means of Geometry, Vol. I, Gakujutsu Bunken Fukyu-Kai, Tokyo.
[24]I. A. Kunin, Teoriya uprugikh sred s mikrostrukturoi

(Theory of Elastic Media with Microstructure), Nauka, Moscow (1975).
[25]A. G. Iosif'yan and N. P. Konopleva, Tezisy dokladov Sovetskoĭ gravitatsionnoĭ konferentsii (Abstracts of Contributions to the Soviet Conf. on Gravitation), Erevan (1972), p. 232.
[26]N. P. Konopleva, Izv. Vyssh. Uchebn. Zaved. Radiofiz. 19, 1025 (1976). [Radiophys. Quantum Electron.].
[27]G. Y. Rainich, Trans. Am. Math. Soc. 27, 106 (1925).
[28]J. A. Wheeler, Geometrodynamics, Academic Press, New York (1962).
[29]E. Newman and R. Penrose, J. Math. Phys. 3, 566 (1962).
[30]E. Cremmer and J. Scherk, Nucl. Phys. B108, 409 (1976).
[31]S. Coleman, Commun. Math. Phys. 55, 113 (1977); S. Coleman and L. Smarr, Commun. Math. Phys. 56, 1 (1977).
[32]N. P. Konopleva, in: Problemy teorii gravitatsii i élementarnykh chastits (Problems of the Theory of Gravitation and Elementary Particles), No. 3, Atomizdat, Moscow (1970), p. 103.
[33]G. de Rham, Variétés Différentiables, Hermann, Paris (1955).
[34]N. P. Konopleva and G. A. Sokolik, in: (Problemy teorii gravitatsii i élementarnykh chastits (Problems of the Theory of Gravitation and Elementary Particles), Atomizdat, Moscow (1966), p. 22.
[35]G. F. Laptev, "Differential geometry of embedded manifolds" [in Russian], Tr. Mosk. Mat. Ob. 2, 275 (1953); in: Tr. 13-go Mat. S'ezda (Proc. of the 13th Mathematical Congress), Vol. 3, USSR Academy of Sciences, Moscow (1958), p. 409.
[36]E. Cartan, Geometriya grupp Li i simmetricheskie prostranstva (Geometry of Lie Groups and Symmetric Spaces) [translated from the French], Izd. Inostr. Lit., Moscow (1949).
[37]Yu. G. Lumiste, in: Tr. geometricheskogo seminara (Proc. of the Seminar on Geometry), Vol. 1, Moscow (1966), p. 191.
[38]L. E. Evtushik, Izv. Vuzov, Ser. Mat., No. 2(81), 32 (1969).
[39]N. P. Konopleva, in: Tezisy dokladov III Mezhvuzovskoĭ nauchnoĭ konferentsii po problemam geometrii (Abstracts of Contributions to the 3rd Inter-University Scientific Conf. on Problems of Geometry), Kazan State University (1967).
[40]J. L. Synge, Relativity: The General Theory, North-Holland, Amsterdam (1960).
[41]A. Z. Petrov, Novye metody v obshcheĭ teorii otnositel'nosti (New Methods in the General Theory of Relativity), Nauka, Moscow (1966).
[42]T. Eguchi, Phys. Rev. D 13, 1561 (1976).

[43]D. A. Alekseevskiǐ, Funktsional'. Analiz i Ego Prilozhen. **2**, No. 2, 1 (1968).
[44]C. A. Uzes, Ann. Phys. (N.Y.), **50**, 534 (1968).
[45]T. Yoneya, J. Math. Phys. **18**, 1759 (1977).
[46]G. Toulouse and M. Kleman, J. Phys. Lett. **37**, 149 (1976).
[47]L. S. Pontryagin, *Nepreryvnye gruppy* (*Continuous Groups*), Nauka, Moscow (1973).
[48]T. T. Wu and C. N. Yang, Phys. Rev. D **14**, 437 (1976); Nucl. Phys. **B107**, 365 (1976).
[49]G. 't Hooft, Nucl. Phys. **B79**, 276 (1974).
[50]A. M. Polyakov, Pis'ma Zh. Eksp. Teor. Fiz. **20**, 430 (1974) [JETP Lett. **20**, 194 (1974)].
[51]G. 't Hooft, Phys. Rev. Lett. **37**, 8 (1976); R. Jackiw and C. Rebbi, Phys. Rev. Lett. **37**, 172 (1976).
[52]F. Lund and T. Regge, Phys. Rev. D **14**, 1524 (1976).
[53]L. D. Faddeev and L. A. Takhtadzhyan, Teor. Mat. Fiz. **21**, 160 (1974).
[54]L. D. Faddeev, Pis'ma Zh. Eksp. Teor. Fiz. **21**, 141 (1975).
[55]J. A. Schouten and D. J. Struik, *Einführung in die Neueren Methoden der Differentialgeometrie*, Noordhoff, Groningen-Batavia (1935).
[56]N. P. Konopleva and G. A. Sokolik, Dokl. Akad. Nauk SSSR **177**, 302 (1967).[Sov. Phys. Dokl. **12**, 1016 (1968)].
[57]N. P. Konopleva, Vestn. Mosk. Univ., Ser. Fiz., No. 3, 73 (1965).
[58]C. N. Yang, Phys. Rev. Lett. **33**, 445 (1974).
[59]S. W. Hawking, Phys. Lett. **60A**, 81 (1977).
[60]Yu. B. Rumer, *Issledovanie po 5-optike* (*Investigations into 5-Optics*), Gostekhizdat, Moscow (1956).
[61]V. A. Fock, *Primenenie ideǐ Lobachevskogo v fizike* (*Application of Lobachevskiǐ's Ideas in Physics*), GTTI, Moscow (1950).
[62]I. E. Tamm, *Osnovy teorii élektrichestva* (*Elements of the Theory of Electricity*), Gostekhizdat, Moscow (1946).

CHAPTER IV. QUANTIZATION OF GAUGE FIELDS

§15. BASIC IDEAS OF THE CONSTRUCTION OF THE QUANTUM THEORY OF GAUGE FIELDS

The geometrical character of gauge fields again becomes apparent when their quantum theory is constructed.

Attempts to quantize geometrical fields by the standard techniques lead to difficulties and contradictions, which are already encountered in the problem of covariant quantization of the electromagnetic field, which has the simplest geometrical structure. Here the difficulties can be circumvented by quantizing the electromagnetic field by the method of Fermi, using an indefinite metric (see, for example, Refs. 1—3). However, it has been found that an uncritical transfer of Fermi's method, which is justified in quantum electrodynamics, to more complicated systems can lead to a violation of the unitarity of the theory. This was discovered for the first time by Feynman[4] in 1963 for the examples of the Yang—Mills field theory and the theory of the gravitational field. Feynman showed how to eliminate the difficulties which he discovered. He demonstrated that the unitarity of a diagram having the form of a closed loop can be restored by subtracting from it another diagram, which also has the form of a loop and which describes the propagation of a fictitious particle.

Feynman's method did not admit a direct generalization to more complicated diagrams. The solution of the problem for arbitrary diagrams was given in 1967 by DeWitt[5] and also by Faddeev and Popov,[6,7] using essentially different approaches. The two approaches are unified by application of the method of path integration, which provides a scheme of covariant perturbation theory for gauge fields.

We shall outline the idea which is fundamental throughout the scheme for constructing the quantum theory which follows.

Fields obtained from one another by gauge transformations (for example, A_μ and $A_\mu + \partial_\mu \Lambda$ in electrodynamics) describe one and the same physical (geometrical) situation and are therefore physically (geometrically) indistinguishable. This suggests that classes of fields obtained from one another by gauge transformations should become the fundamental objects of the theory. Thus, in electrodynamics a single class of fields A_μ includes all fields of the form

$A_\mu + \partial_\mu \Lambda$.

Application of the method of path (functional) integration helps us to construct a theory in which the fundamental objects are classes. This method makes it possible to write quantities of physical interest as integrals "over all fields" with weight $\exp(iS/\hbar)$, where S is the classical action of the system, and $\hbar$ is Planck's constant.* In the path-integral formalism, we can obtain a theory in which the fundamental objects are classes if it is possible to write a path integral as an integral over all classes. This can be done if, for example, we take an integration over a surface in the manifold of all fields which intersects each class once. Then each class will have precisely one representative on this surface. The resulting measure of integration on such surfaces changes when the form of the surface is changed, but all physical results should be independent of the choice of surface.

The modification of the path integral required for the theory of gauge fields is explained for the example of quantization of finite-dimensional mechanical systems in §§16—18. The exposition in §16 and §18 follows the work of Faddeev.[8,9] The application of the path-integral method in field theory is discussed in §19. Next (see §20) we construct the modification of the path integral required in the theory of gauge fields. The central problems here are the choice of measure in the functional space and the transition in the path integral from one gauge to another. Here we write the integrals for the Green's functions and give the perturbation theory for their calculation. The chapter is concluded with a discussion of specific examples. In §21 it is demonstrated how the method of path integration can be used to obtain well-known results of quantum electrodynamics without resorting to an indefinite metric. In §22 we give a detailed discussion of the quantization of fields of Yang—Mills type. In particular, with this example, we give a realization of the scheme for constructing the Green's functions, the S matrix, and perturbation theory, and we consider the transition from one gauge to another. In §23 we solve the problem of covariant quantization of Einstein's gravitational field.

The Hamiltonian form of the theory of gravitation is studied in §24. Here we construct a scheme of canonical quantization of the gravitational field, and we use the path-integral formalism to discuss the relation between covariant quantization and canonical quantization. In §25 we

---

*We shall henceforth employ a system of units with $\hbar = c = 1$, which is the natural system in relativistic quantum theory.

discuss attempts to construct a unified gauge-invariant theory of the weak and electromagnetic interactions. In §26 we discuss the description of vortex-like excitations in quantum field theory.

The quantum theory of gauge fields is a rapidly developing subject, and many interesting problems cannot be covered in the volume of this book. Particular problems which lie beyond the scope of our exposition include the renormalization of gauge theories,[10,11] the derivation of the Slavnov—Ward identities,[12] gauge theories in quark models, and the quantization of supergauge and chiral theories.

§16. MECHANICAL SYSTEMS AND PHASE SPACE

The quantization of classical mechanical systems using the path integral is one of the most convenient of the known methods of quantization and is applicable to situations in which the generally accepted canonical quantization encounters difficulties. We shall first consider the method of path integration as applied to the quantization of mechanical systems with a finite number of degrees of freedom. This makes the discussions and conclusions more concise and lucid. In addition, their general character becomes clearer. Later, we give the generalization to a field theory describing systems with an infinite number of degrees of freedom.

A mechanical system is determined by a Lagrangian

$$L(q, \dot{q}), \tag{16.1}$$

depending on the point q of a coordinate manifold M and the generalized velocity $\dot{q}$ (here $\dot{q}$ specifies a point of the tangent space V_q to the manifold M at the point q). Let n be the dimension of the manifold M. On such a manifold, we can introduce generalized coordinates $q^1, q^2, \ldots, q^n$.

In many cases, and in particular in the problem of quantization, it is convenient to go over from the Lagrangian formalism to the Hamiltonian formalism. For this purpose, one introduces canonical momenta $p_1, \ldots, p_n$ defined by the equations

$$p_i = \partial L/\partial \dot{q}^i, \; i = 1, \ldots, n. \tag{16.2}$$

The transition from the velocities $\dot{q}^i$ to the momenta p_i corresponds to the transition from the tangent space V_q to the cotangent space V_q^* (see, for example, the book of Mackey[13]). The manifold M, together with the cotangent space V_q^* defined at each of its points q, determines the phase space of the mechanical system Γ. The case which is of most interest to us is that in which the relations (16.2) are not solvable for $\dot{q}$. This is the case if, for example, the determinant

$$\det \| \partial^2 L/\partial \dot{q}^i \partial \dot{q}^k \| \tag{16.3}$$

is identically equal to zero. The Lagrangian $L(q, \dot{q})$ in this case is said to be singular. Lagrangians of such interesting fields as the electromagnetic and gravitational field are singular in this sense.*

Let us consider the transition from the Lagrangian formalism to the Hamiltonian formalism for a singular Lagrangian of the form

$$l(\xi, \dot{\xi}) = \sum_{\alpha=1}^{N} f_\alpha(\xi) \dot{\xi}^\alpha - \Phi(\xi). \tag{16.4}$$

The generalized velocities appear here linearly. Any nonsingular Lagrangian can be reduced to the form (16.4) by doubling the number of dynamical variables. In fact, it is easy to verify that in the case of a nonsingular Lagrangian the equations of motion for the Lagrangian

$$l(q, v, \dot{q}, \dot{v}) = \sum_{i=1}^{n} \frac{\partial L(q; v)}{\partial v^i} (\dot{q}^i - v^i) + L(q, v) \tag{16.5}$$

are equivalent to the usual equations of motion for the Lagrangian $L(q, \dot{q})$.

For a singular Lagrangian, the transition to the form (16.4) is not so automatic. The point is that the equations $\partial l/\partial v^i = 0$ reduce to the system

$$\sum_{i=1}^{n} \frac{\partial^2 L}{\partial v^i \partial v^j} (\dot{q}^i - v^i) = 0, \tag{16.6}$$

which is equivalent to the equations $\dot{q}^i = v$ only for a nonzero determinant (16.3), i.e., in the case of a nonsingular Lagrangian. It can be shown, however, that this equivalence also holds for a singular Lagrangian. In examples from field theory, the Lagrangian can be written from the outset in the form (16.4).

We shall reduce the equations of motion for the Lagrangian (16.4) to Hamiltonian form. We begin with the observation that, as is well known from the theory of differential forms of first order (see, for example, Ref. 14), it is possible to find a change of variables $\xi \to (q, p, z)$,

$$q = (q^1, \ldots, q^n),\ p = (p_1, \ldots, p_n),\ z = (z^1, \ldots, z^r),\ 2n + r = N, \tag{16.7}$$

such that the form $\omega = \sum_\alpha f_\alpha d\xi^\alpha$ which enters into the

---

*More precisely, both electromagnetic and gravitational fields are infinite-dimensional analogs of a finite-dimensional mechanical system with determinant (16.3) identically equal to zero.

Lagrangian (16.4) takes the canonical form $\omega = \sum_i p_i dq^i + dS$, apart from an additive term — the total differential dS, the addition of which, as is well known, does not affect the equations of motion. The number of pairs of canonical variables is equal to half of the rank of the antisymmetric matrix

$$\Omega_{\alpha\beta} = (\partial f_\alpha/\partial \xi^\beta - \partial f_\beta/\partial \xi^\alpha). \tag{16.8}$$

The variables of the type z are absent if this matrix is invertible. In this case, we say that the Lagrangian (16.4) is regular. It has the explicitly Hamiltonian form

$$\sum_{i=1}^{n} p_i \dot{q}^i - H(q, p). \tag{16.9}$$

In the general case, the Lagrangian in terms of the variables q, p, and z takes the form

$$l = \sum_{i=1}^{n} p_i \dot{q}^i - \Phi(p, q, z). \tag{16.10}$$

The equations of motion corresponding to the Lagrangian (16.10) contain not only the canonical equations

$$\dot{q}^i = \partial\Phi/\partial p_i; \; \dot{p}_i = -\partial\Phi/\partial q^i \tag{16.11}$$

but also equations of the form

$$\partial\Phi/\partial z^a = 0, \; a = 1, \ldots, r. \tag{16.12}$$

In the regular case, the latter equations are absent, and the problem of reducing the equations of motion to Hamiltonian form is solved as soon as the substitution (16.7) has been found.

It is natural to attempt to use the equations (16.12) to eliminate the variables of the type z. This can be done if

$$\det \| \Phi_{ab} \| \neq 0; \; \Phi_{ab} = \partial^2\Phi/\partial z^a \partial z^b. \tag{16.13}$$

In this case, substitution of the resulting values $z^a = z^a(q, p)$ into the equations (16.11) does not change their Hamiltonian form if the Hamiltonian is taken to be

$$H(q, p) = \Phi(q, p, z(q, p)). \tag{16.14}$$

Indeed, we have, for example,

$$\frac{\partial H}{\partial p} = \left(\frac{\partial \Phi}{\partial p} + \frac{\partial \Phi}{\partial z} \frac{\partial z}{\partial p} \right) \Bigg|_{z = z(q, p)}, \tag{16.15}$$

and the second term in the parentheses on the right-hand side vanishes as a consequence of (16.12). But if the condition (16.13) is not satisfied, we can use the equations (16.12) to express the variables z in terms of q, p, and m

parameters λ, where $m<r$, and $r-m$ is equal to the rank of the matrix Φ_{ab}. We introduce the notation

$$\widetilde{\Phi}(q, p, \lambda)=\Phi(q, p, z(q, p, \lambda)). \tag{16.16}$$

The matrix $\partial^2\widetilde{\Phi}/\partial\lambda^a\partial\lambda^b$ vanishes identically, since otherwise it would be possible to eliminate further variables of the type z from the equations (16.12). Thus, the parameters λ appear linearly in $\widetilde{\Phi}(q, p, z)$, and in terms of the new variables the Lagrangian takes the form

$$l=\sum_{i=1}^{n} p_i \dot{q}^i - H(q, p) - \sum_{a=1}^{m} \lambda_a \varphi^a(q, p). \tag{16.17}$$

It is natural to refer to the variables λ_a, $a=1, \ldots, m$ as Lagrange multipliers and to interpret the coefficients $\varphi^a(q, p)$ which accompany them as constraints imposed on the dynamical variables. The constraint equations

$$\varphi^a(q, p)=0, \; a=1, \ldots, m \tag{16.18}$$

make it possible to eliminate m variables q and p by expressing them in terms of the remaining variables. Then the Lagrangian (16.17) is reduced to the form (16.4) but with a smaller number of variables (n instead of N).

This process of eliminating superfluous variables can be repeated until the Lagrangian takes the Hamiltonian form (16.9). The process of elimination presupposes an explicit solution of the constraint equations of the type (16.18), which in practice is frequently difficult. Therefore it is useful to have a formalism that does not require explicit solution of the constraint equations.

It is natural to regard the constraints, i.e., the functions $\varphi^a(q, p)$, as independent and irreducible, in the sense that the constraint equations (16.18) determine in the phase space Γ a surface M of dimension $2n-m$, an arbitrary function f which vanishes on M being a linear combination of the constraints

$$f=\sum_a c_a(q, p)\, \varphi^a(q, p) \tag{16.19}$$

with, in general, variable coefficients $c_a(q, p)$.

Let us consider the situation, which at first sight appears to be a special case, when the constraints φ^a and the Hamiltonian H satisfy supplementary conditions

$$\{\varphi^a, \varphi^b\}=\sum_c c_c^{ab}\, \varphi^c; \tag{16.20}$$

$$\{H, \varphi^a\}=\sum_b c_b^a\, \varphi^b, \tag{16.21}$$

where c_c^{ab} and c_b^a are certain functions of q and p, and $\{f, g\}$ is the Poisson bracket:

$$\{f, g\} = \sum_{i=1}^{n} \left(\frac{\partial f}{\partial p_i} \frac{\partial g}{\partial q^i} - \frac{\partial f}{\partial q^i} \frac{\partial g}{\partial p_i} \right). \tag{16.22}$$

In other words, we assume that the Poisson brackets of the constraints with one another and with the Hamiltonian vanish on M. Looking ahead, we point out that these are the conditions satisfied by the constraints in the theory of gauge fields.

If the conditions (16.20) are to be satisfied, m must not exceed n.

The equations of motion for the Lagrangian (16.17) consist of the canonical equations

$$\dot{q}^i = \frac{\partial H}{\partial p_i} + \sum_a \lambda_a \frac{\partial \varphi^a}{\partial p_i}; \; \dot{p}_i = -\frac{\partial H}{\partial q_i} - \sum_a \lambda_a \frac{\partial \varphi^a}{\partial q^i} \tag{16.23}$$

and the conditions (16.18).

The conditions (16.20) and (16.21) guarantee that the equations (16.18) are satisfied for arbitrary functions $\lambda_a(t)$ if they are satisfied for the initial conditions. In other words, a trajectory which begins on M does not leave this surface.

It is natural to assume that the observable quantities are not all functions on the manifold M, but only those whose dependence with time is unaffected by the arbitrariness in the choice of $\lambda_a(t)$. This requirement is satisfied by functions $f(q, p)$ which obey the conditions

$$\{f, \varphi^a\} = \sum_b d_b^a \varphi^b. \tag{16.24}$$

Indeed, in the equations of motion

$$\dot{f} = \{H, f\} + \sum_a \lambda_a \{\varphi^a, f\} \tag{16.25}$$

for such functions, the terms depending on λ_a vanish on M.

The function $f(q, p)$, given on M and satisfying the conditions (16.24), does not depend essentially on all the variables. The conditions (16.24) can be regarded as a system of m first-order differential equations on M, for which the equations (16.20) play the role of integrability conditions. Therefore the function f is uniquely determined by its values on a submanifold of initial conditions for this system, which has dimension $(2n - m) - m = 2(n-m)$. It is convenient to take this submanifold to be the surface Γ^* defined by the equations

$$\chi_a(q, p) = 0, \ a = 1, \ldots, m, \qquad (16.26)$$

which are called the supplementary conditions. The functions χ_a must satisfy the condition

$$\det \| \{\chi_a, \varphi^b\} \| \neq 0, \qquad (16.27)$$

since it is only in this case that Γ^* can serve as the initial surface for the equations (16.24). It is also convenient to assume that the quantities χ_a commute with one another*:

$$\{\chi_a, \chi_b\} = 0. \qquad (16.28)$$

In this case, we can introduce canonical variables on the manifold Γ^*. Indeed, if the condition (16.27) is satisfied, then by a canonical transformation in Γ we can transform to new variables in which χ_a take the simple form

$$\chi_a(q, p) = p_a, \qquad (16.29)$$

where the quantities p_a $(a = 1, \ldots, m)$ form part of the canonical momenta of the new system of variables. We write q^a for their conjugate coordinates, and q^* and p^* for the remaining canonical variables. In the new variables, the condition (16.27) can be written in the form

$$\det \| \partial\varphi^a/\partial q^b \| \neq 0, \qquad (16.30)$$

so that the equations (16.18) can be solved for q^a. As a result, the surface Γ^* is given in Γ by the equations

$$p_a = 0, \quad q^a = q^a(q^*, p^*), \qquad (16.31)$$

which mean that the constraint equations (16.18) can be solved for q^a, where q^* and p^* play the role of independent variables on Γ^*. These variables are canonical variables. The Poisson bracket of any functions f and g satisfying the equations (16.25) can be calculated from the formula

$$\{f, g\} = \sum \left(\frac{\partial f^*}{\partial p^*} \frac{\partial g^*}{\partial q^*} - \frac{\partial f^*}{\partial q^*} \frac{\partial g^*}{\partial p^*} \right), \qquad (16.32)$$

where

$$f^* = f(q^a(q^*, p^*), q^*, 0, p^*); \ g^* = g(q^a(q^*, p^*), q^*, 0, p^*). \qquad (16.33)$$

To verify Eq. (16.32), it is convenient to calculate the Poisson brackets in noncanonical coordinates $\eta = (\varphi^a, q^*, p_a, p^*)$. This gives

*Here and in what follows, the commutator of two functions f and g in phase space is defined as the Poisson bracket $\{f, g\}$ given in (16.22). We say that the functions commute if their Poisson bracket is equal to zero.

$$\{f, g\} = \sum_{\alpha, \beta} \{\eta^\alpha, \eta^\beta\} (\partial f/\partial \eta^\alpha)(\partial g/\partial \eta^\beta). \qquad (16.34)$$

As a consequence of the conditions (16.20) and (16.24), a number of terms on the right-hand side of (16.34) vanish, and the result is therefore equal to the right-hand side of (16.32), where $f^* = f(\eta)\mid_{p_a = \varphi^a = 0}$. We stress once more that the conclusion that q^* and p^* are canonical variables is essentially related to the condition (16.28).

Thus, we have two methods of describing the observable quantities in our system. In the first case, the observables are functions on M (more precisely, classes of functions on Γ) satisfying the equations (16.24). The Poisson bracket is defined as the value on M of the Poisson bracket in Γ. In the second method, we must choose supplementary conditions χ_a, solve Eqs. (16.18) and (16.26), and construct the function f^* in accordance with (16.33). It can be shown that this procedure does not depend on the choice of supplementary conditions, since a change in χ_a which preserves the conditions (16.27) and (16.28) reduces to a canonical transformation in Γ.

As we have already mentioned, in practice it is frequently difficult to solve constraint equations of the type (16.18) and more convenient to work with the first method of describing the observables. On the other hand, in the description according to the second method, we are dealing with ordinary phase space and can make use of the familiar equations of mechanics. Thus, to verify the correctness of any particular formula in the first method of description of the observables, it is sufficient to verify that it reduces to the usual formula when we go over to the second method, as described above. This is how we shall proceed in what follows in working with path integrals.

§17. PATH INTEGRAL IN QUANTUM MECHANICS

In 1948 Feynman introduced and studied a path integral over trajectories in the configuration space of a mechanical system.[15] For application to the theory of gauge fields, it is more convenient to use an expression for the path integral obtained by Feynman in 1951, in which the integration is taken over trajectories in phase space.[16]

Consider a one-dimensional mechanical system with Hamiltonian $H(q, p)$, where q is the coordinate $(-\infty < q < \infty)$, and p is the canonical conjugate momentum. The canonical quantization of this system consists in the replacement of the coordinate q and the momentum p by operators $\hat{q}$ and $\hat{p}$ according to the rule

$$q \to \hat{q} \equiv q; \quad p \to \hat{p} = -\mathrm{i}\partial/\partial q \tag{17.1}$$

(we recall that we are using a system of units with $\hbar = 1$). The operators act in the Hilbert space of wave functions $\psi(q)$. Imposing on the functions $\psi(q)$ the normalization condition

$$\int_{-\infty}^{+\infty} |\psi(q)|^2 \, dq = 1, \tag{17.2}$$

we can regard the square of the modulus $|\psi(q)|^2 = \rho(q)$ as the probability density of finding the particle at the point q.

The evolution of the state of the system in time is determined by the Schrödinger equation

$$\mathrm{i}\partial\psi/\partial t = \hat{H}\psi, \tag{17.3}$$

in which the energy operator $\hat{H}$ is obtained from the classical Hamiltonian $H(q, p)$ by replacing q and p by operators according to the rule (17.1). The formal solution of Eq. (17.3) can be written in the form

$$\psi(t) = \hat{U}(t, t_0)\psi(t_0), \tag{17.4}$$

where the evolution operator

$$\hat{U}(t, t_0) = \exp(\mathrm{i}(t_0 - t)\hat{H}) \tag{17.5}$$

is an exponential function of the energy operator $\hat{H}$.

The method of path integration makes it possible to represent the matrix element of the evolution operator in the form of an average over trajectories in phase space of the expression

$$\exp(\mathrm{i}S[t_0, t]), \tag{17.6}$$

where

$$S[t_0, t] = \int_{t_0}^{t} \{p(\tau)\dot{q}(\tau) - H[q(\tau), p(\tau)]\} \, d\tau \tag{17.7}$$

is the classical action for a trajectory $(q(\tau), p(\tau))$ $(t_0 \leqslant \tau \leqslant t)$ in phase space. The average over trajectories is the F e y n m a n p a t h i n t e g r a l. The path integral is usually defined as a limit of a finite-dimensional integral. We shall give one of the possible definitions.

We divide the interval $[t_0, t]$ into N equal parts by means of the points $\tau_1, \ldots, \tau_{N-1}$. Consider on the interval $[t_0, t]$ functions $p(\tau)$ which are constant on the intervals

$$[t_0, \tau), \ (\tau_1, \tau_2), \ \ldots, \ (\tau_{N-1}, t], \tag{17.8}$$

and continuous functions $q(\tau)$ which are linear on the

intervals (17.8). We fix the values of the function $q(\tau)$ at the end points of the interval $[t_0, t]$ by putting

$$q(t_0) = q_0;\ q(t) = q. \tag{17.9}$$

A trajectory $(q(\tau), p(\tau))$ is determined by the values of the piecewise linear function $q(\tau)$ at the points $\tau_1, \ldots, \tau_{N-1}$ (we denote them by $q_1, \ldots, q_{N-1}$) and by the values of the piecewise constant function $p(\tau)$ on the intervals (τ_k, τ_{k+1}). We denote these values by $p_1, \ldots, p_N$.

Consider the finite-dimensional integral

$$(2\pi)^{-N} \int dp_1 dq_1, \ldots dq_{N-1}\, dp_N \exp[\mathrm{i}S(t_0, t)] \equiv J_N(q_0, q;\ t_0, t), \tag{17.10}$$

where $S[t_0, t]$ is the action (17.7) for the trajectory $(q(\tau), p(\tau))$ described just above and determined by the parameters $q_1, \ldots, q_N, p_1, \ldots, p_N$. The basic assertion is that the limit of the integral (17.10) as $N \to \infty$ is equal to the matrix element of the evolution operator:

$$\lim_{N \to \infty} J_N(q_0, q;\ t_0, t) = \langle q | \exp(\mathrm{i}(t_0 - t)\hat{H})\, q_0 \rangle. \tag{17.11}$$

It is easy to verify this assertion in cases when the Hamiltonian H depends only on the coordinate or only on the momentum. If $H = H(q)$ (i.e., H depends only on the coordinate), the classical action for the trajectory $(q(\tau), p(\tau))$ described above has the form

$$\int_{t_0}^{t} (p\dot{q} - H(q))d\tau = p_1(q_1 - q_0) + p_2(q_2 - q_1) + \ldots +$$
$$p_N(q - q_{N-1}) - \int_{t_0}^{t} H(q(\tau))d\tau. \tag{17.12}$$

Integrating with respect to the momenta in (17.10), we obtain a product of δ functions

$$\delta(q_1 - q_0)\delta(q_2 - q_1) \ldots \delta(q - q_{N-1}), \tag{17.13}$$

which enables us to take the expression $\exp[-\mathrm{i}\int_{t_0}^{t} H(q(\tau)d\tau)]$ to be equal to $\exp[\mathrm{i}(t_0 - t)H(q_0)]$ and to remove it from the integral sign. A further integration with respect to the coordinates $q_1, \ldots, q_{N-1}$ removes all the δ functions but one, leading to the result

$$\delta(q_0 - q)\exp[\mathrm{i}(t_0 - t)H(q_0)], \tag{17.14}$$

which is equal to the matrix element of the evolution operator.

If $H = H(p)$ (with a dependence only on the momentum), the action takes the form

$$\int_{t_0}^{t} (p\dot{q} - H)d\tau = p_1(q_1 - q_0) + p_2(q_2 - q_1) + \ldots$$

$$+ p_N (q - q_{N-1}) - \int_{t_0}^{t} H (p (\tau)) d\tau. \qquad (17.15)$$

Integrating in (17.10) first with respect to the coordinates $q_1, \ldots, q_{N-1}$ and then with respect to all the momenta $p_1, \ldots, p_N$, we obtain the expression

$$(1/2\pi) \int dp \exp [ip (q - q_0) + i (t_0 - t) H (p)], \qquad (17.16)$$

which is equal to the matrix element of the evolution operator for the Hamiltonian $\hat{H} = H (\hat{p})$.

The proof of Eq. (17.11) becomes more complicated if the Hamiltonian has a nontrivial dependence on the coordinates and momenta. In this case, the expression (17.10) before the limit is taken is not equal to its limit — the matrix element of the evolution operator. A formula analogous to (17.11) for the evolution operator of an equation of parabolic type has been proved, for example, by Evgrafov.[17] For the Schrödinger equation, a proof is known only in the case when the operator $\hat{H}$ is a sum of a function of the coordinates and a function of the momenta:

$$H = H_1 (q) + H_2 (p). \qquad (17.17)$$

It is Hamiltonians of the type (17.17) that are used in nonrelativistic quantum mechanics.

The path integral, defined as the limit of the expression (17.10) as $N \to \infty$, will be denoted by

$$\int_{q(t_0)}^{q(t)} \exp (iS [t_0, t]) \prod_{\tau} dp (\tau) dq (\tau) / 2\pi. \qquad (17.18)$$

This notation is convenient, although it does not reflect the fact that in the expression (17.10) before the limit is taken the number of integrations with respect to the momenta is one more than the number with respect to the coordinates.

We note that the path integral, defined by Eq. (17.11) as the limit of a finite-dimensional integral, depends on the method of approximating the trajectory $(q (\tau), p (\tau))$. This is due to the fact that we do not have a natural ordering prescription when the arguments of the function $H (q, p)$ are replaced by operators $\hat{q}$ and $\hat{p}$ which do not commute with one another. However, the operators which have physical meaning correspond, as a rule, to functions in which the replacement of the arguments by noncommuting operators leads to a unique result. This is the case for the energy operator of nonrelativistic quantum mechanics, which is equal to the sum of a quadratic function of the momenta and a function of the coordinates. In such cases, the path integral also leads to a unique result.

We shall generalize the path-integral formalism to systems with any finite number of degrees of freedom.

The action of a mechanical system with n degrees of freedom has the form

$$S[t_0, t] = \int \left(\sum_{i=1}^{n} p_i \dot{q}^i - H(q, p) \right) d\tau. \qquad (17.19)$$

Here q^i is the i-th canonical coordinate, p_i is its conjugate canonical momentum, and $H(q, p) \equiv H(q^1, \ldots, q^n; p_1, \ldots, p_n)$ is the Hamiltonian.

By definition, the path integral for the matrix element of the evolution operator is the limit of the finite-dimensional integral obtained from (17.10) by the substitutions

$$(2\pi)^{-N} \to (2\pi)^{-Nn}; \; dq_n \to \prod_{i=1}^{n} dq_k^i; \; dp_k \to \prod_{i=1}^{n} dp_{i,k}, \qquad (17.20)$$

where q_k^i is the value of the i-th coordinate at the point τ_k $(k = 1, \ldots, N-1)$, and $p_{i,k}$ is the value of the i-th momentum on the interval (τ_{k-1}, τ_k). It is necessary here to assume fixed values of all the coordinates $q^1, \ldots, q^n$ at both end points of the time interval $[t_0, t]$.

The path integral defined in this way will be denoted by

$$\int_{q(t_0)=q_0}^{q(t)=q} \exp(iS) \prod_t \prod_{i=1}^{n} dq^i(t)\, dp_i(t)/2\pi. \qquad (17.21)$$

§18. QUANTIZATION OF SYSTEMS WITH CONSTRAINTS

In the preceding section, we considered the quantization of finite-dimensional mechanical systems with a Hamiltonian action of the form (17.19) by means of the path integral. Field theory can be regarded as an infinite-dimensional analog of a mechanical system with constraints. The quantization of a finite-dimensional system with constraints requires modification of the path integral.

We shall exhibit the form of the path integral for a finite-dimensional mechanical system with constraints, specified by the canonical variables q and p, the Hamiltonian $H(q, p)$, and constraints $\varphi^a(q, p)$ satisfying the conditions (16.20) and (16.21) (see §16). We take supplementary conditions $\chi_a(q, p)$ such that the relations (16.27) and (16.28) are satisfied. The basic assertion is that the matrix element of the evolution operator is given by the path integral

$$\int \exp\left\{ i \int_{t_0}^{t} \left(\sum_{i=1}^{n} p_i \dot{q}^i - H(q, p) \right) d\tau \right\} \prod_\tau d\mu(q(\tau), p(\tau)), \qquad (18.1)$$

in which the measure of integration is defined by the formula

$$d\mu(t) = (2\pi)^{m-n} \det \| \{\chi_a, \varphi^b\} \| \prod_a \delta(\chi_a)\,\delta(\varphi_a) \prod_{i=1}^{n} dq^i(t)\,dp_i(t). \quad (18.2)$$

To prove this, we transform the integral (18.1) with the measure (18.2) to the integral (17.21), in which the integration is taken over the trajectories in the physical phase space Γ^*. For this purpose, we go over to the coordinates q^a, q^*, p_a, and p^* described in §16. The integral (18.1) is then transformed into an integral with a different measure:

$$d\tilde{\mu} = (2\pi)^{m-n} \det \| \partial\varphi^a/\partial q^b \| \prod_a \delta(p_a)\,\delta(\varphi^a) \prod_{j=1}^{n} dq^j\,dp_j, \quad (18.3)$$

which can be rewritten as

$$\prod_a \delta(p_a)\,\delta(q^a - q^a(q^*, p))\,dq^a\,dp_a \prod_{j=1}^{n-m} dq^{*j}\,dp_j^*/2\pi. \quad (18.4)$$

The integration with respect to q^a and p_a is removed by the δ functions. As a result, the integral takes the form

$$\int \exp\left\{ i \int_{t_0}^{t} \left(\sum_j p_j^* \dot{q}^{*j} - H^*(q^*, p^*) \right) d\tau \right\} \prod_\tau \prod_{j=1}^{n-m} dq^{*j}\,dp_j^*/2\pi, \quad (18.5)$$

which is identical with (17.21). We can therefore regard Eqs. (18.1) and (18.2) as proved.

We note that the integral (18.1) can be rewritten in the form

$$\int \exp\left\{ i \int_{t_0}^{t} \left(\sum_i p_i \dot{q}^i - H - \sum_a \lambda_a \varphi^a \right) d\tau \right\} \prod_\tau \det \| \{\chi_a, \varphi^b\} \| (2\pi)^{m-n} \times$$

$$\prod_a \delta(\chi^a) \prod_{i=1}^{n} dq^i\,dp_i \prod_b \Delta\tau\,d\lambda_b/2\pi. \quad (18.6)$$

The symbols $\prod_b \Delta\tau\,d\lambda_b/2\pi$ indicate that the expression before the limit is taken involves integrals with respect to the variables $\lambda_b(\tau_i)$ (where τ_i are the points at which the interval $[t_0, t]$ is partitioned) of the form

$$\int \exp\left\{ -i \sum_{i,a} \lambda_a(\tau_i)\,\varphi^a(q(\tau_i); p(\tau_i))\,\Delta\tau \right\} \prod_{i,b} \Delta\tau\,d\lambda_b/2\pi. \quad (18.7)$$

The expression (18.7) is equal to the product of δ functions

$$\prod_{i,a} \delta\,[\varphi^a(q(\tau_i);\ p(\tau_i))]. \quad (18.8)$$

This means that we can carry out the integration with

respect to λ_b in the integral (18.6) and return to the integral (18.1).

We shall now show that the path integral (18.1) does not depend on the choice of supplementary conditions. Let χ_a be an infinitesimal change in these conditions. Apart from a linear combination of constraints, we can represent $\delta\chi_a$ as the result of an infinitesimal canonical transformation in Γ, whose generator is a linear combination of constraints. Indeed, $\delta\chi_a$ can be represented in the form

$$\delta\chi_a = \{\Phi, \chi_a\} + \sum_b c_{ab}\,\varphi^b, \tag{18.9}$$

where

$$\Phi = \sum_a h_a\,\varphi^a, \tag{18.10}$$

and for h_a we can take the solution of the system of equations

$$\sum_b \{\chi_a, \varphi^b\}\,h_b = -\delta\chi_a. \tag{18.11}$$

By virtue of the condition (16.27), this system has a unique solution. With the canonical transformation described above, the constraints are replaced by their linear combinations

$$\delta\varphi^a = \sum_b A^a_b\,\varphi^b, \tag{18.12}$$

where $A^a_b = \{h_b, \varphi^a\} - \sum_c h_c c^{ac}_b$. The quantities entering into the integral (18.1) and the measure (18.2) are changed as follows:

$$\chi_a \to \chi_a + \delta\chi_a; \quad \varphi^a \to \varphi^a + \sum_b A^a_b\,\varphi^b; \quad H \to H;$$

$$\prod_a \delta(\varphi^a) \to \prod_a \delta(\varphi^a + \delta\varphi^a) = \left(1 + \sum_a A^a_a\right)^{-1} \prod_a \delta(\varphi^a);$$

$$\det\|\{\chi_a, \varphi^b\}\| \to \det\|\{\chi_a + \delta\chi_a, \varphi^b + \delta\varphi^b\}\| =$$

$$\det\|\{\chi_a + \delta\chi_a, \varphi^b\}\| \det\left\| \frac{\partial(\varphi^a + \delta\varphi^a)}{\partial\varphi^b} \right\| = \det\|\{\chi_a + \delta\chi_a, \varphi^b\}\| \left(1 + \sum_a A^a_a\right).$$

As a result of the canonical transformation, the measure of integration differs from the measure (18.2) only by the substitution $\chi_a \to \chi_a + \delta\chi_a$. This proves that the integral (18.1) is independent of the choice of supplementary conditions.

We shall further generalize the resulting path integrals for finite-dimensional mechanical systems to field theory, describing systems with an infinite number of degrees of freedom.

§19. PATH INTEGRAL AND PERTURBATION THEORY IN QUANTUM FIELD THEORY

A field theory can be regarded as a theory of a mechanical system with an infinite number of degrees of freedom. Gauge fields them become infinite-dimensional analogs of mechanical systems with constraints.

The path integral in a field theory can be constructed in different ways. One possibility is to begin with the action of the field, written in Hamiltonian form, and construct a path integral over the phase space of the system with infinitely many degrees of freedom. A second possibility is to begin with the action, not written in explicitly Hamiltonian form, and consider a path integral over all fields, which makes it possible to construct a manifestly relativistic theory. In the Hamiltonian approach, relativistic invariance is frequently not obvious and requires a special proof.

The method of integration over all fields can be explained and justified in the case when it is possible to transform the path integrals that have been obtained to integrals of Hamiltonian form.

Let us consider the definition and rules for working with path integrals for the example of the theory of a real scalar field with action

$$S=\int d^4x\left(\frac{1}{2}g_0^{\mu\nu}\frac{\partial\varphi}{\partial x^\mu}\frac{\partial\varphi}{\partial x^\nu}-\frac{m^2}{2}\varphi^2-\frac{g}{3!}\varphi^3\right). \tag{19.1}$$

Here $\varphi(x)$ are field functions depending on the point $x=(x^0, x^1, x^2, x^3)$ of a pseudo-Euclidean space V_4, and $g_0^{\mu\nu}$ is the diagonal Minkowski tensor [1, −1, −1, −1]. The action is a sum of a functional S_0 which is quadratic in the field φ, giving the action of a free field, and the integral of $-(g/3!)\varphi^3$, which describes a self-action with coupling constant g. The factor 1/3! multiplying $g\varphi^3$ is chosen for convenience.

A finite-dimensional approximation is frequently used in the determination of the path integral over all fields.

We choose in the space V_4 a large cubical volume V, divided into N^4 equal small cubes v_i $(i=1, \ldots, N^4)$. We approximate the function $\varphi(x)$ in the volume V by a function which is constant in the volumes v_i, and the first derivatives $\partial\varphi/\partial x_\mu$ by the finite differences

$$[\varphi(x_\nu+\delta_{\mu\nu}\Delta l)-\varphi(x_\nu)]/\Delta l, \tag{19.2}$$

where Δl is the length of the edge of the cube v_i. The approximating piecewise continuous function $\varphi(x)$ is defined by its values in the volumes v_i.

Consider the finite-dimensional integral

$$\int \exp(\mathrm{i}S) \prod_{\substack{i=1 \\ x\in v_i}}^{N^4} n(x)\, d\varphi(x) \tag{19.3}$$

with respect to the values of the function $\varphi(x)$ in the volumes v_i. Here S is the action integral for the approximating function $\varphi(x)$, with the approximation (19.2) for its first derivatives; $n(x)$ is a factor which is independent of $\varphi(x)$, chosen in such a way that when $V \to \infty$ and $v_i \to 0$ the integral (19.3) has the asymptotic form $\exp(cV)$ with a constant c that is independent of V. Usually, $n(x) = a(\Delta l)^{\alpha}$, with constant values (independent of x) of the parameters a and α.

Finite-dimensional integrals of the type (19.3) appear in the expressions before taking the limit in the definition of the path integrals encountered in field theory. We shall write Green's functions in the form of path integrals.

Green's functions are averages of products of two or more field functions with weight $\exp(\mathrm{i}S)$. For example, the two-point function is defined by the formula

$$G(x, y) \equiv -\mathrm{i}\langle \varphi(x)\varphi(y)\rangle =$$

$$-\mathrm{i} \lim_{\substack{V\to\infty \\ v_i\to 0}} \frac{\int \exp(\mathrm{i}S)\,\varphi(x)\,\varphi(y) \prod\limits_{\substack{i=1 \\ x\in v_i}}^{N^4} n(x)\, d\varphi(x)}{\int \exp(\mathrm{i}S) \prod\limits_{\substack{i=1 \\ x\in v_i}}^{N^4} n(x)\, d\varphi(x)}. \tag{19.4}$$

We shall use the following notation for the limit on the right-hand side of this formula:

$$\int \exp(\mathrm{i}S)\varphi(x)\varphi(y)\prod_x n(x)d\varphi(x) / \int \exp(\mathrm{i}S)\prod_x n(x)d\varphi(x). \tag{19.5}$$

The Green's functions are regarded as known if we know the generating functional

$$Z[\eta] = \int \exp(\mathrm{i}S + \mathrm{i}\int \eta(x)\varphi(x)d^4x)\prod_x n(x)d\varphi(x) / \int \exp(\mathrm{i}S)\prod_x n(x)d\varphi(x). \tag{19.6}$$

In particular, the two-point Green's function is given by the formula

$$G(x, y) = \mathrm{i}\frac{\delta}{\delta\eta(x)}\frac{\delta}{\delta\eta(x)} Z[\eta]\,|_{\eta=0}. \tag{19.7}$$

In the theory of a free field, it is easy to calculate

the functional $Z[\eta]$. To do this, in the integration with respect to φ in the numerator of Eq. (19.6) we make the displacement

$$\varphi(x) \to \varphi(x) + \varphi_0(x), \tag{19.8}$$

choosing $\varphi_0(x)$ to satisfy the condition of cancellation of the terms linear in φ in the exponential function. This leads to the following equation for $\varphi_0(x)$:

$$-(\Box + m^2)\varphi_0(x) = -\eta(x). \tag{19.9}$$

The solution to this equation can be expressed in terms of the Green's function $D(x, y)$ of the operator $(-\Box - m^2)$ by the formula

$$\varphi_0(x) = -\int D(x, y)\eta(y)d^4y. \tag{19.10}$$

The Green's function is the solution to the equation

$$(-\Box_x - m^2)D(x, y) = \delta(x - y) \tag{19.11}$$

with a δ function on the right-hand side.

After the displacement (19.8), the integral in the numerator of the right-hand side of (19.6) reduces to the integral in the denominator, multiplied by the factor

$$\exp\{(-i/2)\int \eta(x)D(x, y)\eta(y)d^4xd^4y\}, \tag{19.12}$$

which gives the value of the generating functional in the case of a free field. The two-point function in the theory of a free field [which we shall denote by $G_0(x, y)$], calculated according to Eq. (19.7), is the Green's function of the operator $(-\Box - m^2)$:

$$G_0(x, y) = D(x, y). \tag{19.13}$$

This function is not determined uniquely by Eq. (19.11), but only to within an additive term — the solution of the homogeneous equation $(-\Box - m^2)f = 0$.

However, there exists a most natural choice of the function $D(x, y)$, for which there are many arguments. We shall give one such argument.

The expression $\exp(iS)$ is an oscillating functional of $\varphi(x)$. Let us consider, instead of this functional, the functional $\exp(iS_\varepsilon)$, where S_ε is the complex action

$$S_\varepsilon = (1/2)\int \varphi(-\Box - m^2 + i\varepsilon)\varphi d^4x, \tag{19.14}$$

depending on a non-negative parameter and chosen in such a way that the functional $\exp(iS_\varepsilon)$ is smaller than unity in absolute value and vanishes when $\int\varphi^2 d^4x \to \infty$.

Unique results are obtained if we use the "corrected" action S_ε in the determination of the Green's functions and then go to the limit $\varepsilon \to +0$ in the results. In particular,

$D(x, y)$ becomes the limit of the Green's function of the operator $(-\Box - m^2 + i\varepsilon)$. The latter is uniquely determined. It depends on the difference $(x-y)$ and is given by the formula

$$D(x-y) = (2\pi)^{-4}\int \exp(ik(x-y))/(k^2 - m^2 + i\varepsilon). \qquad (19.15)$$

The limit of this function as $\varepsilon \to +0$ is called the causal or Feynman Green's function and is denoted by $D_F(x-y)$.

Thus, for the average of the product of two fields $\varphi(x)$ and $\varphi(y)$ in the theory of a free field, we have

$$\langle \varphi(x)\varphi(y)\rangle = iD_F(x-y). \qquad (19.16)$$

The average of the product of any odd number of functions φ is obviously equal to zero. For the average of an even number of functions, it is easy to derive the assertion known as Wick's theorem.

The average of a product of an even number of functions, $\varphi(x_1)\ldots\varphi(x_{2n})$, is equal to the sum of the products of all possible pairwise averages. For example,

$$\langle \varphi(x_1)\varphi(x_2)\varphi(x_3)\varphi(x_4)\rangle = \langle \varphi(x_1)\varphi(x_2)\rangle\langle \varphi(x_3)\varphi(x_4)\rangle + \langle \varphi(x_1)\varphi(x_3)\rangle\langle \varphi(x_2)\varphi(x_4)\rangle + \langle \varphi(x_1)\varphi(x_4)\rangle\langle \varphi(x_2)\varphi(x_3)\rangle. \qquad (19.17)$$

We shall obtain a proof for the average of $2n$ functions by differentiating the functional (19.12) $2n$ times and putting $\eta = 0$.

Wick's theorem is used to construct the formal perturbation theory and its associated diagrammatic technique. We shall construct the perturbation theory for a scalar field with the Lagrangian (19.1). We represent the functional $\exp(iS)$ in the form

$$\exp(iS) = \exp(iS_0)\exp(iS_1), \qquad (19.18)$$

where S_0 is the action of a free field, and the term

$$S_1 = -(g/3!)\int \varphi^3(x)d^4x \qquad (19.19)$$

describes the self-action. The perturbation theory is based on a series expansion of $\exp(iS_1)$ with respect to g under the path-integral sign in the form

$$\exp(iS_1) = \sum_{n=0}^{\infty} \frac{(-ig)^n}{n!\,(3!)^n}\int \varphi^3(x_1)\ldots\varphi^3(x_n)\,d^4x_1\ldots d^4x_n \qquad (19.20)$$

with a subsequent term-by-term integration of the resulting series. For example, for the two-point Green's function, we obtain a representation in the form of a quotient of two series:

$$G(x, y) = -\mathrm{i}\,\frac{\sum_{n=0}^{\infty}\frac{(-\mathrm{i}g)^n}{n!\,(3!)^n}\int \exp(\mathrm{i}S_0)\,\varphi(x)\,\varphi(y)}{\sum_{n=0}^{\infty}\frac{(-\mathrm{i}g)^n}{n!\,(3!)^n}\int \exp(\mathrm{i}S_0)} \rightarrow$$

$$\frac{\int \varphi^3(x_1)\ldots\varphi^3(x_n)\,d^4x_1\ldots d^4x_n\prod_x n(x)\,d\varphi(x)}{\int \varphi^3(x_1)\ldots\varphi^3(x_n)\,d^4x_1\ldots d^4x_n\prod_x n(x)\,d\varphi(x)}. \qquad (19.21)$$

Dividing the numerator and denominator on the right-hand side of (19.21) by the integral

$$\int \exp(\mathrm{i}S_0)\prod_x n(x)d\varphi(x), \qquad (19.22)$$

the problem is reduced to the calculation of averages of the type

$$\langle\varphi^3(x_1)\ldots\varphi^3(x_n)\rangle_0 \equiv \frac{\int \exp(\mathrm{i}S_0)\,\varphi^3(x_1)\ldots\varphi^3(x_n)\prod_x n(x)\,d\varphi(x)}{\int \exp(\mathrm{i}S_0)\prod_x n(x)\,d\varphi(x)} \qquad (19.23)$$

in the denominator of the right-hand side of (19.21) and to the calculation of averages

$$\langle\varphi(x)\varphi(y)\varphi^3(x_1)\ldots\varphi^3(x_n)\rangle_0 \qquad (19.24)$$

in the numerator.

Here we require Wick's theorem, which represents an average $\langle\ldots\rangle_0$ in the form of a sum of all possible pairwise averages, making it possible to calculate each term of the series in (19.21). Feynman pointed out that a graph, or diagram, can be associated with each term of the series in question. Perturbation theory, in which a diagram is associated with each term of the series, has become known as the d i a g r a m m a t i c t e c h n i q u e.[18] In the theory of a scalar field with self-action under consideration, we can construct the diagrams as follows.

We associate the average (19.23) with a diagram in the form of n points (each with three outgoing lines), representing the points $x_1, \ldots, x_n$ in a pseudo-Euclidean space V_4. Such a diagram for the case $n = 4$ has the form

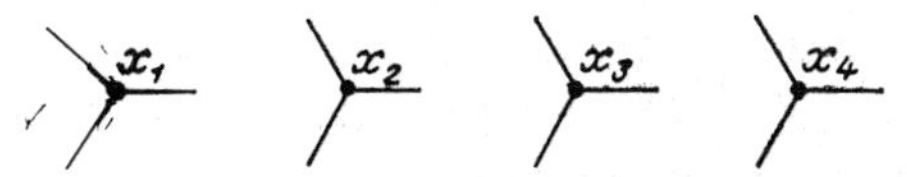

(19.25)

We associate the average (19.24) with a diagram obtained

from the corresponding diagram for the average (19.23) by adding two points (each with a single outgoing line), representing the points x and y in V_4. For example, for $n = 4$, we obtain

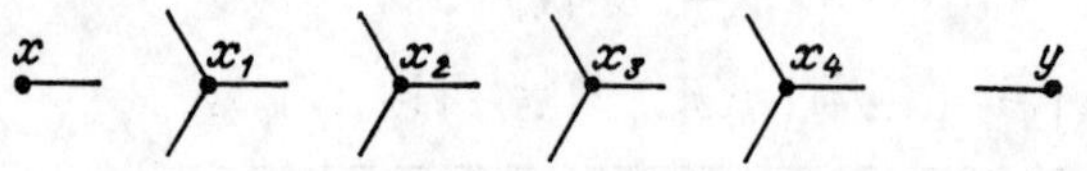

(19.26)

We shall call these diagrams *prediagrams* to distinguish them from the diagrams that will be introduced below. Prediagrams have symmetry with respect to permutation of the outgoing lines at each point. Therefore we can speak of the symmetry group G_n of an n-point prediagram of order

$$R_n = n!(3!)^n. \tag{19.27}$$

The symmetry of the prediagrams reflects the symmetry of the averages (19.23) and (19.24) corresponding to them, which are unchanged under permutation of the arguments $x_1, \ldots, x_n$ and under permutation within the average in each triplet of field functions $\varphi(x_i)\varphi(x_i)\varphi(x_i) = \varphi^3(x_i)$.

Note that the expression R_n^{-1} [together with $(-ig)^n$] appears in the series of (19.21) as a factor in front of the averages $\langle \ldots \rangle_0$.

According to Wick's theorem, the averages (19.23) and (19.24) are sums of products of all possible pairwise averages. We associate a diagram with each method of forming pairwise averages by drawing a line between each pair of points x_i and x_j of the prediagram if there is an average $\langle \varphi(x_i)\varphi(x_j) \rangle$ among the pairwise averages. The number of lines is equal to the number of pairs, i.e., half the number of averaged field functions.

All diagrams arising from the prediagram (19.25) are represented in (19.28), and all those arising from the prediagram (19.26) are represented in (19.29)*:

---

*No allowance is made in (19.28) and (19.29) for diagrams containing subdiagrams linked to the principal part of a single line (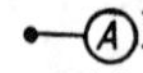). In the operator formalism, this means that the interaction is written in the form of a normal product $(g/3!) : \varphi^3 :$.

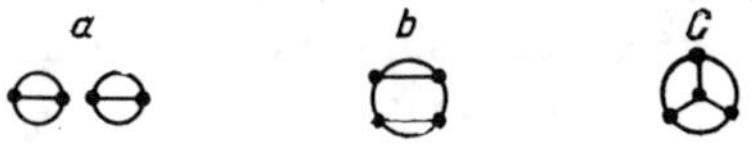

(19.28)

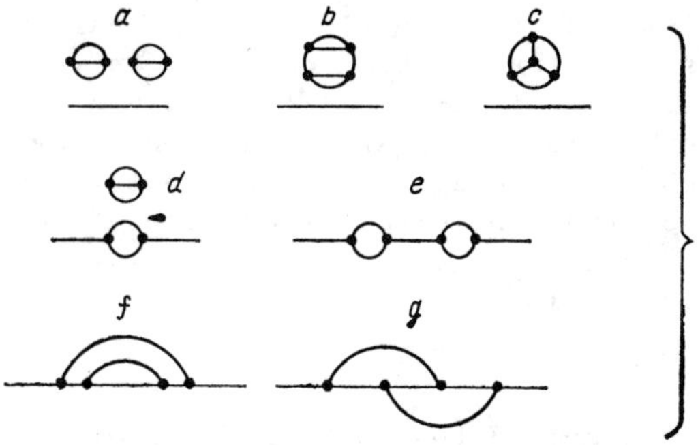

(19.29)

The expression corresponding to a diagram is obtained by integrating the product of pairwise averages with respect to $x_1, \ldots, x_n$ and multiplying the result by $(-ig)^n R_n^{-1}$ and by the number of ways in which this diagram is obtained from the prediagram. It is easy to see that this number is equal to the ratio $R_n/r_{n,d}$ of the order R_n of the symmetry group of the prediagram to the order $r_{n,d}$ of the symmetry group of the diagram obtained from the prediagram by drawing lines between its vertices. As a result, the numerical factor in front of the integral with respect to $x_1, \ldots, x_n$ becomes equal to $(-ig)^n r_{n,d}^{-1}$.

The rules of correspondence which are obtained can be reformulated as follows. We associate each line joining the points x_i and x_j with a Green's function $D_F(x_i - x_j)$, which differs from the average $\langle \varphi(x_i)\varphi(x_j)\rangle_0$ by a factor i, and we associate a vertex with the coupling constant g:

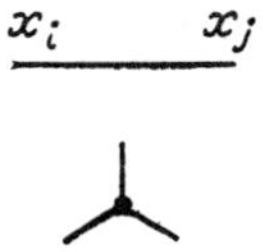

$D_F(x_i - x_j)$

g

(19.30)

The expression for a diagram is obtained by integrating the product of the expressions corresponding to the elements of the diagram — the vertices and lines — with respect to the coordinates of the vertices and multiplying the result by $i^{l-n-1} r_{n,d}^{-1}$, where l is the number of lines of the diagram, n is the number of its vertices, and $r_{n,d}$ is the order of the symmetry group.

The presence of the symmetry factor $r_{n,d}$ is not always clearly noted in the literature. This is perhaps due to the fact that in quantum electrodynamics (the only theory in which higher-order diagrams must be taken into account for a comparison with experiment) this factor is equal to unity for all diagrams except the vacuum diagrams, which need not be considered for the description of physical effects.

To calculate the contributions of the various diagrams to the Green's function, we need consider only the connected diagrams, i.e., those in which it is possible to pass from any vertex of the diagram to any other vertex by moving along the lines of the diagram. To prove this, we take into account the fact that the diagrams corresponding to the series in the denominator of Eq. (19.21) give a total contribution of the form

$$\exp \sum_i D_i^c, \tag{19.31}$$

where $\sum D_i^c$ is the sum of the contributions of all the connected vacuum diagrams (without external lines). Equation (19.31) follows from the fact that a diagram consisting of n_1 connected components of the first kind, n_2 connected components of the second kind, etc., has as a symmetry factor the expression

$$r^{-1} = \prod_i ((n_i!)\, r_i^{n_i})^{-1}, \tag{19.32}$$

where r_i is the order of the symmetry group of the connected component of type i. The factors $(n_i!)^{-1}$ reflect the symmetry of the diagram with respect to permutations of identical components and lead to the exponential function (19.31). The remaining step is to note that the sum of the contributions of the diagrams in the numerator of (19.21) reduces to the sum of the contributions of the connected diagrams, multiplied by the factor (19.31).

In textbooks on quantum field theory (see, for example, Refs. 1—3), the diagrammatic technique is usually constructed by means of an operator method. Its derivation by means of path integrals, which we have given here for the example of a real scalar field, seems more natural. In fact, Feynman arrived at his diagrams via the path integral.

For concrete calculations, the diagrammatic technique in momentum space is more convenient. This representation is obtained by taking the Fourier transform $\tilde{\varphi}(k)$ of the field functions $\varphi(x)$ defined by

$$\varphi(x) = (2\pi)^{-4} \int \exp(ikx)\tilde{\varphi}(k)d^4k \tag{19.33}$$

and considering as the Green's functions averages of the form

$$\langle \tilde{\varphi}(k_1)\ldots\tilde{\varphi}(k_n)\rangle. \tag{19.34}$$

The expressions corresponding to the elements of the diagrams — the vertices and lines — take the form

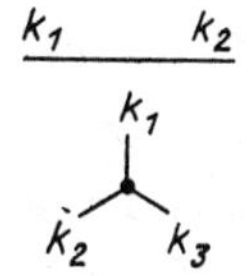

$$\left.\begin{array}{c} \delta(k_1+k_2)(k_1^2-m^2+i\varepsilon)^{-1} \\ g\delta(k_1+k_2+k_3) \end{array}\right\} \tag{19.35}$$

The contribution of a specific diagram in the momentum diagrammatic technique is obtained by integrating the product of the expressions corresponding to its elements in accordance with (19.35) with respect to all internal momenta and multiplying the result by $r_{n,d}^{-1}(i/(2\pi)^4)^{l-n-1}$, where n is the number of vertices, l is the number of lines, and $r_{n,d}$ is the order of the symmetry group of the diagram.

Note that the Green's functions in momentum space contain as a factor the δ function $\delta(\sum_i k_i)$, which ensures conservation of 4-momentum:

$$G(k_i) = M(k_i)\delta(\sum_i k_i). \tag{19.36}$$

Knowing the Green's functions, we can calculate the elements of the S matrix according to the formula

$$S(k_1,\ldots,k_n) = \lim_{k_i^2 \to m_i^2} M(k_i)\left\{\prod_{i=1}^{n}(k_i^2-m_i^2)\,\theta(\pm k_i^0)\,|2k_i^0|^{-1/2}(2\pi)^{-3/2}\right\}. \tag{19.37}$$

The proof of this formula is not given here; it can be found in many textbooks on quantum field theory.

An explanation and justification of the path integrals over all fields in quantum field theory can be given in the case when it is possible to transform them to integrals in Hamiltonian form, which are a generalization to field theory of the integrals obtained above in the quantization of finite-dimensional mechanical systems.

Returning to the example of a real scalar field, we write in Hamiltonian form the path integral

$$\int \exp(iS)\prod_x n(x)d\varphi(x). \tag{19.38}$$

To do this, we consider the integral

$$\int \exp(iS[\varphi,\pi])\prod_x n(x)d\varphi(x)d\pi(x), \tag{19.39}$$

where the expression

$$S[\varphi, \pi] = \int (\pi\partial_0\varphi - \pi^2/2 - (\nabla\varphi)^2/2 - m^2\varphi^2/2 - g\varphi^3/3!)d^4x \quad (19.40)$$

is the same as the action (19.1) with $\pi(x)$ replaced by $\partial_0\varphi(x)$. The action (19.40) has Hamiltonian form. The corresponding Hamiltonian is

$$H = \int d^3x\, (\pi^2/2 + (\nabla\varphi)^2/2 + m^2\varphi^2/2 + g\varphi^3/3!), \quad (19.41)$$

where the functions $\varphi(x)$ and $\pi(x)$ have the meaning of the density of the coordinate and its conjugate momentum. We shall show that the integral (19.39) with respect to the variables $\varphi(x)$ and $\pi(x)$ reduces to the integral (19.38) over all fields. For this, we note that the integral with respect to π in Eq. (19.39) can be taken in explicit form if we make the displacement

$$\pi(x) \to \pi(x) + \partial_0\varphi(x), \quad (19.42)$$

after which the integral becomes a product of the integral (19.38) with respect to φ and an integral with respect to π:

$$\int \exp[-(i/2)\int \pi^2(x)d^4x] \prod_x d\pi(x), \quad (19.43)$$

which reduces to a product of normalization factors. In the equations for the Green's functions — the averages of products of several fields — the integrals of the type (19.43) appear in the numerator and in the denominator and therefore cancel.

Thus, we have succeeded in reducing the path integral in the theory of a real scalar field to Hamiltonian form by artificially introducing an integral with respect to a new variable — the canonical momentum. This prescription will be used later to prove the Hamiltonian character of specific systems of quantum field theory.

The scheme of path integration over all fields provides a method of quantizing Bose fields. In the operator formalism, this quantization reduces to the replacement of the field functions by operators with Bose commutation relations.

The quantization of Fermi fields can be realized by means of a path integral with respect to anticommuting variables (for more details, see the book of Berezin[19]). For this purpose, the following basic facts are needed.

An integral over Fermi fields (over an infinite Grassmann algebra with an involution) is defined as the limit of the integral over an algebra with unity and a finite even number of generators x_i and x_i^* $(i = 1, 2, \ldots n)$ obeying the commutation relations

$$x_i x_j + x_j x_i = 0;\; x_i^* x_j^* + x_j^* x_i^* = 0;\; x_i x_j^* + x_j^* x_i = 0. \quad (19.44)$$

Any element of the algebra $f(x, x^*)$ is a polynomial of the form

$$f(x, x^*) = \sum_{a_i, b_i = 0, 1} c_{a_1,\dots,a_n, b_1,\dots, b_n} x_1^{a_1} \dots x_n^{a_n} (x_1^*)^{b_1} \dots (x_n^*)^{b_n} \quad (19.45)$$

with coefficients $c_{a_1,\dots,a_n, b_1,\dots,b_n}$ from the field of complex numbers. By virtue of the commutation relations (19.44) with $i=j$, we have $x_i^2 = (x_i^*)^2 = 0$, so that powers of the generators higher than the first vanish. We need consider only the ordering of the factors adopted in (19.45), since any other order can be reduced to this one by means of the commutation relations (19.44).

We define an involution operation acting on the element (19.45) according to the rule

$$f \to f^* = \sum_{a_i b_i = 0, 1} \bar{c}_{a_1,\dots, a_n, b_1,\dots, b_n} x_n^{b_n} \dots x_1^{b_1} (x_n^*)^{a_n} \dots (x_1^*)^{a_1}. \quad (19.46)$$

On this algebra, we can introduce the integral

$$\int f(x, x^*) dx^* dx \equiv \int f(x_1, \dots, x_n, x_1^*, \dots, x_n^*) dx_1^* dx_1 \dots dx_n^* dx_n, \quad (19.47)$$

which is determined by the formulas

$$\int dx_i = 0; \int dx_i^* = 0; \int x_i dx_i = 1; \int x_i^* dx_i^* = 1 \quad (19.48)$$

and by the requirement that the symbols dx_i and dx_i^* anticommute with each other and with the generators if we impose the natural linearity condition

$$\int (c_1 f_1 + c_2 f_2) dx^* dx = c_1 \int f_1 dx^* dx + c_2 \int f_2 dx^* dx. \quad (19.49)$$

On integration of the sum (19.45), the only nonzero contribution comes from the term with $a_i = b_i = 1$ for all $i = 1, 2, \dots, n$.

The following two equations will be useful in what follows:

$$\int \exp(-x^* A x) dx^* dx = \det A; \quad (19.50)$$

$$\int \exp(-x^* A x + \eta^* x + x^* \eta) dx^* dx / \int \exp(-x^* A x) dx^* dx =$$

$$\exp(\eta^* A^{-1} \eta), \quad (19.51)$$

where $x^* A x = \sum_{i, k} a_{ik} x_i^* x_k$ is the quadratic form in the generators x_i and x_i^* corresponding to the matrix A. The expressions $\eta^* x = \sum_i \eta_i^* x_i$ and $x^* \eta = \sum_i x_i^* \eta_i$ are linear forms in the generators x_i and x_i^*, whose coefficients η_i and η_i^* anticommute with each other and with the generators. The elements η_i and η_i^*, together with the generators x_i and x_i^*, can be regarded as generators of a larger algebra. The expression

$\eta^* A^{-1} \eta$ in Eq, (19.51) is the quadratic form of the matrix A^{-1}, which is the inverse of the matrix A.

The exponential function in the integrands of Eqs. (19.50) and (19.51) is determined by the series expansion, in which only the first few terms of the expansion are non-zero by virtue of the commutation relations (19.44).

It is easy to prove Eq. (19.50) by noting that only the n-th term of the expansion of the exponential function contributes to the integral. Equation (19.51) can be proved by making the displacement $x \to x + \tilde{\eta}$ and $x^* \to x^* + \tilde{\eta}$, which destroys the linear form with respect to x and x^* in the exponential integrand.

A detailed proof and some generalizations of Eqs. (19.50) and (19.51) can be found in the monograph of Berezin.[19]

§20. QUANTUM THEORY OF GAUGE FIELDS

We shall now outline the general procedure for quantizing gauge fields in the formalism of the path integral over all fields. We recall that a gauge field is a connection of a fiber space, whose base is space-time V_4 and whose fiber is a finite-dimensional space of representations of some group G_0. We denote the gauge field by A and its components by A_μ^a, where $\mu = 0, 1, 2, 3$ is a space-time index and a is an "isotopic" index. The gauge group G is a direct product of the groups G acting at each point of V_4:

$$G = \prod_x G_0(x). \qquad (20.1)$$

Let Ω be an element of the gauge group, which is a function on V_4 with a value in G_0. We denote by A^Ω the result of the action of the element Ω on the field A. The set of fields A^Ω, where A is fixed and Ω runs over the gauge group, is usually called the *orbit of the gauge group*.

As we have seen, the quantization of a field with action S reduces to an average of the functional $\exp(iS)$ over all fields. In the theory of gauge fields, the action $S[A]$ is gauge-invariant, i.e., it is the same for all fields obtained from one another by means of gauge transformations:

$$S[A^\Omega] = S[A]. \qquad (20.2)$$

In other words, the action is a functional on classes of fields obtained from one another by means of gauge transformations. In this situation, it is natural to consider the problem of going over from an integration over all fields to an integration over classes of fields. In what follows, we outline one of the possible approaches to this problem.

To construct the path integral, it is necessary to choose a measure in the manifold of all fields A. The simplest measure is

$$d\mu[A]=\prod_x\prod_{\mu,a} dA_\mu^a(x), \tag{20.3}$$

where the symbol $\prod_x$ was defined in §19. We call this measure l o c a l. In concrete examples from the theory of gauge fields, the measure (20.3) has the property of gauge invariance:

$$d\mu\,[A^\Omega]=d\mu\,[A]. \tag{20.4}$$

Invariance of the action $S\,[A]$ and of the measure $d\mu\,[A]$ with respect to gauge transformations $A\to A^\Omega$ has the consequence that the corresponding path integral

$$\int \exp\,(iS\,[A])d\mu[A] \tag{20.5}$$

becomes proportional to the "volume of the orbit," i.e., to the path integral

$$\int \prod_x d\Omega\,(x) \tag{20.6}$$

with respect to the gauge group G. Here $\prod_x d\Omega\,(x)$ is the invariant measure on the group G, which is equal to a product on the groups G_0 acting at each point of space-time V_4.

The approach to integration over classes outlined here consists in the explicit separation of this factor for the path integral. Such a separation can be realized in several ways. The idea of one of them is to go over from the integral (20.5) to an integral over a surface in the manifold of all fields, which has a single intersection with the orbits of the gauge group.

Let the equation of the surface be

$$f\,(A)=0. \tag{20.7}$$

The equation $f\,(A^\Omega)=0$ for any $A\,(x)$ must have a unique solution for $\Omega\,(x)$.

We introduce a functional $\Delta_f[A]$, defining it by the condition

$$\Delta_f\,[A]\int \prod_x \delta[f\,(A^\Omega\,(x))]d\Omega\,(x)=1. \tag{20.8}$$

This is an integration with respect to the gauge group G of the infinite-dimensional δ function $\prod_x\delta\,(f\,(A^\Omega\,(x)))$. Such a δ function is a functional defined by specification of the rule for integrating it with other functionals. In what follows, we shall give several specific examples of the calculation of integrals of the type (20.8). We note that the functional $\Delta_f[A]$ is gauge-invariant, i.e.,

$$\Delta_f[A^\Omega] = \Delta_f[A]. \qquad (20.9)$$

To separate the factor (20.6) from the path integral (20.4), we insert the left-hand side of Eq. (20.8) (which is equal to unity) under the integral sign and make the change of variable $A^\Omega \to A$. The measure $d\mu[A]$ and the functionals $S[A]$ and $\Delta_f[A]$ are invariant under this change of variable. The integral (20.4) reduces to a product of the group volume and the integral

$$\int \exp(iS[A])\Delta_f[A]\,[\prod_x \delta(f(A))]\,d\mu[A]. \qquad (20.10)$$

It is this integral that will be the starting point in the quantum theory of gauge fields.

We shall show that the integral (20.10), which formally depends on the choice of the surface $f(A) = 0$, is actually invariant with respect to the choice of surface. To prove this, we insert "another unity" under the integral sign in (20.10):

$$1 = \Delta_g[A] \int \prod_x \delta(g(A(x)))d\Omega(x), \qquad (20.11)$$

where $g(A) = 0$ is the equation of another surface which, like the surface $f(A) = 0$, has a single intersection with the orbits of the group.

Changing the order of integration with respect to A and Ω, then making the displacement $A^\Omega \to A$, and finally, again changing the order of integration with respect to A and Ω, we transform the integral (20.10) to the form

$$\int \exp(iS[A])\Delta_g[A]\,\{\prod_x \delta(g(A))\}d\mu[A]. \qquad (20.12)$$

This procedure enables us to transform in a path integral from one surface to another, or, as we might say, from one gauge to another. In particular, this method is convenient in going over from the Hamiltonian form of the path integral to an integral in a relativistic gauge. Below, we shall consider this transition for specific examples.

It is possible to give a more general method than that described just above for separating the volume of the gauge group from the path integral. Consider a gauge-noninvariant functional $F[A]$. We define a functional $\Phi[A]$ by the equation

$$\Phi[A] \int F[A^\Omega]\prod_x d\Omega(x) = 1. \qquad (20.13)$$

This functional is gauge-invariant. Of course, we must require that the path integral on the left-hand side of Eq. (20.13) actually exists. Inserting the left-hand side of

(20.13) into the integral (20.5) and then making the displacement $A^{\Omega} \to A$, we obtain the product of the group volume (20.6) and the integral

$$\int \exp (iS[A])\Phi[A]F[A]d\mu[A]. \qquad (20.14)$$

The integral (20.10) introduced above is a particular case of (20.14). The independence of the integral (20.14) on the choice of functional is proved in the same way as the independence of the integral (20.10) on the choice of the surface $f(A) = 0$.

We define the Green's functions in the theory of gauge fields as averages of products of field functions at different points of space-time V_4. The generating functional for the Green's functions has the form

$$Z[\eta] = \frac{\int \exp\{iS[A] + i\int \eta A d^4 x\} F[A]\,\Phi[A]\,d\mu[A]}{\int \exp(iS[A])\,F[A]\,\Phi[A]\,d\mu[A]}, \qquad (20.15)$$

where $S[A]$ is the action of the field A, $d\mu[A]$ is a local gauge-invariant measure, the functionals F and Φ are defined above, and $\int \eta A d^4 x$ is the linear functional

$$\int \Big(\sum_{\mu,\, a} \eta_a^{\mu}(x)\, A_{\mu}^{a}(x)\Big)\, d^4 x, \qquad (20.16)$$

where $\eta_a^{\mu}(x)$ are arbitrary test functions.

The Green's functions — variational derivatives of the functional (20.15) — depend on the choice of gauge, i.e., on the choice of the functional $F[A]$. Physical results obtained by averaging the gauge-invariant functionals do not depend on the choice of gauge.

We point out a characteristic difference between perturbation theory in the theory of gauge fields and the perturbation theory developed in §19 for the theory of a scalar field. Suppose that the action S contains a small parameter ε. The action S_0 corresponding to zero value of the parameter ε will be assumed to be a quadratic form in the field functions. This is the case in the majority of examples from field theory.

In constructing the perturbation theory in the path-integral formalism in §19, we expanded the functional $\exp(iS)$ in a series of the form

$$\exp(iS) = \exp(iS_0)\exp(iS_1) = \exp(iS_0) \sum_{n=0}^{\infty} a_n \varepsilon^n / n! \qquad (20.17)$$

in powers of ε. The resulting functional series was then integrated term by term, where the individual terms of the series of perturbation theory were calculated by means of

Wick's theorem, which follows from the fact that S_0 is a nondegenerate quadratic form.

In the theory of gauge fields, we have under the path-integral sign not only the functional $\exp(iS)$, but also a product of functionals $F\Phi$. In perturbation theory, this product is also expanded in a series in ε. If F is a δ functional, it is difficult to expand it in a series in ε. Therefore we must choose the equation of the surface $f(A) = 0$ in such a way that it does not contain the parameter ε. This applies, for example, to the equations which distinguish the Lorentz and Coulomb gauges in electrodynamics and in Yang—Mills theory. In the theory of gravitation, equations with analogous properties are, for example, conditions of harmonicity.

We note that the function $\Delta_f[A]$ depends on the parameter ε even when the equation $f(A) = 0$ does not contain this parameter. It is then necessary to expand the functional $\Delta_f[A]\exp(iS[A])$ in a series of the form

$$\exp(iS[A]) \sum_{n=0}^{\infty} (\varepsilon^n/n!) b_n[A]. \tag{20.18}$$

In the general case, the product

$$F_0\Phi_0 \exp(iS_0) \equiv M_0, \tag{20.19}$$

where F_0 and Φ_0 are the principal terms of the functionals F and Φ, must be such that Wick's theorem holds in the integration with M_0 by products of fields. This property is satisfied if M_0 is an exponential function of a nondegenerate quadratic form or a product of an exponential with a degenerate quadratic form and a δ functional corresponding to the flat surface orthogonal to the null directions of this quadratic form.

The construction of the perturbation theory resulting from the expansion (20.18) will be considered in more detail for specific examples. The rule formulated just above, which ensures the validity of Wick's theorem, will be satisfied in these cases.

§21. QUANTUM ELECTRODYNAMICS

We shall show how the general scheme of quantization of gauge fields works by means of specific examples.* We

*In this section and in the next section, we shall write vector indices as subscripts without distinguishing between covariant and contravariant components. Repeated Greek indices indicate a summation with allowance for the pseudo-Euclidean metric, and repeated Latin indices indicate a summation over the values 1, 2, 3. For example,

begin with the electromagnetic field, which has the simplest geometrical structure.

The action of the free electromagnetic field

$$S = 1/4 \int (\partial_\mu A_\nu - \partial_\nu A_\mu)^2 d^4x \qquad (21.1)$$

is invariant with respect to the Abelian group of gauge transformations

$$A_\mu(x) \to A_\mu(x) + \partial_\mu \lambda(x). \qquad (21.2)$$

It follows from the preceding section that the quantization of gauge fields is achieved by means of the path integral of the functional

$$\Phi F \exp(iS), \qquad (21.3)$$

where S is the action of the system, F is an arbitrary gauge-noninvariant functional, and the functional Φ^{-1} is the average of F with respect to the gauge group.

The local measure of integration, which in this case has the form

$$d_\mu[A] = \prod_x \prod_{\mu=0}^{3} dA_\mu(x), \qquad (21.4)$$

is obviously gauge-invariant. It remains to choose the functional $F[A]$. The most convenient functionals for the further construction of perturbation theory have the form

$$\left.\begin{aligned} F_1[A] &= \prod_x \delta(\partial_\mu A_\mu(x)); \\ F_2[A] &= \prod_x \delta(\operatorname{div} \mathbf{A}(x)); \\ F_3[A] &= \exp\left\{(-i/2d_l) \int (\partial_\mu A_\mu)^2 d^4x\right\}. \end{aligned}\right\} \qquad (21.5)$$

The functionals F_1 and F_3 lead to manifestly relativistic quantization, and the use of F_2 is convenient in going over to a Hamiltonian theory. The corresponding gauge-invariant functionals are given by the equations

$$\left.\begin{aligned} \Phi_1^{-1}[A] &= \int \prod_x \delta(\partial_\mu (A_\mu(x) + \partial_\mu \lambda(x))) d\lambda(x); \\ \Phi_2^{-1}[A] &= \int \prod_x \delta(\operatorname{div} \mathbf{A}(x) + \nabla\lambda(x))\, d\lambda(x); \\ \Phi_3^{-1}[A] &= \int \exp\{-(i/2d_l) \int (\partial_\mu (A_\mu(x) + \partial_\mu \lambda(x)))^2 d^4x\} \prod_x d\lambda(x). \end{aligned}\right\} \qquad (21.6)$$

$k^2 \equiv k_\mu k_\mu = k_0^2 - \mathbf{k}^2$, $\partial_\mu A_\mu = \partial_0 A_0 - \partial_1 A_1 - \partial_2 A_2 - \partial_3 A_3$, and $k_i k_i = k_1^2 + k_2^2 + k_3^2$; the symbol $\delta_{\mu\nu}$ denotes the Minkowski tensor (with nonzero components $\delta_{00} = -\delta_{11} = -\delta_{22} = -\delta_{33} = 1$).

All these functionals are in fact independent of the field $A_\mu(x)$, and this can be seen by making the displacement $\lambda \to \lambda - \Box^{-1}\partial_\mu A_\mu$ in the first and third cases, and $\lambda \to \lambda - \Delta^{-1}\mathrm{div}\,\mathbf{A}$ in the second case. Therefore, apart from a constant (infinite) factor, we can take

$$\Phi_1 = \Phi_2 = \Phi_3 = 1. \tag{21.7}$$

The form of the path integral is now determined in all three cases.

The use of the functional F_2 implies an integration over fields satisfying the equation

$$\mathrm{div}\,\mathbf{A} = 0. \tag{21.8}$$

This is the well-known condition of the Coulomb gauge. We shall show how an integral with the functional F_2 can be transformed to an integral which is explicitly of Hamiltonian form. Such a transformation was made in §19 for a real scalar field by introducing an integral over an auxiliary field. In our case, our objective can be achieved by means of a path integral of the form

$$\int \exp(iS[A_\mu, F_{\mu\nu}]) \prod_x \delta(\mathrm{div}\,\mathbf{A}(x)) \prod_\mu dA_\mu(x) \prod_{\mu<\nu} dF_{\mu\nu}(x) \tag{21.9}$$

with the action

$$S[A_\mu, F_{\mu\nu}] = \int((1/4)F_{\mu\nu}F_{\mu\nu} - (1/2)F_{\mu\nu}(\partial_\mu A_\nu - \partial_\nu A_\mu))d^4x, \tag{21.10}$$

depending not only on the vector $A_\mu(x)$, but also on the antisymmetric tensor $F_{\mu\nu}(x)$. In classical theory, $F_{\mu\nu}$ is the intensity of the electromagnetic field:

$$F_{\mu\nu} = \partial_\mu A_\nu - \partial_\nu A_\mu. \tag{21.11}$$

In going over to quantum theory, we take the functions A_μ and $F_{\mu\nu}$ to be independent and integrate with respect to them as independent variables. The integral with respect to $F_{\mu\nu}$ in (21.9) can be evaluated exactly. To do this, it is sufficient to make the displacement

$$F_{\mu\nu} \to F_{\mu\nu} + \partial_\mu A_\nu - \partial_\nu A_\mu, \tag{21.12}$$

which transforms the integral (21.9) into a product of an integral with respect to A_μ and an integral with respect to $F_{\mu\nu}$ of the form

$$\int \exp((i/4)\int F_{\mu\nu}F_{\mu\nu}d^4x) \prod_x \prod_{\mu<\nu} dF_{\mu\nu}(x). \tag{21.13}$$

This integral is simply a normalization constant.

We rewrite the action (21.10) in three-dimensional notation:

$$\int [\mathbf{E}\partial_0\mathbf{A} - (1/2)\,\mathbf{E}^2 + (1/2)\,\mathbf{H}^2 - (\mathbf{H}, \mathrm{curl}\,\mathbf{A}) + A_0\,\mathrm{div}\,\mathbf{E}]\,d^4x, \quad (21.14)$$

where

$$E_i = F_{0i};\ H_1 = F_{23};\ H_2 = F_{31};\ H_3 = F_{12}. \quad (21.15)$$

We evaluate the integral with respect to $\mathbf{H}$ in (21.9), which reduces to the substitution $\mathbf{H} \to \mathrm{curl}\,\mathbf{A}$ in the action (21.14). We then integrate with respect to A_0, and this gives the functional

$$\prod_x \delta\,(\mathrm{div}\,\mathbf{E}\,(x)). \quad (21.16)$$

We obtain

$$\int \exp\,(iS\,[\mathbf{A}, \mathbf{E}]) \prod_x \delta\,(\mathrm{div}\,\mathbf{A}\,(x))\delta(\mathrm{div}\,\mathbf{E}\,(x)) \prod_{i=1}^{3} dA_i\,(x) dE_i\,(x) \quad (21.17)$$

with the action of Hamiltonian form

$$\int [\mathbf{E}\partial_0\mathbf{A} - (1/2)\,\mathbf{E}^2 - (1/2)\,(\mathrm{curl}\,\mathbf{A})^2]\,d^4x. \quad (21.18)$$

The integral (21.17) is an analog of the integrals considered in §18 in the quantization of finite-dimensional mechanical systems with constraints. Here the role of the constraints is played by $\mathrm{div}\,\mathbf{E}$, and the role of the supplementary condition is played by the equation of the Coulomb gauge (21.8). The transverse components (in the three-dimensional sense) of the vectors $\mathbf{A}$ and $\mathbf{E}$ giving the vector potential and the electric field intensity can be regarded as independent variables.

This scheme of quantization of the electromagnetic field makes it possible to construct a formalism for quantum electrodynamics without introducing an indefinite metric.

The need to modify the path integral in quantum electrodynamics was noted by Bialynicki-Birula,[20] who was apparently the first to consider a path integral with the δ functional $\prod_x \delta\,(\partial_\mu A_\mu)$ (but without indicating the possible appearance of an additional factor of the type Δ_f in going over to a field theory with a non-Abelian gauge group).

The Lagrangian of spinor quantum electrodynamics,

$$L\,(x) = \bar{\psi}\,(x)(i\gamma_\mu\,(\partial_\mu - ieA_\mu\,(x)) - m)\psi\,(x) - (1/4)(\partial_\mu A_\nu\,(x) - \partial_\nu A_\mu\,(x))^2; \quad (21.19)$$

where γ_μ are the Dirac matrices, contains not only the electromagnetic field $A_\mu\,(x)$, but also the four-component spinors $\psi\,(x)$ and $\bar{\psi}\,(x)$ describing the Fermi electron–positron field. The trilinear term $e\bar{\psi}\gamma_\mu\psi A_\mu$ describes the interaction of the electromagnetic field with the electron–

positron field. The Lagrangian (21.19) is invariant with respect to the Abelian group of gauge transformations

$$\left.\begin{array}{l} A_\mu(x) \to A_\mu(x) + \partial_\mu \lambda(x); \\ \psi(x) \to \exp(\mathrm{i}e\lambda(x))\,\psi(x);\ \bar\psi(x) \to \bar\psi(x)\exp(-\mathrm{i}e\lambda(x)). \end{array}\right\} \qquad (21.20)$$

In the scheme of path integration, we must take the components of the spinors $\psi_\alpha(x)$ and $\bar\psi_\alpha(x)$ to be mutually anticommuting elements of a Grassmann algebra and integrate the functional $\exp(\mathrm{i}S)$ with respect to the measure

$$\prod_x \left(\prod_\mu dA_\mu(x) \prod_\alpha d\bar\psi_\alpha(x)\, d\psi_\alpha(x) \right) \equiv \prod_x dA\, d\bar\psi\, d\psi. \qquad (21.21)$$

We now outline the construction of perturbation theory in the parameter e — the coefficient in the trilinear term of interaction of the electromagnetic field with the electron—positron field. The functionals $F_1, F_2,$ and F_3 do not depend on the parameter e, so that the problem reduces to term-by-term integration of the series

$$\exp(\mathrm{i}e\Delta S) = \sum_{n=0}^{\infty} \frac{(\mathrm{i}e)^n}{n!} (\Delta S)^n. \qquad (21.22)$$

The form of the Green's functions depends on the choice of the functional $F[A]$. We shall find the generating functionals for the unperturbed functions corresponding to zero value of the parameter e. With the choice $F = F_1$, we must calculate the integral

$$Z_0[\eta] = \frac{\int \exp \mathrm{i}\left(S_0 + \int (\bar\eta\psi + \bar\psi\eta + \eta_\mu A_\mu)\, d^4x\right) \prod_x \delta(\partial_\mu A_\mu)\, dA\, d\bar\psi\, d\psi}{\int \exp(\mathrm{i}S_0) \prod_x \delta(\partial_\mu A_\mu)\, dA\, d\bar\psi\, d\psi}, \qquad (21.23)$$

where $\bar\eta$, η, and η_μ are the sources of the fields ψ, $\bar\psi$, and A_μ. The integration in (21.23) is taken with respect to the fields $A_\mu(x)$ satisfying the Lorentz condition

$$\partial_\mu A_\mu(x) = 0. \qquad (21.24)$$

Therefore the integrals containing the functional $F_1[A]$ will be called *integrals in the Lorentz gauge*. The functional (21.23) can be calculated by the standard procedure of making the displacements

$$\psi \to \psi + \psi_0;\ \bar\psi \to \bar\psi + \bar\psi_0;\ A_\mu \to A_\mu + A_\mu^{(0)}, \qquad (21.25)$$

which eliminates the terms linear in $\bar\psi$, ψ, and A_μ in the exponential function of the integrand in the numerator of

(21.23). For the δ functional $F_1[A]$ to remain unchanged under the displacements (21.25), it is sufficient to choose the field $A_\mu^{(0)}(x)$ to be transverse, i.e., to satisfy the condition (21.24). The expression for the functional $Z_0[\eta]$ is found to be equal to

$$\exp[-i\int\bar{\eta}(x)G(x-y)\eta(y)d^4xd^4y - (i/2)\int\eta_\mu(x)D_{\mu\nu}^{tr}(x-y)\eta_\nu(y)d^4xd^4y], \tag{21.26}$$

where

$$\left.\begin{aligned} G(x-y) &= (2\pi)^{-4}\int d^4p\exp(ip(x-y))(\hat{p}+m)/(p^2-m^2+i0); \\ D_{\mu\nu}^{tr}(x-y) &= (2\pi)^{-4}\int d^4k\exp(ik(x-y))(-k^2\delta_{\mu\nu}+k_\mu k_\nu)/(k^2+i0)^2. \end{aligned}\right\} \tag{21.27}$$

Here $(p, x-y) = p_\mu(x-y)_\mu$ and $\hat{p} = \gamma_\mu p_\mu$.

The need for the substitution $k^2 \to k^2 + i0$ in the integrands for the Green's functions was explained in §19. The validity of Wick's theorem follows from Eq. (21.26), and this leads to a diagrammatic technique in which each electron (respectively, photon) line corresponds to a function G (respectively, $D_{\mu\nu}^{tr}$), and each vertex corresponds to a coupling constant e.

Use of the functionals F_2 and F_3 results in analogous schemes of perturbation theory, which differ only in the form of the photon Green's function. The integrals with the functional F_2 will be called *integrals in the Coulomb gauge*. The unperturbed Green's function in the Coulomb gauge is given by the equations

$$\left.\begin{aligned} D_{\mu\nu}^q(x-y) &= (2\pi)^4\int d^4k\exp(i(k, x-y))\tilde{D}_{\mu\nu}^q(k); \\ \tilde{D}_{00}^q(k) &= \mathbf{k}^{-2};\ \tilde{D}_{0i}^q(k) = 0;\ \tilde{D}_{ij}^q(k) = (k^2+i0)^{-1}(\delta_{ij}-k_ik_j/\mathbf{k}^2). \end{aligned}\right\} \tag{21.28}$$

The unperturbed function $D_{\mu\nu}$ for the functional F_3 has the form

$$(k^2-i0)^{-2}(-k^2\delta_{\mu\nu}+(1-d_l)k_\mu k_\nu). \tag{21.29}$$

This expression is particularly simple when $d_l = 1$. The corresponding Green's function $\delta_{\mu\nu}(-k^2-i0)^{-1}$ is usually called the *function in the Feynman gauge*.

Thus, we have obtained three schemes of perturbation theory, which differ in the form of the photon Green's functions. These schemes are well known and lead to identical results for the physical quantities.

The method of path integration is also convenient in deriving exact relations that follow from gauge invariance. As an illustration, we shall derive the Ward identity, and also the relation for the Green's functions in different gauges.

Consider the electron Green's function in the Lorentz gauge:

$$G_L(x-y) = -i\langle\psi(x)\bar{\psi}(y)\rangle_L \equiv$$

$$-i\frac{\int \exp(iS)\,\psi(x)\,\bar{\psi}(y)\prod_x \delta(\partial_\mu A_\mu)\,dAd\bar{\psi}d\psi}{\int \exp(iS)\prod_x \delta(\partial_\mu A_\mu)\,dAd\bar{\psi}d\psi}. \tag{21.30}$$

A rotation of the spinor fields $\psi(x)\to\exp\{iec(x)\psi(x)\}$ and $\bar{\psi}(x)\to\bar{\psi}(x)\exp\{-iec(x)\}$ in the integral of the numerator of (21.30) leads to the appearance in the integrand of the factor

$$\exp i\,e\{c(x)-c(y)+\int c(z)\partial_\mu j_\mu(z)d^4z\}, \tag{21.31}$$

where $j_\mu=\bar{\psi}\gamma_\mu\psi$. Differentiating with respect to $c(z)$ and then setting $c\equiv 0$, we obtain the equation

$$G_L(x-y)(\delta(x-z)-\delta(y-z)) = i\langle\psi(x)\bar{\psi}(y)\partial_\mu j_\mu(z)\rangle_L, \tag{21.32}$$

from which, after the transition to the momentum representation, there follows the Ward identity[25]

$$G^{-1}(\hat{p})-G^{-1}(\hat{q}) = (p-q)_\mu\Gamma_\mu(p,q), \tag{21.33}$$

which relates the electron Green's function $G(\hat{p})$ to the irreducible vertex part $\Gamma_\mu(p,q)$. Clearly, this identity holds for any gauge of the photon function, since the change of variables in the path integral in its derivation affected only the spinor fields.

We now consider the transition from the Coulomb gauge to the Lorentz gauge for the case of the single-electron function:

$$G_R(x-y) = -i\langle\psi(x)\bar{\psi}(y)\rangle_R \equiv$$

$$-i\frac{\int \exp(iS)\,\psi(x)\,\bar{\psi}(y)\prod_x \delta(\mathrm{div}\,\mathbf{A})\,dAd\bar{\psi}d\psi}{\int \exp(iS)\prod_x \delta(\mathrm{div}\,\mathbf{A})\,dAd\bar{\psi}d\psi}. \tag{21.34}$$

In the numerator and denominator on the right-hand side of this equation, we insert the integral $\int\prod_x\delta(\partial_\mu A_\mu-\Box\lambda)d\lambda(x)$, which, as was shown above, does not depend on $A_\mu(x)$, and we then make the transformation (21.20), which leaves the action unchanged. Then the δ function $\delta(\partial_\mu A_\mu-\Box\lambda)$ becomes $\delta(\partial_\mu A_\mu)$, and $\delta(\mathrm{div}\,\mathbf{A})$ becomes $\delta(\mathrm{div}\,\mathbf{A}+\Delta\lambda)$. In the numerator, there appears a factor $\exp\{ie(\lambda(x)-\lambda(y))\}$, in which we can replace $\lambda(x)$ by the solution $c(x)$ of the equation

$$\Delta c(x)+\mathrm{div}\,\mathbf{A}(x)=0, \tag{21.35}$$

i.e., by the function

$$(1/4\pi)\int|\mathbf{x}-\mathbf{z}|^{-1}\delta(x_0-z_0)\mathrm{div}\,\mathbf{A}d^4z\equiv\int l_i(x-z)A_i(z)d^4z, \quad (21.36)$$

where $l_i(x-z)=\delta(x_0-z_0)(\partial/\partial x_i)(4\pi|\mathbf{x}-\mathbf{z}|)^{-1}$. The integrals with respect to $\lambda(x)$ in the numerator and denominator then cancel. The resulting equation

$$G_R(x-y)=-i\langle\psi(x)\bar{\psi}(y)\exp\{ie\int(l_i(x-z)-l_i(y-z))A_i(z)d^4z\}\rangle_L \quad (21.37)$$

expresses the Green's function of the electron in the Coulomb gauge after expansion in powers of e in the form of a series in Green's functions

$$\left\langle\psi(x)\bar{\psi}(y)\prod_{k=1}^{n}A_{i_k}(z_k)\right\rangle_L$$

in the Lorentz gauge.

§22. YANG—MILLS FIELDS

The Yang—Mills field theory[22] is the simplest example of a theory with a non-Abelian gauge group.

It is convenient to describe the Yang—Mills vector field associated with a simple compact Lie group G by matrices $B_\mu(x)$ with values in the Lie algebra of this group:

$$B_\mu(x)=\sum_{a=1}^{n}b_\mu^a(x)\tau_a. \quad (22.1)$$

Here τ_a are linearly independent matrices in the adjoint representation of the Lie algebra, normalized by the conditions

$$\mathrm{tr}\,\tau_a\tau_b=-2\delta_{ab}; \quad (22.2)$$

n is the number of parameters of the group, and $b_\mu^a(x)$ is a numerical function with a vector index μ and an isotopic index a. As is well known, in the adjoint representation the latter index can also be used to label the matrix elements, so that

$$(B_\mu)_{ab}=(\tau_c)_{ab}b_\mu^c=t_{abc}b_\mu^c, \quad (22.3)$$

where t_{abc} are the structure constants of the group, which are antisymmetric in all three indices.

The Lagrangian of the Yang—Mills field,

$$\mathscr{L}(x)=1/8\,\mathrm{tr}\,F_{\mu\nu}F_{\mu\nu}, \quad (22.4)$$

where

$$F_{\mu\nu}=\partial_\nu B_\mu-\partial_\mu B_\nu+\varepsilon[B_\mu,B_\nu], \quad (22.5)$$

is invariant with respect to the gauge transformations

$$B_\mu \to \Omega B_\mu \Omega^{-1} + (1/\varepsilon)\,\partial_\mu \Omega \Omega^{-1} \tag{22.6}$$

with the matrix Ω acting in the adjoint representation of the group.

For quantization in the path-integral formalism, it is convenient to make use of the analogs of the functionals $F[A]$ employed in §21 for quantum electrodynamics. Here they have the form

$$\left.\begin{aligned} F_1[B] &= \prod_x \delta(\partial_\mu B_\mu(x)) \equiv \prod_x \prod_a \delta(\partial_\mu b_\mu^a(x)); \\ F_2[B] &= \prod_x \delta(\operatorname{div} \mathbf{B}(x)) = \prod_x \prod_a \delta(\operatorname{div} \mathbf{b}^a(x)); \\ F_3[B] &= \exp\left((i/4d_l)\int \operatorname{tr}(\partial_\mu B_\mu(x))^2 d^4x\right) = \\ &= \exp\left((-i/2d_l)\int \sum_a (\partial_\mu b_\mu^a(x))^2 d^4x\right). \end{aligned}\right\} \tag{22.7}$$

The functionals F_1 and F_2 select from among all fields those that satisfy the conditions

$$f_L[B] = \partial_\mu B_\mu = 0 \text{ for } F_1;\quad f_R[B] = \operatorname{div} \mathbf{B} = 0 \text{ for } F_2. \tag{22.8}$$

Each of these equations is a matrix equation and actually constitutes n supplementary conditions (according to the number of parameters of the group G).

We shall denote the factor Φ corresponding to the functional F_1 by Δ_L. This factor appears in the path integral in front of the δ functional of $\partial_\mu B_\mu$, and it is therefore sufficient to know its value only for transverse fields ($\partial_\mu B_\mu = 0$). In this case, the entire contribution to the integral

$$\Delta_L^{-1}[B] = \int \prod_x \delta(\partial_\mu B_\mu^\Omega(x)) d\Omega(x)) \tag{22.9}$$

comes from the neighborhood of the unit element, in which we can make the substitution

$$\Omega(x) = 1 + \varepsilon u(x), \tag{22.10}$$

where $u(x)$ is an element of the Lie algebra, and retain in $\partial_\mu B_\mu^\Omega$ only the terms linear in u:

$$\partial_\mu B_\mu^\Omega = \partial_\mu (B_\mu + \varepsilon [u, B_\mu] + \partial_\mu u) = \Box u - \varepsilon [B_\mu, \partial_\mu u] \equiv \hat{A} u, \tag{22.11}$$

where $\Box = \hat{A}_0$ is the d'Alembertian operator. In place of the matrices $u(x)$, we introduce the column

$$u(x) = \sum_{a=1}^{n} \tau_a u_a(x), \tag{22.12}$$

on which the operator $\hat{A}$ acts according to the rule

$$(\hat{A}u)_a = (\Box u - \varepsilon[B_\mu, \partial_\mu u])_a \equiv (\Box\delta_{ac} - \varepsilon\,(B_\mu)_{ac}\partial_\mu)u_c = \Box u_a - \varepsilon t_{abc}b_\mu^b\partial_\mu u_c. \tag{22.13}$$

The integral (22.9) can be written in the form

$$\Delta_L^{-1}[B] = \int \prod_x \prod_a \delta\left((\hat{A}u)_a\right) du_a. \tag{22.14}$$

Formally, $\Delta_L[B]$ is the determinant of the operator $\hat{A}$. Extracting the trivial (infinite) factor $\det \Box$, we can then expand the logarithm of $\Delta_L[B]$ in a series in ε:

$$\ln \Delta_L[B] = \ln(\det \hat{A}/\det \hat{A}_0) = \mathbf{Tr}\ln(1 - \varepsilon\,\Box^{-1}B_\mu\,\partial_\mu) = -\sum_{n=2}^{\infty}(\varepsilon^n/n)\int d^4x_1\ldots d^4x_n\,\mathrm{tr}\,[B_\mu(x_1)\ldots B_{\mu_n}(x_n)] \times \partial_{\mu_1}D(x_1 - x_2)\ldots\partial_{\mu_n}D(x_n - x_1). \tag{22.15}$$

Here $D(x)$ is the Feynman Green's function of the d'Alembertian operator (19.17). The symbol Tr in (22.15) and in subsequent equations denotes the trace in the operator sense, in contrast to the symbol tr, which denotes the trace of a matrix.

We shall denote the corresponding factor in the Coulomb gauge by Δ_R. Analogous calculations lead to the equation

$$\ln \Delta_R[B] = \mathbf{Tr}\ln(1 - \varepsilon\Delta^{-1}B_i\,\partial_i) = -\sum_{n=2}^{\infty}(\varepsilon^n/n)\int d^4x_1\ldots d^4x_n\,\mathrm{tr}\,(B_{i_1}(x_1)\ldots B_{i_n}(x_n)) \times \partial_{i_1}\tilde{D}(x_1 - x_2)\ldots\partial_{i_n}\tilde{D}(x_n - x_1), \tag{22.16}$$

where

$$\tilde{D}(x) = -(2\pi)^{-4}\int (d^4k/\mathbf{k}^2)\exp(\mathrm{i}\,(kx)) = -\,\delta(x_0)(4\pi|\,\mathbf{x}\,|)^{-1}. \tag{22.17}$$

The indices $i_1, \ldots, i_n$ in (22.16) take the values 1, 2, 3. The factor $\Phi_3[B]$ is given by the formula

$$\Phi_3^{-1}[B] = \int \exp\{(\mathrm{i}/4d_l)\int \mathrm{tr}(\partial_\mu B_\mu^\Omega(x))^2 d^4x\}\prod_x d\Omega(x). \tag{22.18}$$

For the integral on the right-hand side of (22.18), it is not possible to obtain a closed expression analogous to Eqs. (22.15) and (22.16), which represent the factors Δ_L and Δ_R in the form of determinants. As we shall show below, this does not interfere with the development of a simple scheme of perturbation theory in this case.

We shall first construct a perturbation theory in the Lorentz gauge. This results from a series expansion in ε of the functional

$$\Delta_L[B]\exp(\mathrm{i}S[B]) = \exp(\mathrm{i}S + \ln\Delta_L[B]). \tag{22.19}$$

It is convenient to interpret the expression $\ln \Delta_L$ as addition to the action S. The term of order n in the expansion of $\ln\Delta_L$ in a series in ε leads to the appearance in the diagrams of a vertex with n outgoing lines. The explicit expression for this term which follows from (22.15) suggests an interpretation of the vertex as a loop with n outgoing lines, along which there propagates a fictitious scalar particle interacting with the vector field according to the law $\sim\varepsilon \operatorname{tr}(\varphi B_\mu \partial_\mu \varphi)$. This statement can be given a precise meaning by writing the determinant in the form of an integral with respect to the anticommuting variables η^a and $\bar{\eta}^a$:

$$\det(\Box - \varepsilon B_\mu \partial_\mu) = \int \exp\{i \int L(B_\mu, \bar{\eta}, \eta) d^4x\} \prod_x \prod d\bar{\eta}^a(x) d\eta^a(x), \quad (22.20)$$

where

$$L(B_\mu, \bar{\eta}, \eta) = -{}^1/_2 \operatorname{tr} \bar{\eta}(\Box - \varepsilon B_\mu \partial_\mu)\eta = \bar{\eta}^a \Box \eta^a - \varepsilon t_{abc} b^c_\mu \bar{\eta}^a \partial_\mu \eta^b. \quad (22.21)$$

Equation (22.20) is an infinite-dimensional analog of (19.50). It shows that our system can be regarded as a system of Bose fields $b^a_\mu(x)$ interacting with one another and with the scalar Fermi fields $\eta^a(x)$ and $\bar{\eta}^a(x)$.

The construction of the perturbation theory and the diagrammatic technique is analogous in many respects to that outlined in §22 for quantum electrodynamics. The elements of the diagrammatic technique in the Yang—Mills theory are lines of two types, corresponding to transverse vector particles and fictitious scalar particles, and also vertices describing the interaction of the vector particles with the scalar particles and with each other.

We shall represent the vector particles by solid lines, and the fictitious scalar particles by dashed lines. The elements of the diagrams are vertices and lines of the form

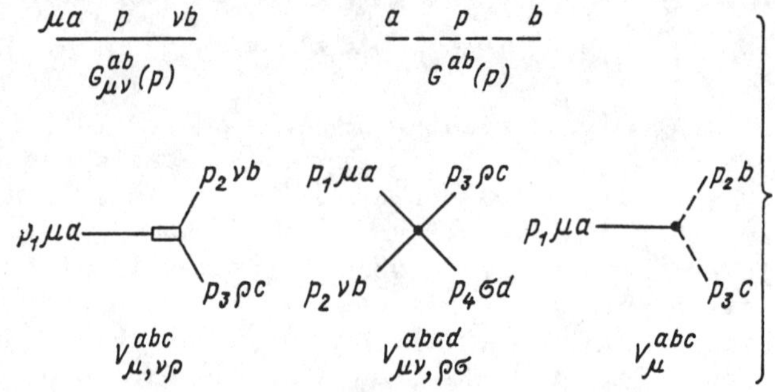

(22.22)

We write the expressions for the diagrammatic elements given in (22.22) in the momentum representation:

$$\left.\begin{aligned}
&G^{ab}_{\mu\nu}(p) = -\delta_{ab}(p^2\delta_{\mu\nu} - p_\mu p_\nu)(p^2 + i0)^{-2};\\
&G^{ab}(p) = -\delta_{ab}(p^2 + i0)^{-1};\\
&V^{abc}_{\mu,\nu\rho} = i\varepsilon t_{abc}(p_{1\nu}\delta_{\mu\rho} - p_{1\rho}\delta_{\mu\nu});\\
&V^{abcd}_{\mu\nu,\rho\sigma} = \varepsilon^2 t_{abe} t_{cde}(\delta_{\mu\rho}\delta_{\nu\sigma} - \delta_{\mu\sigma}\delta_{\nu\rho});\\
&V^{abc}_{\mu} = (i\varepsilon/2)\, t_{abc}(p_3 - p_2)_\mu.
\end{aligned}\right\} \qquad (22.23)$$

To find the contribution of a given diagram, we must take the product of the expressions corresponding to all its elements, integrate with respect to the independent 4-momenta, sum over the independent discrete indices, and multiply the result by

$$r^{-1}\,(i/(2\pi)^4)^{l-v-1}\,(-2)^s, \qquad (22.24)$$

where v is the number of vertices of the diagram, l is the number of its internal lines, s is the number of closed loops of fictitious scalar particles, and r is the order of the symmetry group of the diagram. Note that $l - v - 1 = c$ is the number of independent loops of the diagram.

We shall show that in the perturbation theory developed here the transverse Green's function $G^{ab}_{\mu\nu}$ can, without altering the physical results, be replaced by a function with an arbitrary longitudinal part:

$$G^{ab}_{\mu\nu}(d_l p) = -\delta_{ab}\,(p^2 + i0)^{-2}(p^2\delta_{\mu\nu} + (d_l - 1)p_\mu p_\nu). \qquad (22.25)$$

The original proof of this fact, given by DeWitt,[5] was cumbersome. Here we shall present a proof proposed by 't Hooft.[10] Consider a family of gauge conditions of the form

$$\partial_\mu B_\mu(x) - c(x) = 0. \qquad (22.26)$$

The average of a gauge-invariant functional X with respect to the fields with the gauge condition (22.26) can be written in the form

$$\langle X\rangle = \frac{\int X[B]\exp(iS[B])\,\Delta_c[B]\prod_x \delta(\partial_\mu B_\mu - c)\,dB}{\int \exp(iS[B])\,\Delta_c[B]\prod_x \delta(\partial_\mu B_\mu - c)\,dB}, \qquad (22.27)$$

where $\Delta_c[B]$ is a factor corresponding to the condition (22.26). The numerator and denominator on the right-hand side of (22.27) are independent of c. Using this fact, we rewrite the expression (22.27) in the form

$$\langle X\rangle = \frac{\int \exp\{(i/4d_l)\int \mathrm{tr} c^2 d^4x\}\, dc \int X[B] \exp(iS[B])\,\Delta_c[B] \prod_x \delta(\partial_\mu B_\mu - c)\, dB}{\int \exp\{(i/4d_l)\int \mathrm{tr}\, c^2 d^4x\}\, dc \int \exp(iS[B])\,\Delta_c[B] \prod_x \delta(\partial_\mu B_\mu - c)\, dB} \tag{22.28}$$

i.e., as a quotient of path integrals with respect to the variable $c(x) = \sum_a \tau_a c^a(x)$. Integrating with respect to c, we obtain for the average $\langle X\rangle$ the expression

$$\langle X\rangle = \frac{\int X[B]\tilde{\Delta}[B] \exp\{i(S[B] + (1/_4 d_l)\int \mathrm{tr}(\partial_\mu B_\mu)^2 d^4x)\} \prod_x dB}{\int \tilde{\Delta}[B] \exp\{i(S[B] + (1/_4 d_l)\int \mathrm{tr}(\partial_\mu B_\mu)^2 d^4x)\} \prod_x dB}. \tag{22.29}$$

The factor

$$\tilde{\Delta}[B] = \Delta_c[B]|_{c=\partial_\mu B_\mu} \tag{22.30}$$

is defined by the equation

$$\Delta_c^{-1}[B] = \int \prod_x \delta(\partial_\mu B_\mu^\Omega - c) d\Omega(x). \tag{22.31}$$

The calculation of the factor $\Delta_c[B]$ is analogous to the calculation of Δ_L in (22.9). The entire contribution to the integral (22.31) comes from the neighborhood of the unit element, in which we can make the substitution (22.10) and retain in $(\partial_\mu B_\mu - c)$ only the terms linear in u:

$$\partial_\mu B_\mu^\Omega - c = \partial_\mu(\varepsilon[u, B_\mu] + \partial_\mu u) = \tilde{\hat{A}}. \tag{22.32}$$

The operator $\tilde{\hat{A}}$ acts on the column $u(x)$ defined above by Eq. (22.12) according to the rule

$$(\tilde{\hat{A}}u)_a = \Box\, u_a - \varepsilon t_{abc}\, \partial_\mu(b_\mu^b u_c); \tag{22.33}$$

the operator $\tilde{\hat{A}}$ is conjugate to $\hat{A}$, and the determinants of the operators $\hat{A}$ and $\tilde{\hat{A}}$ are therefore equal:

$$\tilde{\Delta}[B] = \det \tilde{\hat{A}} = \det \hat{A} = \Delta_L[B]. \tag{22.34}$$

The perturbation-theory calculation of $\langle X\rangle$ as a quotient of path integrals in (22.29) leads to a diagrammatic technique with the Green's function (22.25) for the vector particle. As before, the contribution of the factor $\Delta_L[B] = \tilde{\Delta}[B]$ is interpreted as the contribution of additional diagrams describing the interaction of the vector particles with the fictitious scalar particles.

The perturbation theory developed here is not the only possible one. A different form of perturbation theory and the diagrammatic technique arises in the so-called *first-order formalism*. This formalism is obtained if we write the Lagrangian (22.4) in the form

$$L(x) = -(1/8)\mathrm{tr}\, F_{\mu\nu}F_{\mu\nu} + (1/4)\mathrm{tr}\, F_{\mu\nu}(\partial_\nu B_\mu - \partial_\mu B_\nu + \varepsilon [B_\mu, B_\nu]) \tag{22.35}$$

and integrate with respect to B_μ and $F_{\mu\nu}$ as independent variables. The term "first-order formalism" means that only first derivatives appear in the Lagrangian (22.35).

Using the Lorentz gauge, we obtain a path integral of the form

$$\int \exp(\mathrm{i}S[B, F])\Delta_L[B]\prod_x \delta(\partial_\mu B_\mu)dBdF, \tag{22.36}$$

where the expression

$$dB(x)\,dF(x) = \prod_a \prod_\mu db^a_\mu(x) \prod_{\mu<\nu} df^a_{\mu\nu}(x), \tag{22.37}$$

like $B(x)$, is gauge-invariant.

If we expand the functional $\Delta_L[B]\exp(\mathrm{i}S[B, F])$ in a series in ε under the integral sign in (22.36), we obtain a new variant of the diagrammatic perturbation theory involving three lines, corresponding to the functions $\langle BB\rangle$, $\langle BF\rangle$, and $\langle FF\rangle$, and one vertex, describing the trilinear interaction εFBB. The elements of the diagrammatic technique have the form

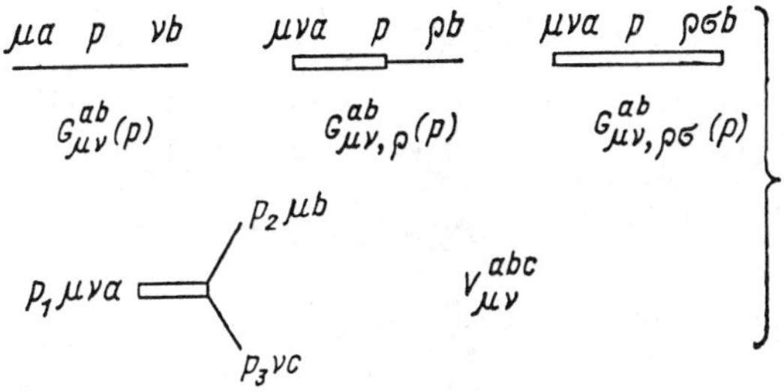

(22.38)

where the field B is represented by a single line, and the field F by a double line. The expressions corresponding to the elements (22.38) are given by the formulas

$$\left.\begin{aligned}
&G^{ab}_{\mu\nu} = \delta_{ab}(-p^2\delta_{\mu\nu} + p_\mu p_\nu)(p^2 + \mathrm{i}0)^{-2};\\
&G^{ab}_{\mu\nu,\rho} = \mathrm{i}\delta_{ab}(p_\nu\delta_{\mu\rho} - p_\mu\delta_{\nu\rho})(p^2 + \mathrm{i}0)^{-1};\\
&G^{abcd}_{\mu\nu,\rho\sigma} = \delta_{ab}(\delta_{\mu\rho}\delta_{\nu\sigma} - \delta_{\mu\sigma}\delta_{\nu\rho} - (p^2 + \mathrm{i}0)^{-1}\\
&\times(\delta_{\mu\rho}p_\nu p_\sigma + \delta_{\nu\sigma}p_\mu p_\rho - \delta_{\mu\sigma}p_\nu p_\rho - \delta_{\nu\rho}p_\mu p_\sigma));\\
&V^{abc}_{\mu\nu} = \varepsilon t_{abc}.
\end{aligned}\right\} \tag{22.39}$$

The lines and vertices describing the propagation of the fictitious scalar particles and their interaction with the vector particles remain the same as in the second-order formalism, since the factor Δ_L, which depends only on B and not on F, remains the same.

The first-order formalism is convenient for the transition to canonical quantization. We shall consider this transition on the basis of an integral with respect to B_μ in the Coulomb gauge with sources which are the generating functional for the Green's functions:

$$Z[\eta]=\frac{\int \exp\{iS[B, F]+i\int(\eta_{\mu a}b_{\mu a}+1/2\eta_{\mu\nu a}f_{\mu\nu a})d^4x\}\Delta_R[B]\prod_x\delta(\operatorname{div}\mathbf{B})dBdF}{\int \exp(iS[B, F])\Delta_R\prod_x\delta(\operatorname{div}\mathbf{B})\,dBdF}. \tag{22.40}$$

As in electrodynamics, we take the dynamical variables to be the transverse components, in the three-dimensional sense, of the fields B_i and $F_{0i}(i = 1, 2, 3)$. We shall assume that the sources occur with the chosen dynamical variables, i.e., that the following conditions are satisfied:

$$\eta_{0a}=\eta_{ika}=\partial_i\eta_{0ia}=\partial_i\eta_{ia}=0. \tag{22.41}$$

In three-dimensional notation, the Lagrangian (22.35) takes the form

$$L(x)=\operatorname{tr}\{-(1/8)F_{ik}F_{ik}+(1/4)F_{0i}F_{0i}+(1/4)F_{ik}(\partial_kB_i-\partial_iB_k+$$
$$\varepsilon[B_i, B_k]-(1/2)F_{0i}\partial_0B_i-(1/2)B_0(\partial_iF_{0i}-\varepsilon[B_i, F_{0i}])\}. \tag{22.42}$$

Owing to the absence of sources associated with B_0 and F_{ik}, we can integrate with respect to these variables in (22.38), and this leads to the appearance of the δ functional

$$\prod_x\delta(\partial_iF_{0i}-\varepsilon[B_i, F_{0i}]) \tag{22.43}$$

and to the replacement of F_{ik} by

$$H_{ik}\equiv\partial_kB_i-\partial_iB_k+\varepsilon[B_i, B_k] \tag{22.44}$$

in the integral with respect to the remaining variables B_i and F_{0i}. We insert in the integral (22.40) the factor

$$\int\prod_x\delta(\Delta c+\partial_iF_{0i})dc(x)), \tag{22.45}$$

which is actually independent of F_{0i}, and we then make the displacement $F_{0i}\to F_{0i}-\partial_ic$. The functional $\prod_x\delta(\Delta c+\partial_iF_{0i})$ then becomes $\prod_x\delta(\partial_iF_{0i})$, and $\prod_x\delta(\partial_iF_{0i}-\varepsilon[B_i, F_{0i}])$ becomes $\prod_x\delta(\Delta c-\partial_iF_{0i}-\varepsilon[B_i, \partial_ic]+\varepsilon[B_i, F_{0i}])$, which is equal to $\prod_x\delta(\Delta c-\varepsilon[B_i, \partial_ic]+\varepsilon[B_i, F_{0i}])$, since $\partial_iF_{0i}=0$.

Let $c_0(x)$ be the solution of the equation

$$\Delta c - \varepsilon [B_i, \partial_i c] = -\varepsilon [B_i, F_{0i}], \tag{22.46}$$

which can be expressed in terms of the B-dependent Green's function:

$$c_0 (x) = -\varepsilon \int D (x, y; B)[B_i (y), F_{0i} (y)]d^3y. \tag{22.47}$$

Making the displacement $c \to c + c_0$, we find the functional $\prod_x \delta (\Delta c - \varepsilon[B_i, \partial_i c])$, and the function $c(x)$ can be set equal to zero everywhere, except for the argument of this δ functional. The integral

$$\int \prod_x \delta (\Delta c - \varepsilon[B_i, \partial_i c])dc (x) \tag{22.48}$$

cancels with the factor $\Delta_R[B]$. As a result, the functional (22.40) is reduced to the form

$$Z [\eta] = \frac{\int \exp \left\{iS [B_i, F_{0i}] + i \int \left(\eta_i^a b_i^a + \eta_{0i}^a f_{0i}^a\right) d^4 x\right\} \prod_x \delta (\partial_i B_i) \delta (\partial_i F_{0i}) \, dBdF}{\int \exp \{iS [B_i, F_{0i}]\} \prod_x \delta (\partial_i B_i) \delta (\partial_i F_{0i}) \, dBdF}, \tag{22.49}$$

where

$$S [B_i, F_{0i}] = \int dx_0 \left(\int f_{0i}^a \partial_0 b_i^a d^3x - H\right); \tag{22.50}$$

$$H = \int d^3 x \left(\tfrac{1}{4} h_{ik}^a h_{ik}^a + \tfrac{1}{2} f_{0i}^a f_{0i}^a + \tfrac{1}{2} \partial_i c_0^a \partial_i c_0^a\right). \tag{22.51}$$

In these equations, $S [B_i, F_{0i}]$ is the action corresponding to the Hamiltonian H, where the transverse fields in b_i and f_{0i} have the meaning of canonically conjugate coordinates and momenta.

As was shown in §19 for the example of a scalar field, the formalism of the path integral with respect to the canonically conjugate coordinates and momenta is equivalent to canonical quantization. When applied to a system with the Hamiltonian (22.51), canonical quantization reduces to the replacement of the functions b_i^a and f_{0i}^b, in terms of which h_{ik}^a and c_0^a are expressed, by the operators $\hat{b}_i^a(x)$ and $\hat{f}_{0j}^b (y)$ obeying the commutation relations

$$[\hat{b}_i^a (\mathrm{x}), f_{0j} (\mathrm{y})] = i\delta_{ab}\delta_{ij}^{\mathrm{tr}} (\mathrm{x} - \mathrm{y}) \equiv$$

$$(i\delta_{ab}/(2\pi)^3) \int d^3k \exp (i (\mathbf{k}, \mathrm{x} - \mathrm{y}))(\delta_{ij} - k_i k_j/\mathbf{k}^2). \tag{22.52}$$

The Hamiltonian (22.51) becomes a self-adjoint and positive definite energy operator. This quantization of the Yang–Mills field was proposed by Schwinger.[23] Thus, the path-integral formalism leads to Schwinger canonical quantization. We stress that the presence of the factor $\Delta_R[B]$ in the original integral (22.40) was essential for its reduction to manifestly Hamiltonian form.

Let us consider the construction of the S matrix for the Yang—Mills field. It is natural to begin with the path integral in the Coulomb gauge. It is in this gauge that the integral reduces to an integral with respect to canonical variables. The unitarity of the S matrix is obvious here, at any rate when its elements are calculated by perturbation theory.

The element of the S matrix describing the transition of m incoming particles into $(n-m)$ outgoing particles can be expressed in terms of the Fourier transform $G^{a_1\ldots a_n}_{i_1\ldots i_n}$ $(p_1, \ldots p_n)$ of the Green's function in the Coulomb gauge:

$$\langle b^{a_1}_{i_1}(x_1)\ldots b^{a_n}_{i_n}(x_n)\rangle_R \equiv \frac{\int \exp(iS)\,\Delta_R[B]\,b^{a_1}_{i_1}(x_1)\ldots b^{a_n}_{i_n}(x_n)\prod_x \delta(\operatorname{div}\mathbf{B}(x))\,dB(x)}{\int \exp(iS)\,\Delta_R[B]\prod_x \delta(\operatorname{div}\mathbf{B}(x))\,dB(x)} \tag{22.53}$$

by means of the formula

$$S^{a_1\ldots a_n}_{i_1\ldots i_n}(p_1,\ldots,p_n) = \lim_{p_k^2\to 0}\left(\prod_{k=1}^n Z_R^{-1/2}\,u_k\,(\hat{e}_k)_{i_k j_k}\right) G^{a_1\ldots a_n}_{j_1\ldots j_n}(p_1,\ldots,p_n), \tag{22.54}$$

where

$$(\hat{e}_k)_{ij} = \delta_{ij} - (p_k)_i(p_k)_j/\mathbf{p}_k^2 \tag{22.55}$$

is the (transverse) polarization operator, and the factors u_k are equal to $p_k^2\theta(p_k^0)|2p_k^0|^{-1/2}(2\pi)^{-3/2}$ for incoming particles and $p_k^2\theta(-p_k^0)\,|2p_k^0|^{-1/2}\,(2\pi)^{-3/2}$ for outgoing particles. Finally, Z_R is the residue as $p^2\to 0$ of the full single-particle Green's function in the Coulomb gauge, under the assumption that as $p^2 \to 0$ the Green's function has the form

$$G^{ab}_{ij} = [Z_R\delta_{ab}/(p^2 + i0)](\delta_{ij} - p_i p_j/\mathbf{p}^2) \tag{22.56}$$

(apart from infrared singularities, of course).

The expressions for the elements of the S matrix in the Coulomb gauge are not manifestly relativistically invariant. We transform them by going over to the relativistic Lorentz gauge.

In the path integrals in the numerator and denominator, we insert the factor

$$\Delta_L[B]\int \prod_x \delta\,((\partial_\mu B^\Omega_\mu)\,d\Omega(x), \tag{22.57}$$

which is equal to unity. We then make the displacements $B^\Omega \to B$ and $B \to B^{\Omega^{-1}}$. This gives an integral over the gauge group,

$$\int \left(B^{\Omega^{-1}}(x_1)\right)^{a_1}_{i_1}\ldots\left(B^{\Omega^{-1}}(x_n)\right)^{a_n}_{i_n}\prod_x \delta\left(\operatorname{div}\mathbf{B}^{\Omega^{-1}}\right)d\Omega(x), \tag{22.58}$$

from which we can remove the product $(B^{\Omega^{-1}}(x_1))^{a_1}_{i_1} \ldots (B^{\Omega^{-1}}(x_n))^{a_n}_{i_n}$. The remaining integral with respect to Ω cancels with $\Delta_R[B]$.

The expression (22.53) for the Green's function takes the form

$$\frac{\int \exp(iS)\, \Delta_L[B] \left(B^{\Omega^{-1}}(x_1)\right)^{a_1}_{i_1} \cdots \left(B^{\Omega^{-1}}(x_n)\right)^{a_n}_{i_n} \prod_x \delta(\partial_\mu B_\mu)\, dB}{\int \exp(iS)\, \Delta_L[B] \prod_x \delta(\partial_\mu B_\mu)\, dB}. \quad (22.59)$$

The matrix Ω^{-1} in (22.59) is chosen to satisfy the condition of three-dimensional transversality of the field $\mathbf{B}^{\Omega^{-1}}$ ($\mathrm{div}\, \mathbf{B}^{\Omega^{-1}} = 0$). The expansion of $B^{\Omega^{-1}}$ in powers of ε gives the series

$$B^{\Omega^{-1}}(x) = B^{\mathrm{tr}} - (\varepsilon/2)[\Delta^{-1}\mathrm{div}\, \mathbf{B},\, B + B^{\mathrm{tr}}]^{\mathrm{tr}} + \ldots \quad (22.60)$$

Here we have given the first two terms of the expansion; the superscript tr denotes the three-dimensional transverse part of the corresponding vector. The integral (22.59) can be calculated by perturbation theory by first expanding the function $B^{\Omega^{-1}}$ in the series (22.60), where each term of the expansion, depending on a product of m fields B, is associated with a vertex having m outgoing lines.

Thus, the transition to the Lorentz gauge for the Green's functions is rather complicated. However, in constructing the S matrix, it is sufficient to know the Green's function only on the mass shell (all $p_k^2 \to 0$). The factors u_k then vanish, and the transition to the Lorentz gauge reduces to insertions at the external ends:

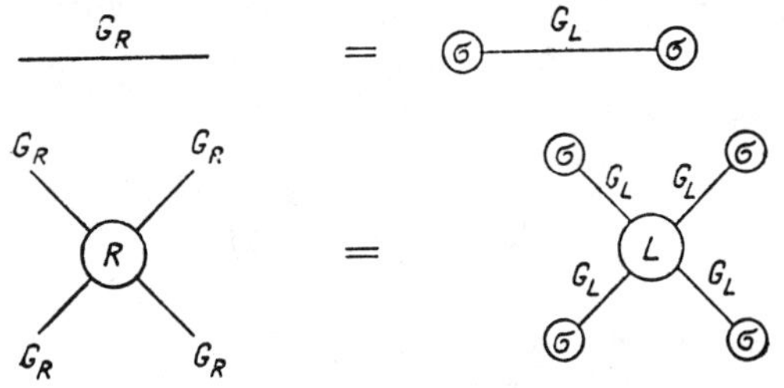

(22.61)

This insertion (which we denote by σ) receives contributions from all subdiagrams beginning with a vertex generated by the expansion (22.60) and ending with a vertex connected to the rest of the diagram by only a single line. It is in diagrams with this structure that the vanishing of the factors u_k is compensated by poles of the single-particle functions G_L in the Lorentz gauge (with residues Z_L). The contribution of all the remaining diagrams vanishes on the mass shell, and the Green's function (22.59) in the Coulomb gauge differs from the corresponding function in the Lorentz gauge by a factor σ^n. Comparing the single-particle functions (the first of the diagrammatic equations (22.61)), it follows that $\sigma = (Z_R/Z_L)^{1/2}$, i.e., the result is expressed in terms of the ratio of residues. Consequently, on the mass shell we can go over to the Green's functions in the Lorentz gauge and replace Z_R by Z_L, and thus write the S-matrix element in a manifestly Lorentz-invariant form.

We conclude this section with a discussion of the corrections of second-order perturbation theory to the Green's function. These are of interest because they demonstrate a situation which is opposite to the well-known zero-charge situation in quantum electrodynamics.[1,2]

The Green's function of the photon in the transverse gauge for $k^2 \gg m^2$ (where m is the electron mass) has the form

$$D^{\mathrm{tr}}_{\mu\nu} = (k^2 + i0)^{-1}(k^2\delta_{\mu\nu} - k_{\mu\nu} - k_\mu k_\nu)(k^2 + i0 - P)^{-1} \simeq$$

$$(k^2 + i0)^{-2}(k^2\delta_{\mu\nu} - k_\mu k_\nu)(1 - (e^2/12\pi^2)\ln(-k^2/m^2))^{-1}, \quad (22.62)$$

if we restrict the self-energy part to the single-loop diagram of second order, taking its asymptotic form for $k^2 \gg m^2$. The approximation (22.62) has an unphysical ("ghost") pole for $(e^2/12\pi^2)\ln(-k^2/m^2) \approx 1$. The existence of such a pole in the exact Green's function of the photon would lead to serious contradictions with a number of general principles of the theory.

For the Yang–Mills field, the formula for the Green's function at large k^2 has the form

$$G^{ab}_{\mu\nu} = \delta_{ab}(k^2 + i0)^{-2}(k^2\delta_{\mu\nu} - k_\mu k_\nu)(1 + (11\varepsilon^2/12\pi^2)\ln(-k^2/M^2))^{-1}, \quad (22.63)$$

where M^2 is a renormalization constant. In this formula, there is a plus sign in front of the logarithmic term in the denominator, and no difficulties involving an unphysical pole or zero-charge arise. This fact was discussed in Refs. 24 and 25. More recently, it has been shown that the situation remains unchanged when allowance is made for higher-order diagrams of perturbation theory.

Equation (22.63) is readily proved in the "first-order formalism," where it is necessary to calculate the following diagrams:

$$\Sigma^{ab}_{\mu\nu} = \quad \text{[diagram]} + \text{[diagram]}$$

$$\Sigma^{ab}_{\mu\nu,\rho} = \quad \text{[diagram]}$$

$$\Sigma^{ab}_{\mu\nu,\rho\sigma} = \quad \text{[diagram]}$$

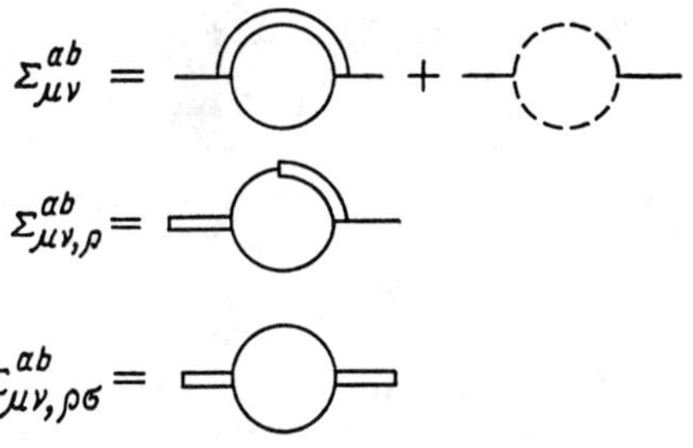

(22.64)

These diagrams include internal lines of three types for the Yang–Mills field, and dashed lines corresponding to fictitious scalar particles.

Expressions for the self-energy parts of (22.63) can be calculated from the rules formulated above and are found to have the following form:

$$\Sigma^{ab}_{\mu\nu}(k) = (\delta_{ab}\,\varepsilon^2/12\pi^2)\{(\delta_{\mu\nu}k^2 - k_\mu k_\nu)\ln(-k^2/k_0^2)$$
$$+ (ak^2 + b)\delta_{\mu\nu} + ck_\mu k_\nu\};$$
$$\Sigma^{ab}_{\mu\nu,\sigma}(k) = (\delta_{ab}\,\varepsilon^2\,3i/16\pi^2)(k_\mu\delta_{\nu\sigma} - k_\nu\delta_{\mu\sigma})\{\ln(-k^2/k_0^2) + d\};$$
$$\Sigma^{ab}_{\mu\nu,\sigma\rho}(k) = (\delta_{ab}\,\varepsilon^2/16\pi^2)\{(\delta_{\mu\rho}\delta_{\nu\sigma} - \delta_{\mu\sigma}\delta_{\nu\rho})\{\ln(-k^2/k_0^2) + e\}$$
$$+ (2(k^2+i0))^{-1}(k_\mu k_\rho\delta_{\nu\sigma} + k_\nu k_\sigma\delta_{\mu\rho} - k_\mu k_\sigma\delta_{\nu\rho} - k_\nu k_\rho\delta_{\mu\rho})\}.$$

(22.65)

Here k_0 is a fixed 4-momentum with $k_0^2 > 0$, and a, b, c, d, and e are renormalization constants. Actually, one can uniquely determine the first derivatives of $\Sigma^{ab}_{\mu\nu,\rho\sigma}(k)$, the second derivatives of $\Sigma^{ab}_{\mu\nu,\sigma}(k)$, and the third derivatives of $\Sigma^{ab}_{\mu\nu}(k)$. For the transversality of $\Sigma^{ab}_{\mu\nu}(k)$, we require $b = 0$ and $a = -c$. Knowing the self-energy parts (22.65), we calculate the Green's function in second-order perturbation theory according to the formula

$$G^{ab}_{\mu\nu} + G^{ac}_{\mu,\sigma}\Sigma^{cd}_{\sigma\rho}G^{db}_{\rho\nu} + G^{ac}_{\mu\sigma}\Sigma^{cd}_{\sigma,\rho\lambda}G^{db}_{\rho\lambda,\nu} + G^{ac}_{\mu,\rho\lambda}\Sigma^{cd}_{\rho\lambda,\sigma}G^{db}_{\sigma\nu} +$$
$$G^{ac}_{\mu,\rho\lambda}\Sigma^{cd}_{\rho\lambda,\sigma\tau}G^{db}_{\sigma\tau,\nu} = (k^2+i0)^{-2}(k^2\delta_{\mu\nu} - k_\mu k_\nu)\{1 - (11\varepsilon^2/12\pi^2) \times$$
$$\ln(-k^2/k_0^2) + f\} \approx (k^2+i0)^{-2}(k^2\delta_{\mu\nu} - k_\mu k_\nu)\{1 + (11\varepsilon^2/12\pi^2) \times$$
$$\ln(-k^2/k_0^2) + f\}^{-1},$$

(22.66)

where the renormalization constant f is a linear combination of a, d, and e ($f = a + 9d + 3e$). The Green's function $G^{ab}_{\mu\nu,\sigma}(k)$ and the self-energy part $\Sigma^{ab}_{\mu\nu,\sigma}(k)$ differ in sign from $G^{ab}_{\sigma,\mu\nu}(k)$ and $\Sigma^{ab}_{\sigma,\mu\nu}(k)$, respectively. This completes the proof of Eq. (22.63).

§23. QUANTIZATION OF THE GRAVITATIONAL FIELD

The quantum theory of gravity is constructed largely in the hope that the gravitational field may become a natural "physical regularizer" which eliminates infinities from quantum field theory. The gravitational field can be regarded as a special case of a gauge field, and its quantization is achieved in accordance with the general scheme of quantization of gauge fields. Here we shall consider only gravitational fields which have no singularities (such as "black holes") and are asymptotically flat at infinity. The gauge transformations are coordinate transformations which do not affect the spatial infinity, and the symmetry group is the Poincaré group.

The specific features of the quantization of the gravitational field are related mainly to its self-action. Therefore we shall consider primarily the "free" self-interacting gravitational field.

Of the various methods of parametrizing the gravitational field, two are most widely used: the metric-tensor method, and the moving-tetrad method. We shall give a formulation of these methods.

In the metric-tensor formalism, the gravitational field is described by the potentials $g_{\mu\nu}(x)$ and the Christoffel symbols $\Gamma^{\rho}_{\mu\nu}(x)$. The latter can be treated either as independent dynamical quantities (the Palatini formalism) or as functions of $g_{\mu\nu}$:

$$\Gamma^{\rho}_{\mu\nu} = (1/2)g^{\rho\sigma}(\partial_{\mu}g_{\nu\sigma} + \partial_{\nu}g_{\mu\sigma} - \partial_{\sigma}g_{\mu\nu}). \tag{23.1}$$

The contravariant matrix $g^{\mu\nu}$ is inverse to $g_{\mu\nu}$, and g is the determinant of the matrix $g_{\mu\nu}$.

For an asymptotically flat gravitational field, the space-time manifold is topologically equivalent to a 4-dimensional Euclidean space and can be parametrized by global coordinates x_{μ} $(-\infty < x_{\mu} < \infty;\ \mu = 0, 1, 2, 3)$. We shall assume that these coordinates are consistent with the conditions at the spatial infinity, so that

$$g_{\mu\nu} = \eta_{\mu\nu} + O(r^{-1});\ \Gamma^{\rho}_{\mu\nu} = O(r^{-1}), \tag{23.2}$$

where $r = \sqrt{(x^1)^2 + (x^2)^2 + (x^3)^2}$, and $\eta_{\mu\nu}$ is the Minkowski tensor with signature (+---).

The action functional, which has the form

$$S = (2\varkappa^2)^{-1}\int\{-\Gamma^{\rho}_{\mu\rho}\partial_{\nu}(\sqrt{-g}\,g^{\mu\nu}) + \Gamma^{\rho}_{\mu\nu}\partial_{\rho}(\sqrt{-g}\,g^{\mu\nu}) + \sqrt{-g}\,g^{\mu\nu}(\Gamma^{\rho}_{\mu\sigma}\Gamma^{\sigma}_{\rho\nu} - \Gamma^{\rho}_{\mu\nu}\Gamma^{\sigma}_{\rho\sigma})\}\,d^4x, \tag{23.3}$$

where $\varkappa$ is Newton's constant, is invariant with respect to the group of coordinate transformations acting on the quantities $g^{\mu\nu}$ and $\Gamma^{\rho}_{\mu\nu}$ according to the rules

$$\left.\begin{aligned} \delta g^{\mu\nu} &= -\eta^{\lambda}\partial_{\lambda} g^{\mu\nu} + g^{\mu\lambda}\partial_{\lambda}\eta^{\nu} + g^{\nu\lambda}\partial_{\lambda}\eta^{\mu}; \\ \delta\Gamma^{\rho}_{\mu\nu} &= -\eta^{\lambda}\partial_{\lambda}\Gamma^{\rho}_{\mu\nu} - \Gamma^{\rho}_{\mu\lambda}\partial_{\lambda}\eta^{\lambda} - \Gamma^{\rho}_{\nu\lambda}\partial_{\mu}\eta^{\lambda} \\ &+ \Gamma^{\lambda}_{\mu\nu}\partial_{\lambda}\eta^{\rho} - \partial_{\mu}\partial_{\nu}\eta^{\rho}. \end{aligned}\right\} \quad (23.4)$$

Here we have written equations for the infinitesimal transformations; η^{μ} are the infinitesimal components of the vector field which generates the coordinate transformations

$$\delta x^{\mu} = \eta^{\mu}(x). \quad (23.5)$$

Variation of the action (23.3) with respect to $\Gamma^{\rho}_{\mu\nu}$ leads to equations whose solutions are the functions (23.1). In this sense, $\Gamma^{\rho}_{\mu\nu}$ can be regarded as independent variables.

Substituting in (23.3) the explicit expression (23.1) for the Christoffel symbols $\Gamma^{\rho}_{\mu\nu}$ in terms of the metric tensor, the action takes the form

$$S = (4\varkappa^2)^{-1}\int(h^{\rho\sigma}\partial_{\rho}h^{\mu\nu}\partial_{\nu}h_{\mu\sigma} - (1/2)h^{\rho\sigma}\partial_{\rho}h^{\mu\nu}\,\partial_{\sigma}h_{\mu\nu} + (1/4)h^{\rho\sigma}\partial_{\rho}\ln h\partial_{\sigma}\ln h)d^4x, \quad (23.6)$$

where for convenience we have used the contravariant density

$$h^{\mu\nu} = \sqrt{-g}\,g^{\mu\nu};\ h = \det h^{\mu\nu}. \quad (23.7)$$

In the moving-tetrad formalism, the gravitational field is described by the tetrad components $e^{\mu a}(x)$ and the coefficients of torsion $\omega_{\mu ab}(x) = -\varphi_{\mu ba}(x)$. The set $e^{\mu a}(x)$ forms a matrix with a positive determinant $e(x)$. The action functional

$$S = (2\varkappa^2)^{-1}\int\{\omega_{\mu ab}\partial_{\mu}(e^{-1}e^{\mu a}e^{\nu b}) - \omega_{\mu ab}\partial_{\nu}(e^{-1}e^{\mu a}e^{\nu b}) + e^{-1}e^{\mu a}e^{\nu b}(\omega_{\mu ac}\omega^{c}_{\nu b} - \omega_{\nu ac}\omega^{c}_{\mu b})\} \quad (23.8)$$

is invariant with respect to coordinate transformations

$$\left.\begin{aligned} \delta e^{\mu a} &= -\eta^{\lambda}\partial_{\lambda}e^{\mu a} + e^{\lambda a}\partial_{\lambda}\eta^{\mu}; \\ \delta\omega_{\mu ab} &= -\eta^{\lambda}\partial_{\lambda}\omega_{\mu cb} - \omega_{\lambda ab}\partial_{\mu}\eta^{\lambda} \end{aligned}\right\} \quad (23.9)$$

and local Lorentz rotations

$$\left.\begin{aligned} \delta e^{\mu a} &= \eta^{a}_{b}e^{\mu b}; \\ \delta\omega_{\mu ab} &= -\eta^{c}_{a}\omega_{\mu cb} + \eta^{c}_{b}\omega_{\mu ac} + \partial_{\mu}\eta_{ab}. \end{aligned}\right\} \quad (23.10)$$

Variation of S with respect to ω leads to equations that enable us to express ω explicitly in terms of e. It is convenient to write the solution in the form

$$\omega_{\mu ab} = e^{c}_{\mu}\omega_{cab} \equiv (1/2)e^{c}_{\mu}(\Omega_{abc} + \Omega_{bca} - \Omega_{cab}), \quad (23.11)$$

where

$$\Omega_{abc} = e_{\mu a}\,\Omega^{\mu}_{bc} \equiv e_{\mu a}\left(e^{\nu}_{b}\,\partial_{\nu}\,e^{\mu}_{c} - e^{\nu}_{c}\,\partial_{\nu}\,e^{\mu}_{b}\right).$$

We can assume that this has already been done, so that S is a functional of only the variables $e^{\mu a}$.

We speak of a first-order formalism if the variables $g_{\mu\nu}$ and $\Gamma^{\rho}_{\mu\nu}$ (or $e^{\mu a}$ and $\omega_{\mu ab}$) are assumed to be independent. But if Γ are expressed in terms of g, and e in terms of ω, we speak of a second-order formalism.

The descriptions of the free gravitational field in terms of $g^{\mu\nu}$ and $e^{\mu a}$ are equivalent. The different number of components — 10 in the first case and 16 in the second — is compensated by the difference in the gauge group, which is parametrized by four functions in the first case and ten in the second. The moving-tetrad formalism is convenient for the description of an interaction with a spinor field.

The equivalence of the first-order and second-order formalisms may disappear when interactions with other fields are switched on. Geometrically, Eq. (23.11) determines a connection without torsion. The minimal interaction of the gravitational field with a spinor field in the first-order formalism leads to the appearance of torsion.[26]

The exposition which follows will be given mainly for the example of a second-order tensor formalism.

The coordinate transformations of the metric tensor form a non-Abelian gauge group depending on four arbitrary functions (Eq. (23.4) gives the infinitesimal transformations). Therefore, in accordance with the general scheme of quantization of gauge fields, to quantize a system with action S we must integrate the functional $\exp(iS)$ over a surface in the manifold of all fields specified by four equations. It is convenient to take these equations to be the De Donder—Fock harmonicity conditions[27]

$$\partial_{\nu}\,(\sqrt{-g}g^{\mu\nu}) = l^{\mu}\,(x), \tag{23.12}$$

where $l^{\mu}(x)$ is a given vector field. The arbitrariness in the choice of $l^{\mu}(x)$ will be convenient for the formal transformations. The harmonicity conditions (23.12) constitute an analog of the Lorentz gauge in electrodynamics and in Yang—Mills theory.

The conditions (23.12) are not general covariant conditions, and it is for this reason that they can serve to parametrize the classes. The analog of the equation $f(A^{\Omega}) = 0$ is a complicated nonlinear equation for the parameters of a coordinate transformation which converts a given metric into a harmonic one. In perturbation theory, this equation has a unique solution.

The local gauge-invariant measure has the form*

$$\prod_x g^{5/2}(x) \prod_{\mu\leqslant\nu} dg^{\mu\nu}(x) = \prod_x h^{-5/2}(x) \prod_{\mu\leqslant\nu} dh^{\mu\nu}(x). \qquad (23.13)$$

To prove the gauge invariance of the measure (23.13), we consider the product $\prod_x$ as a product over "physical" points. The transformation of the metric tensor at one and the same "physical" point is given by the first of the equations (23.4) without the first term, so that

$$\delta g^{\mu\nu} = g^{\mu\lambda}\,\partial_\lambda\eta^\nu + g^{\nu\lambda}\,\partial_\lambda\eta^\mu; \;\; \delta g = -2g\partial_\mu\eta^\mu. \qquad (23.14)$$

Therefore

$$\prod_{\mu\leqslant\nu} d(g^{\mu\nu}+\delta g^{\mu\nu}) = (1+5\partial_\mu\eta^\mu)\prod_{\mu\leqslant\nu} dg^{\mu\nu}; \;\; g+\delta g = (1-2\partial_\mu\eta^\mu)\,g. \qquad (23.15)$$

Hence for a fixed "physical" point there follows invariance of the expression $g^{5/2}\prod_{\mu\leqslant\nu} dg^{\mu\nu}$, which ensures gauge invariance of the measure (23.13).

With the parametrization of the classes given by (23.12) and the measure (23.13), we obtain the following expression for the path integral:

$$\int \exp(iS)\,\Delta_h[g]\prod_x\Big[\prod_\mu \delta(\partial_\nu\, g^{\mu\nu}-l^\mu)\Big(g^{5/2}\prod_{\mu\leqslant\nu} dg^{\mu\nu}\Big)\Big], \qquad (23.16)$$

where in accordance with (20.8) the functional $\Delta_h[g]$ is determined by the equation

$$\Delta_h[g]\int\prod_x\Big[\prod_x \delta(\partial_\nu(h^{\mu\nu})^\Omega) - l^\mu(x)\Big]\,d\Omega(x) = 1 \qquad (23.17)$$

and is expressed in terms of an integral over the gauge group of a δ functional.

Let us calculate this integral. The expression $\Delta_h[g]$ appears in the integral (23.16) only on the surface determined by (23.12). For such $g^{\mu\nu}$, the total contribution to the integral (23.17) comes from an infinitesimal

---

*After the publication of the first edition of this book, there appeared Refs. 28—30, which present arguments in favor of the Leutwyler measure $d\mu = \prod_x (g^{7/2} g^{00} \prod_{\mu\leqslant\nu} dg^{\mu\nu})$ in the theory of gravitation instead of the measure (23.13) used here. It is shown in these same papers that the distinction between the measures gives rise to terms of purely renormalization type, whose role is to eliminate the divergences from perturbation theory. In other respects, the structure of perturbation theory remains unchanged. Therefore, with this in mind, we do not modify the treatment of perturbation theory for the gravitational field.

neighborhood of the unit element of the group. In this neighborhood, the action of the group transformations on $h^{\mu\nu}$ and on the measure $d\Omega$ of the gauge group can be parametrized by means of the infinitesimal functions $\eta^{\mu}(x)$ introduced above in (23.5). In this parametrization,

$$\partial_{\nu}(h^{\mu\nu})^{\Omega} - l^{\mu} = \partial_{\nu}(h^{\nu\lambda}\partial_{\lambda}\eta^{\mu}) - \partial_{\lambda}(\partial_{\nu}h^{\mu\nu}\eta^{\lambda}). \tag{23.18}$$

The measure $d\Omega$ in the unit element has a simple form:

$$d\Omega = \prod_{x}\prod_{\mu} d\eta^{\mu}(x). \tag{23.19}$$

As a result, the integral in (23.17) can be written as follows:

$$\int \prod_{x,\mu} \delta(\partial_{\nu}(h^{\nu\lambda}\partial_{\lambda}\eta^{\mu}) - \partial_{\lambda}(\partial_{\nu}h^{\mu\nu}\eta^{\lambda}))\, d\eta^{\mu}(x). \tag{23.20}$$

Formally, this integral is equal to $(\det \hat{A})^{-1}$, where $\hat{A}$ is an operator acting on the four functions η^{μ} according to the rule

$$(\hat{A}\eta)^{\mu} = \partial_{\nu}(h^{\nu\lambda}\partial_{\lambda}\eta^{\mu}) - \partial_{\lambda}(\partial_{\nu}h^{\mu\nu}\eta^{\lambda}). \tag{23.21}$$

Thus, we find that

$$\Delta_{h}[g] = \det \hat{A}. \tag{23.22}$$

For the formulation of perturbation theory, it is convenient to represent $\det \hat{A}$ as a Gaussian integral over auxiliary fields. These fields must be anticommuting, since the integral must give the first power of the determinant. These requirements are satisfied by the expression

$$\det \hat{A} = \int \exp\left(i \int \bar{\theta}^{\mu}(x) A_{\mu\nu}\theta^{\nu}(x)\, d^{4}x\right) \prod_{x,\mu} d\bar{\theta}^{\mu}(x)\, d\theta^{\mu}(x), \tag{23.23}$$

where $\theta^{\mu}(x)$ and $\bar{\theta}^{\mu}(x)$ are anticommuting classical fields satisfying the relation

$$\theta^{\mu}(x)\theta^{\nu}(y) + \theta^{\nu}(y)\theta^{\mu}(x) = 0 \tag{23.24}$$

and analogous relations for $(\theta, \bar{\theta},)$ and $(\bar{\theta}, \bar{\theta})$. Then the integral (23.16) can be written in the form

$$\int \exp\{iS[g] + i\int \theta^{\mu} A_{\mu\nu}[g]\theta^{\nu}\, d^{4}x\} \prod_{x}\left[\prod_{\mu}\delta(\partial_{\nu}h^{\mu\nu} - e^{\mu}) \times \left(g^{5/2}\prod_{\mu\leqslant\nu} dg^{\mu\nu}\right)\right]\prod_{\mu} d\bar{\theta}^{\mu}\, d\theta^{\mu} \tag{23.25}$$

and can be applied directly for the formulation of perturbation theory. However, we transform it further, making use of the arbitrariness in the choice of l^{μ}. The method of transformation proposed by 't Hooft[10] was explained in §22 for the example of the Yang–Mills field. The integral (23.25), by its very construction, does not depend on l^{μ}.

Therefore we can average it over l^μ with an arbitrary weight. For the weight, we choose an exponential function of a quadratic form in the fields:

$$\exp \{(i\alpha/4) \int l^\mu (x)\eta_{\mu\nu} l^\nu (x) d^4x\}, \tag{23.26}$$

where $\eta_{\mu\nu}$ is the Minkowski tensor. The averaging can be performed explicitly and gives the expression

$$\int \exp \{iS[g] + (i\alpha/4) \int \partial_\rho h^{\mu\rho} \eta_{\mu\nu} \partial_\sigma h^{\nu\sigma} d^4 x + i \int \bar{\theta}^\mu A_{\mu\nu} \theta^\nu d^4 x\} \times$$

$$\prod_x g^{5/2} \left(\prod_{\mu\leqslant\nu} dg^{\mu\nu}\right) \left(\prod_\mu d\bar{\theta}^\mu d\theta^\mu\right), \tag{23.27}$$

which contains a quadratic form of the longitudinal parts of the field h^μ with an arbitrary coefficient α. It follows from the foregoing considerations that the integral is independent of α.

The expression (23.27) leads to a diagrammatic technique of perturbation theory. We shall take the independent variables in the path integral (23.27) to be $h^{\mu\nu} = \sqrt{-g} g^{\mu\nu}$, as well as θ^μ and $\bar{\theta}^\mu$. We put

$$h^{\mu\nu} = \eta^{\mu\nu} + \varkappa u^{\mu\nu} \tag{23.28}$$

and regard $u^{\mu\nu}$ as a tensor field describing the gravitational field. The action functional takes the form

$$S = S_2 + \sum_{n=1}^{\infty} \varkappa^n S_{n+2}, \tag{23.29}$$

where S_2 is a quadratic form, and S_n is a form in the power n and the variables $u^{\mu\nu}$ and their first derivatives.

In many respects, the linearization of (23.12) may be unnatural. It may violate the signature of the metric tensor if $u^{\mu\nu}$ is not sufficiently small. There exist parametrizations which are free of this defect, for example, the exponential parametrization

$$h^{\mu\nu} = \eta^{\mu\lambda}(\exp \varkappa\Phi)^{\lambda\nu}. \tag{23.30}$$

In principle, the expansion (23.29) can also be calculated in this parametrization. Note that the quadratic form S_2 is independent of the choice of parametrization.

Owing to the invariance of the action with respect to the transformations (23.4), the quadratic form

$$S_2 = {}^1/_4 \int (-\eta_{\nu\sigma} \delta^\alpha_\rho \delta^\beta_\mu + {}^1/_2 \eta^{\alpha\beta} \eta_{\mu\rho} \eta_{\nu\sigma} + {}^1/_4 \eta_{\mu\nu} \eta_{\rho\sigma} \eta^{\alpha\beta}) \times$$

$$\partial_\alpha u^{\mu\nu} \partial_\beta u^{\rho\sigma} d^4 x \tag{23.31}$$

is degenerate. It does not contain longitudinal components. By introducing the auxiliary fictitious fields θ^μ and $\bar{\theta}^\mu$, we have ensured that the quadratic form in the exponential function of the integrand in (23.27) becomes nondegenerate.

Thus, we can uniquely determine the operators inverse to the operators of the quadratic forms in $u^{\mu\nu}$, θ^μ, and $\bar{\theta}^\mu$, i.e., the propagators of the particles corresponding to the lines of the diagrams.

We shall use a solid line to represent the propagator of the graviton $\langle h^{\mu\nu} h^{\rho\sigma}\rangle$, and a dashed line to represent the propagator of the fictitious vector particle $\langle \theta^\mu \bar{\theta}^\nu \rangle$. The vertices of the diagram are generated by the forms S_{n+2} of the expansion (23.29), and also by the form $\int \bar{\theta}^\mu A_{\mu\nu} \theta^\nu d^4x$, which includes a trilinear interaction $\sim \bar{\theta} u \theta$ of the graviton with the fictitious vector particle.

The elements of the diagrammatic technique have the form

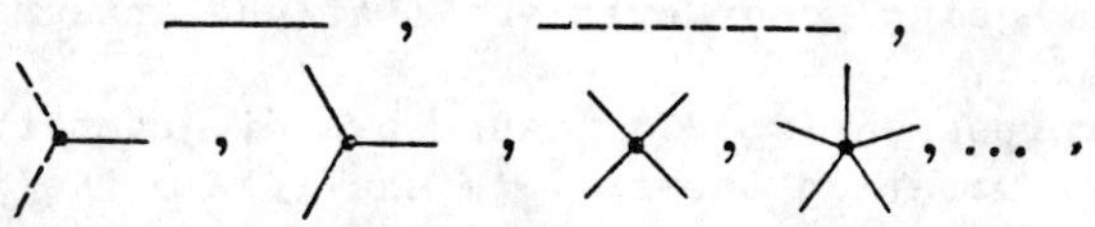

(23.32)

The expression corresponding to the propagator of the graviton is given by the formula

$$G^{\mu\nu,\rho\sigma}(k) = (2/k^2)(\eta^{\mu\rho}\eta^{\nu\sigma} + \eta^{\mu\sigma}\eta^{\nu\rho} + (\alpha^{-1} - 2)\eta^{\mu\nu}\eta^{\rho\sigma}) +$$
$$(2(1-\alpha^{-1})k^{-4})(2k^\mu k^\nu \eta^{\rho\sigma} + 2k^\rho k^\sigma \eta^{\mu\nu} - k^\mu k^\rho \eta^{\nu\sigma} - k^\nu k^\rho \eta^{\mu\sigma} -$$
$$k^\mu k^\sigma \eta^{\nu\rho} - k^\nu k^\sigma \eta^{\mu\rho}), \tag{23.33}$$

which contains the parameter α. The quantity α^{-1} is analogous to the parameter d_l in quantum electrodynamics and in the Yang—Mills theory, and it has the meaning of the coefficient of the longitudinal part of the propagator. The physical results do not depend on the arbitrariness in the choice of the parameter α.

The propagator of the fictitious vector particle has the form

$$G^{\mu\nu} = -\eta^{\mu\nu}/k^2, \tag{23.34}$$

and the vertex for its interaction with gravitons is given by the expression

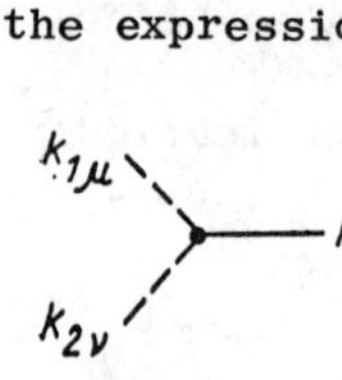

$$k_{3\rho\sigma} = \frac{\varkappa}{2}[k_{1\nu}(\delta^\mu_\sigma k_{3\rho} + \delta^\mu_\rho k_{3\sigma}) - \delta^\mu_\nu(k_{1\rho}k_{2\sigma} + k_{1\sigma}k_{2\rho})], \tag{23.35}$$

with $k_1 + k_2 + k_3 = 0$.

We also give an expression for the third-order vertex, corresponding to linearization of (23.28):

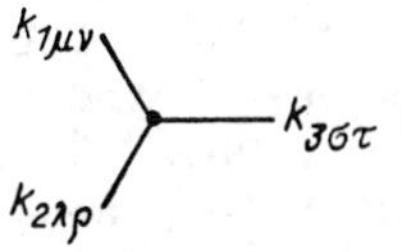

$$= \frac{\varkappa}{32}\left\{\frac{k_1^2}{2}(\eta_{\mu\lambda}\eta_{\rho\sigma}\eta_{\tau\nu} + \eta_{\mu\tau}\eta_{\nu\lambda}\eta_{\rho\sigma} + \right.$$

$$\eta_{\mu\sigma}\eta_{\nu\lambda}\eta_{\rho\tau} + \eta_{\nu\sigma}\eta_{\mu\nu}\eta_{\rho\tau} + \eta_{\mu\tau}\eta_{\lambda\rho}\eta_{\nu\sigma} + \eta_{\mu\sigma}\eta_{\nu\rho}\eta_{\lambda\tau} +$$

$$\eta_{\nu\tau}\eta_{\mu\rho}\eta_{\lambda\sigma} + \eta_{\nu\sigma}\eta_{\mu\rho}\eta_{\lambda\tau}) + k_1^2\eta_{\mu\nu}(\eta_{\rho\sigma}\eta_{\lambda\tau} + \eta_{\rho\tau}\eta_{\lambda\sigma}) +$$

$$(k_{2\mu}k_{3\nu} + k_{2\nu}k_{3\mu})\eta_{\lambda\rho}\eta_{\sigma\tau} + (k_{2\mu}k_{3\nu} + k_{2\nu}k_{3\mu})(\eta_{\lambda\sigma}\eta_{\rho\tau} + \eta_{\lambda\rho}\eta_{\sigma\tau}) -$$

$$k_{1\nu}k_{1\tau}(\eta_{\mu\lambda}\eta_{\rho\sigma} + \eta_{\mu\rho}\eta_{\lambda\sigma}) - k_{1\nu}k_{1\sigma}(\eta_{\mu\lambda}\eta_{\rho\tau} + \eta_{\mu\rho}\eta_{\nu\tau}) - k_{1\mu}k_{1\tau} \times$$

$$(\eta_{\nu\lambda}\eta_{\rho\tau} + \eta_{\nu\rho}\eta_{\lambda\sigma}) - k_{1\mu}k_{1\sigma}(\eta_{\nu\lambda}\eta_{\rho\tau} + \eta_{\nu\rho}\eta_{\lambda\tau}) - k_{2\nu}k_{3\rho}\eta_{\nu\sigma}\eta_{\mu\tau} -$$

$$k_{2\nu}k_{3\lambda}\eta_{\rho\sigma}\eta_{\mu\tau} - k_{2\mu}k_{3\rho}\eta_{\lambda\sigma}\eta_{\nu\tau} - k_{2\mu}k_{3\lambda}\eta_{\rho\sigma}\eta_{\nu\tau} - k_{2\nu}k_{3\rho}\eta_{\lambda\tau}\eta_{\mu\sigma} -$$

$$k_{2\nu}k_{3\lambda}\eta_{\rho\tau}\eta_{\mu\sigma} - k_{2\mu}k_{3\rho}\eta_{\lambda\tau}\eta_{\nu\sigma} - k_{2\mu}k_{3\lambda}\eta_{\rho\tau}\eta_{\nu\sigma} + \text{sum over}$$

permutations of the pairs (μ, ν), (σ, τ), and (λ, ρ).

(23.36)

The contribution of a given diagram is obtained by taking the product of the expressions of the form (23.33)–(23.36) corresponding to its elements, integrating this with respect to the internal momenta, and multiplying the result by

$$r^{-1}(-1)^s(i/(2\pi)^4)^{l-v-1}, \tag{23.37}$$

where r is the order of the symmetry group of the diagram, l is the number of internal lines, v is the number of vertices, and s is the number of loops of fictitious vector particles.

The fictitious vector particles corresponding to these fields are fermions, i.e., they violate the spin–statistics relation. This shows that their role reduces to the subtraction of the contributions from the unphysical degrees of freedom.

In addition to the diagrams described above, perturbation theory contains contributions of renormalization type, proportional to powers of $\delta^{(4)}(0)$. These contributions are produced by the local factor $\prod_x h^{-5/2}(x)$ which occurs in the measure. On linearization of (23.28), we have

$$\prod_{x,\mu\leqslant\nu} dh^{\mu\nu} = \prod_{x,\mu\leqslant\nu} du^{\mu\nu}. \tag{23.38}$$

This factor must be taken into account in the construction of perturbation theory. Formally, its role reduces to the appearance of a contribution to the action of the form

$$\Delta S = (5/2)i\delta^{(4)}(0)\int \ln h(x)d^4x, \tag{23.39}$$

which produces vertices proportional to $\delta^{(4)}(0)$. The appearance of such renormalization terms has been noted in many papers devoted to nonlinear theories (see Refs. 31 and 32). We note that these terms are absent in the exponential parametrization (23.30). Apart from a constant factor, the measure (23.13) here has the simple form

$$\prod_x \prod_{\mu\leqslant\nu} d\Phi^{\mu\nu} \tag{23.40}$$

without any local additions.

Thus, we have obtained a diagrammatic perturbation theory in the formalism of a path integral over the fields $g^{\mu\nu}$ (or $h^{\mu\nu}$). In a number of cases, in particular, when going over to a Hamiltonian formulation, a first-order formalism is more convenient. We are dealing with a first-order formalism if the variables $g^{\mu\nu}$ and $\Gamma^{\rho}_{\mu\nu}$ are regarded as independent. The measure in this case (apart from powers of the volume) has the form

$$g^{15/2} \prod_{\mu\leqslant\nu} dg^{\mu\nu} \prod_{\substack{\mu\leqslant\nu \\ \rho}} d\Gamma^{\rho}_{\mu\nu} = h^{5/2} \prod_{\mu\leqslant\nu} dh^{\mu\nu} \prod_{\substack{\mu\leqslant\nu \\ \rho}} d\Gamma^{\rho}_{\mu\nu}. \tag{23.41}$$

The power of the determinant g in the measure is such that after taking the Gaussian integral with respect to the variables Γ the measure is identical with (23.13).

We shall describe the diagrammatic technique in the first-order formalism. The elements of the diagrams associated with the fictitious vector particles remain unchanged. In addition to the tensor propagator $\langle uu\rangle$ in (23.33), the scheme of perturbation theory includes the propagators $\langle u, \gamma\rangle$ and $\langle\gamma\gamma\rangle$. The three different propagators correspond to the lines

$\langle uu\rangle =$;
$\langle u\gamma\rangle =$;
$\langle\gamma\gamma\rangle =$.

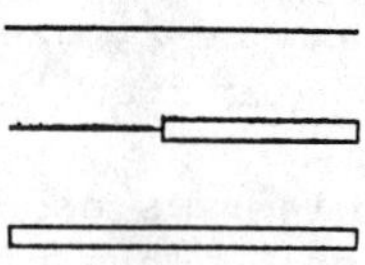

(23.42)

In the momentum representation, they have the form

$$G^{\mu\nu,\rho}_{\sigma\tau}(k) = (i/2)(\eta_{\sigma\alpha}\delta^{\rho}_{\beta}k_{\tau} + \eta_{\tau\alpha}\delta^{\rho}_{\beta}k_{\sigma} - \eta_{\sigma\alpha}\eta_{\tau\beta}k^{\rho})G^{\mu\nu,\alpha\beta}(k) \equiv$$

$$\Omega^{\rho}_{\sigma\tau,\alpha\beta}(k)G^{\alpha\beta,\mu\nu}(k);$$

$$G^{\rho\lambda}_{\sigma\tau,\mu\nu}(k) = {}^1/_4(\delta^{\rho}_{\mu}\delta^{\lambda}_{\sigma}\eta_{\nu\tau} + \delta^{\rho}_{\nu}\delta^{\lambda}_{\sigma}\eta_{\mu\tau} + \delta^{\rho}_{\mu}\delta^{\lambda}_{\tau}\eta_{\nu\sigma} + \delta^{\rho}_{\nu}\delta^{\lambda}_{\tau}\delta_{\mu\sigma}) -$$

$$^1/_6(\delta^{\lambda}_{\nu}\delta^{\rho}_{\tau}\eta_{\mu\sigma} + \delta^{\lambda}_{\mu}\delta^{\rho}_{\tau}\eta_{\nu\sigma} + \delta^{\lambda}_{\nu}\delta^{\rho}_{\sigma}\eta_{\mu\tau} + \delta^{\lambda}_{\mu}\delta^{\rho}_{\sigma}\eta_{\nu\tau}) +$$

$$\Omega^{\lambda}_{\mu\nu,\alpha\beta}(k)\Omega^{\rho}_{\sigma\tau,\gamma\delta}(-k)G^{\alpha\beta,\gamma\delta}(k). \tag{23.43}$$

The only graviton vertex is produced by the trilinear form

$$(\varkappa/2)\int u^{\mu\nu}\left(\gamma^{\rho}_{\mu\sigma}\,\gamma^{\sigma}_{\rho\nu}-\gamma^{\rho}_{\mu\nu}\,\gamma^{\sigma}_{\rho\sigma}\right)d^4x \qquad (23.44)$$

and has the expression

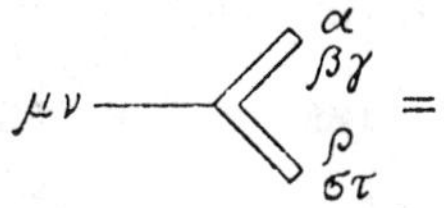

$$=$$

$$\frac{\varkappa}{8}\{2(\delta^{\sigma}_{\mu}\delta^{\tau}_{\nu}+\delta^{\sigma}_{\nu}\delta^{\tau}_{\mu})(\delta^{\beta}_{\rho}\delta^{\gamma}_{\alpha}+\delta^{\beta}_{\alpha}\delta^{\gamma}_{\rho})-\delta^{\sigma}_{\mu}\delta^{\tau}_{\alpha}\delta^{\gamma}_{\nu}\delta^{\beta}_{\rho}-\delta^{\sigma}_{\mu}\delta^{\beta}_{\nu}\delta^{\gamma}_{\rho}\delta^{\tau}_{\alpha}-$$
$$\delta^{\sigma}_{\nu}\delta^{\tau}_{\alpha}\delta^{\beta}_{\mu}\delta^{\gamma}_{\rho}-\delta^{\sigma}_{\nu}\delta^{\tau}_{\alpha}\delta^{\gamma}_{\mu}\delta^{\beta}_{\rho}-\delta^{\tau}_{\mu}\delta^{\sigma}_{\alpha}\delta^{\beta}_{\nu}\delta^{\gamma}_{\rho}-\delta^{\tau}_{\mu}\delta^{\sigma}_{\alpha}\delta^{\gamma}_{\nu}\delta^{\beta}_{\rho}-\delta^{\tau}_{\nu}\delta^{\sigma}_{\alpha}\delta^{\beta}_{\mu}\delta^{\gamma}_{\rho}-$$
$$\delta^{\tau}_{\nu}\delta^{\sigma}_{\alpha}\delta^{\gamma}_{\mu}\delta^{\beta}_{\rho}\}. \qquad (23.45)$$

The renormalization elements proportional to $\delta^{(4)}(0)$ are produced by the local factor $\prod_x h^{5/2}(x)$ in the measure (23.41), whose contribution can be interpreted as an addition to the action of the form

$$\Delta S=-(5/2)i\delta^{(4)}(0)\int \ln h\,(x)d^4x. \qquad (23.46)$$

We have considered in detail the case of the gravitational field in a vacuum. The introduction of an interaction with other fields does not essentially alter the scheme of constructing the perturbation theory. For matter fields with nondegenerate Lagrangians interacting with the gravitational field, no new fictitious particles appear. Such particles and their corresponding diagrams occur only when one includes fields with a larger gauge group than in the theory of gravitation, for example, the electromagnetic field or Yang—Mills field. Without considering this case in detail, we give an expression for the path integral corresponding to interacting electromagnetic and gravitational fields:

$$\left.\begin{aligned}&\int \exp\{iS[g^{\mu\nu},A_\mu]\}\,\Delta[g]\prod_x\delta[\partial_\mu(h^{\mu\nu}A_\nu)]\\&\times\prod_\mu\delta(\partial_\nu h^{\mu\nu})\,g^{5/2}\prod_{\mu\leqslant\nu}dg^{\mu\nu}\prod_\mu dA_\mu;\\&S[g^{\mu\nu},A_\mu]=S_g-{}^1/_4\int(\partial_\mu A_\nu-\partial_\nu A_\mu)\\&\times(\partial_\lambda A_\rho-\partial_\rho A_\lambda)\,g^{\mu\nu}g^{\lambda\rho}\sqrt{-g}\,d^4x,\end{aligned}\right\} \qquad (23.47)$$

where S_g is the action of the free gravitational field, and $\Delta[g]$ is equal to the product of determinants

$$\det\hat{A}\,\det(\partial_\nu(h^{\mu\nu}\partial_\nu)), \qquad (23.48)$$

in which $\hat{A}$ is the operator (23.21). The presence of the nontrivial second factor in this product shows that the unimportant fictitious scalar particle which could have been introduced in the description of the electromagnetic field also interacts with the gravitational field. Thus, covariant perturbation theory for electromagnetic and gravitational fields entails a fictitious neutral scalar particle.

§24. CANONICAL QUANTIZATION OF THE GRAVITATIONAL FIELD

In the preceding section, we have constructed a formalism of covariant quantization of the gravitational field and a relativistic perturbation theory in accordance with the general scheme of quantization of gauge fields outlined in §20. In going over to canonical (operator) quantization, two problems arise: 1) the reduction of the action of the gravitational field to Hamiltonian form; 2) the transformation of the path integral in the form (23.25) to an integral with respect to canonical variables. The first of these problems could have been solved as soon as Einstein had written his gravitational field equations (1916). However, over 40 years elapsed before its solution was given by Dirac in 1958.[33] Apparently, the trouble lay in insufficient attention to the problem, and also in its technical complexity.

We shall consider the transition to a Hamiltonian theory in the path-integral formalism. A Hamiltonian formulation of the classical theory of gravitation was first developed by Dirac.[33] Several variants of this formulation have been obtained by many authors.[32-37,8] In constructing a manifestly Hamiltonian form of Einstein's equations, one encounters a difficult problem — the solution of the constraint equations. We shall consider a generalized Hamiltonian formulation of the theory of gravitation in which it is not necessary to solve the constraint equations, but in which it is sufficient to verify their commutation relations. This generalized formulation is a field-theoretic analog of the formulation developed in §16 for finite-dimensional mechanical systems. We shall show that the action of the gravitational field can be reduced to a form analogous to (16.17) for finite-dimensional systems, where the corresponding constraints and Hamiltonian satisfy the conditions (16.20) and (16.21). We shall follow the method proposed by Faddeev[8] in a form which is especially suitable for the gravitational field. For this purpose, it is convenient to use a first-order formalism. We consider the expression for the action of the gravitational field in the form (23.3) and collect the terms in the Lagrangian density function containing derivatives with respect to the time:

$$(2\varkappa^2)^{-1}(\Gamma^0_{\mu\nu}\partial_0 h^{\mu\nu}-\Gamma^\rho_{\mu\rho}\partial_0 h^{\mu 0})=(2\varkappa^2)^{-1}(\Gamma^0_{ik}\partial_0 h^{ik}+(\Gamma^0_{i0}-\Gamma^k_{ik})\partial_0 h^{i0}-\Gamma^i_{0i}\partial_0 h^{00}). \qquad (24.1)$$

This expression does not contain the variables Γ^μ_{00}, which appear linearly in $L(h, \Gamma)$ and have the meaning of Lagrange multipliers. The factors accompanying Γ^μ_{00} (which we denote by A^{00}_μ) are the constraints. The constraint equations

$$\left.\begin{aligned} A^{00}_0 &= h^{ik}\Gamma^0_{ik}+h^{00}\Gamma^i_{00}+\partial_i h^{i0}=0;\\ A^{00}_i &= 2h^{k0}\Gamma^0_{ik}+h^{00}(\Gamma^0_{i0}-\Gamma^k_{ik})+\partial_i h^{00}\end{aligned}\right\} \qquad (24.2)$$

enable us to express the variables Γ^i_{0i} and $\Gamma^0_{i0}-\Gamma^k_{ik}$ in terms of Γ^0_{ik} and $h^{\mu\nu}$. The terms containing derivatives with respect to the time then take the form

$$(2\varkappa^2)^{-1}(\Gamma^0_{ik}/h^{00})\partial_0(h^{00}h^{ik}-h^{i0}h^{k0}), \qquad (24.3)$$

if we omit the terms

$$\begin{aligned}&(2\varkappa^2 h^{00})^{-1}(\partial_0 h^{00}\partial_i h^{i0}-\partial_i h^{00}\partial_0 h^{i0})=\\ &(1/2\varkappa^2)(\partial_0 \ln h^{00}\partial_i h^{i0}-\partial_i \ln h^{00}\partial_0 h^{i0}),\end{aligned} \qquad (24.4)$$

which vanish after integration by parts.

Equation (24.3) suggests that the natural dynamical variables are

$$q^{ik}=h^{i0}h^{k0}-h^{00}h^{ik};\quad \pi_{ik}=-(1/h^{00})\Gamma^0_{ik}. \qquad (24.5)$$

The variables $\Gamma^\rho_{\mu\nu}$, being different from Γ^0_{ik}, are nondynamical, and they can be eliminated by means of the constraint equations

$$\partial L(h, \Gamma)/\partial\Gamma^\rho_{\mu\nu}=0,\quad \Gamma^\rho_{\mu\nu}\neq\Gamma^0_{ik}. \qquad (24.6)$$

The system (24.6) includes (24.2) together with the equations

$$\left.\begin{aligned}&\partial_k h^{i0}+h^{is}\Gamma^0_{sk}+h^{00}\Gamma^i_{k0}+h^{0s}\Gamma^i_{sk}-h^{i0}\Gamma^s_{ks}=0;\\ &\partial_k h^{ij}+h^{i\sigma}\Gamma^j_{\sigma k}+h^{j\sigma}\Gamma^i_{\sigma k}-h^{ij}\Gamma^\sigma_{k\sigma}=0.\end{aligned}\right\} \qquad (24.7)$$

The solution of the system (24.2) and (24.7), which expresses the "nondynamical" variables Γ^0_{i0}, Γ^k_{i0}, and Γ^k_{ij} in terms of $h^{\mu\nu}$ and Γ^0_{ik}, has the form

$$\left.\begin{aligned}&\Gamma^0_{i0}=\Gamma^s_{is}-\partial_i h^{00}/h^{00}-(h^{0s}/h^{00})\Gamma^0_{is};\\ &\Gamma^k_{i0}=-(1/h^{00})(\partial_i\ h^{k0}+h^{0s}\Gamma^k_{is}-h^{k0}\Gamma^s_{is}+h^{ks}\Gamma^0_{is});\\ &\Gamma^k_{ij}=\overset{*}{\Gamma}{}^k_{ij}+(h^{k0}/h^{00})\Gamma^0_{ij}.\end{aligned}\right\} \qquad (24.8)$$

Here $\overset{*}{\Gamma}{}^k_{ij}$ are three-dimensional connection coefficients generated by the three-dimensional metric g_{ik} $(i, k = 1, 2, 3)$.

We substitute the resulting expressions (24.8) for Γ^0_{i0}, Γ^k_{i0}, and Γ^k_{ij} into the Lagrangian density function $L(\mathrm{h}, \Gamma)$. Neglecting terms of the divergence type, which vanish after integration over the three-dimensional space, and taking into account the asymptotic conditions (23.2), the result of the substitution reduces to the form

$$\frac{1}{2\varkappa^2}(\pi_{ik}(x)\,\partial_0\, q^{ik}(x)) - H(x) - \left(\frac{1}{h^{00}(x)} - 1\right) T_0(x) - \frac{h^{i0}(x)}{h^{00}(x)} T_i(x); \qquad (24.9)$$

$$\left.\begin{aligned} T_0(x) &= q^{ij}\, q^{kl}(\pi_{ik}\,\pi_{jl} - \pi_{ij}\,\pi_{kl}) + g_3 R_3; \\ T_i(x) &= 2(\nabla_i(q^{kl}\,\pi_{kl}) - \nabla_k(q^{kl}\,\pi_{il})); \\ H(x) &= T_0(x) - \partial_i\,\partial_k\, q^{ik}(x). \end{aligned}\right\} \qquad (24.10)$$

Here $q_3 = \det g_{ik}$, and R_3 is the three-dimensional scalar curvature generated by the three-dimensional metric g_{ik} $(i, k = 1, 2, 3)$. The symbol ∇_k in the expressions for the constraints T_i denotes the covariant derivative with respect to the metric g_{ik}.

Arnowitt, Deser, and Misner[34] noted that the canonical variables in the expression for the constraints have a perspicuous geometrical meaning. The functions g^{ik} and π_{ik} serve as coefficients of the first and second quadratic forms of a surface $x^0 = \text{const}$ embedded in four-dimensional space-time with metric $g_{\mu\nu}$ and connection $\Gamma^\rho_{\mu\nu}$. More precisely, q^{ik} is a contravariant metric density with weight +2, and π_{ik} is a covariant metric density with weight -1. The constraints are the relations of Codazzi and Gauss, which are well known in the theory of surfaces (see, for example, Ref. 38).

Equation (24.9) represents the solution to the problem of reducing the action of the gravitational field to a generalized Hamiltonian form analogous to (16.17) for a finite-dimensional system with constraints. It can be verified that the constraints T_μ are in an involution. To write the explicit relations, it is convenient to introduce the quantities

$$T(\eta) = \int T_k(x)\eta^k(x)d^3x; \quad T_0(\varphi) = \int T_0(x)\varphi(x)d^3x. \qquad (24.11)$$

Here η is a vector field and φ is a scalar field, or, more precisely, a scalar density with weight -1. We have the relations

$$\left.\begin{aligned} \{T(\eta_1), T(\eta_2)\} &= T([\eta_1, \eta_2]); \\ \{T(\eta), T_0(\varphi)\} &= T_0(\eta\varphi); \\ \{T_0(\varphi), T_0(\psi)\} &= T(\varphi\eta_\psi - \psi\eta_\varphi), \end{aligned}\right\} \qquad (24.12)$$

where $[\eta_1, \eta_2]$ is the bracket of the vector fields, i.e., the vector field with components $\eta_1^l \partial_l \eta_2^k - \eta_2^l \partial_l \eta_1^k$, $\eta\varphi = \eta^l \partial_l \varphi - \partial_l \eta^l \varphi$, and η_φ is the vector field with components $q^{ik}\partial_k \varphi$. The relations (24.12) constitute a field-theoretic analog of (16.20). The first line in (24.12) shows that the connections $T_k(x)$ $(k = 1, 2, 3)$ have the meaning of the generators of coordinate transformations. The remaining relations do not have a simple group meaning.

Note the divergence $(-\partial_i \partial_k q^{ik})$ in the Hamiltonian density $H(x)$. If the constraint equations $T_\mu = 0$ are satisfied, the Hamiltonian H reduces to a three-dimensional integral of a divergence, i.e., to an integral over an infinitely remote surface. The last integral is determined by the asymptotic form of the functions q^{ik} as $r = |\mathbf{x}| \to \infty$. For an asymptotically flat gravitational field, we have

$$q^{ik} = \delta^{ik}(1 + \varkappa^2 M/2\pi r) + O(r^{-2}), \qquad (24.13)$$

where M is the total mass, which can be calculated by integrating $H(x)$:

$$H = \int H(x)\, d^3x = -(1/2\varkappa^2)\int \partial_i\, \partial_k\, q^{ik}\, d^3x =$$

$$-(1/2\varkappa^2) \lim_{S\to\infty} \oint \partial_k\, q^{ik}\, dS_i = M. \qquad (24.14)$$

Thus, we can assume that $H = \int H(x) d^3x$ has the meaning of the energy of the gravitational field. The integrand

$$H(x) = T_0(x) - \partial_i \partial_k q^{ik}(x), \qquad (24.15)$$

which has the meaning of the energy density, is equal to the sum of two quadratic forms — the quadratic form of the first derivatives of q^{ik}, and the quadratic form of the "momenta" π_{ik} — as is assumed for the energy density of a wave field. In our case, this is the energy of the gravitational field, which has two polarizations in accordance with the usual calculation:

$$2 = 6\,(\text{coordinates}) - 4\,(\text{constraints}). \qquad (24.16)$$

In the weak-field approximation, the Hamiltonian is a quadratic form in the completely transverse components of the linearized field.

It is this fact, and also (24.14) given above, which equates the energy of the gravitational field to the mass, which provide the justification for choosing the action of this field as (23.3), where the derivatives act not on the Christoffel symbols $\Gamma^\rho_{\mu\nu}$, but on the metric tensor $g_{\mu\nu}$.

Once the action of the gravitational field has been reduced to generalized Hamiltonian form, we can construct a Hamiltonian form of the path integral by first choosing supplementary conditions. Frequent use is made of the conditions first proposed by Dirac[30]:

$$\partial_k (q^{-1/3} q^{ik}) = 0,\ i = 1, 2, 3;\ \pi = q^{ik}\pi_{ik}, \tag{24.17}$$

where $q = \det q^{ik}$. These conditions have a simple geometrical meaning: the surface $x^0 = \text{const}$ is minimal, and the coordinates x^1, x^2, x^3 on it are "harmonic" [the equations $\partial_k (q^{-1/3} q^{ik}) = 0$ are the conditions of "three-dimensional harmonicity"].

To prove the equivalence of the canonical and relativistic forms of the path integral, different supplementary conditions are more convenient, namely,

$$\ln q = \Phi(x);\ q^{ik} = 0,\ i \neq k, \tag{24.18}$$

where Φ is a function with asymptotic form c/r at infinity. The relations (16.27) and (16.28) are satisfied for these conditions. The matrix of the Poisson brackets of the conditions (24.18) with the constraints is determined by the equations

$$\left.\begin{aligned}
&(\hat{C}\eta)^0 = \{T_\eta, \ln q - \Phi(x)\} = -\eta^s \partial_s \ln q - 4\partial_s \eta^s + 4\pi\eta^0;\\
&(\hat{C}\eta)^1 = \{T_\eta, q^{23}\} = -\eta^s \partial_s q^{23} + q^{2s}\partial_s \eta^3 + q^{3s}\partial_s \eta^2\\
&-2q^{23}\partial_s \eta^s - 2(\pi^{23} - q^{23}\pi)\eta^0;\\
&(\hat{C}\eta)^2 = \{T_\eta, q^{31}\} = -\eta^s \partial_s q^{31} + q^{3s}\partial_s \eta^1\\
&+q^{1s}\partial_s \eta^3 - 2q^{31}\partial_s \eta^s - 2(\pi^{31} - q^{31}\pi)\eta^0;\\
&(\hat{C}\eta)^3 = \{T_\eta, q^{12}\} = -\eta^s \partial_s q^{12} + q^{1s}\partial_s \eta^2 + q^{2s}\partial_s \eta^1\\
&-2q^{12}\partial_s \eta^s - 2(\pi^{12} - q^{12}\pi)\eta^0,
\end{aligned}\right\} \tag{24.19}$$

where

$$T_\eta = \int (T_0\eta^0 + T_i\eta^i)\, d^3x. \tag{24.20}$$

The matrix C is not degenerate if the curvature of the metric g_{ik} is nonzero.

We introduce the notation

$$\ln q - \Phi = \chi_0;\ q^{23} = \chi_1;\ q^{31} = \chi_2;\ q^{12} = \chi_3. \tag{24.21}$$

The path integral in Hamiltonian form for the gravitational field is given by

$$\int \exp\left\{i\int\left(\pi_{ik}\partial_0 q^{ik} - \frac{h^{0i}}{h^{00}}T_i - \left(\frac{1}{h^{00}} - 1\right)T_0 - H(x)\right)d^4x\right\} \times$$

$$\det\{T_\mu, \chi_a\}\prod_x\left[\prod_{a=0}^{3}\delta(\chi_a)\prod_{i\leqslant k} d\pi_{ik}\, dq^{ik}\, d\,\frac{1}{h^{00}}\prod_{i=1}^{3} d\left(\frac{h^{0i}}{h^{00}}\right)\right]. \tag{24.22}$$

We shall reduce this expression to a form in which the integration is taken only with respect to the field $g^{\mu\nu}$. This will enable us to identify the required measure of integration. For this purpose, we must integrate with respect to the fields π_{ik}. There is dependence on π_{ik} not only in the functional $\exp(iS)$, but also in the determinant $\det\{T_\mu, \chi_a\}$. We write this determinant in the form of a path integral with respect to the anticommuting variables η^μ and $\bar{\eta}^\mu$ $(\mu = 0, 1, 2, 3)$:

$$\det\{T_\mu, \chi_a\} = \int \exp[i\int \bar{\eta}^\mu C_{\mu\nu}(\pi, q)\, \eta^\nu d^4x] \prod_x d\bar{\eta}^\mu(x)\, d\eta^\mu(x). \quad (24.23)$$

The functions π_{ik} appear only in the coefficients $C_{\mu 0}$ of the operator $\hat{C}$, which do not contain derivatives, and they appear linearly. In the integral with respect to π_{ik}, we make the displacement

$$\pi_{ik} \to \pi_{ik} + \pi_{ik}(g), \quad (24.24)$$

where $\pi_{ik}(g)$ is the expression $\pi_{ik} = -(1/h^{00})\Gamma^0_{ik}$ in terms of the metric tensor in accordance with (23.1). With this displacement, the action functional $S[g^{\mu\nu}, \pi_{ik}]$ — an integral of the expression (24.9) — reduces to

$$S[g^{\mu\nu}] - (2\varkappa^2)^{-1} \int (1/h^{00})\, q^{ij} q^{kl} (\pi_{ik}\pi_{jl} - \pi_{ij}\pi_{kl})\, d^4x. \quad (24.25)$$

Here $S[g^{\mu\nu}]$ is the action (23.3), in which the Christoffel symbols $\Gamma^\rho_{\mu\nu}$ are expressed in terms of the metric tensor. The quadratic form $\bar{\eta}^\mu C_{\mu\nu}(\pi_{ik}, q^{ik})\,\eta^\nu$ becomes

$$\bar{\eta}^\mu C_{\mu\nu}[\pi_{ik}(g), q^{ik}]\,\eta^\nu + \bar{\eta}^\mu l_\mu(\pi_{ik})\,\eta^0, \quad (24.26)$$

where $l_\mu(\pi_{ik})$ are linear forms in π_{ik}, whose explicit structure is not required. We now make another displacement in the integral, which eliminates the form $\bar{\eta}^\mu l_\mu(\pi_{ik})\eta^0$ linear in π_{ik}. Instead of this form, we obtain a form quadratic in $\eta^\mu\eta^0$ and containing no derivatives, and therefore equal to zero identically, since $(\eta^0)^2 \equiv 0$. We then take the Gaussian integral with respect to π_{ik}. But the integral with respect to η^μ and $\bar{\eta}^\mu$ can again be written as a determinant of the operator $\hat{C}_1$, which differs from the operator $\hat{C}$ by the fact that the symbols π_{ik} in it are replaced by their expressions in terms of the metric tensor. The action of the operator $\hat{C}_1$ is determined by the equations

$$\left.\begin{aligned}
(\hat{C}_1\eta)^0 &= -\eta^\lambda\partial_\lambda \ln q - 4\partial_s\eta^s - [(h^{0s}/h^{00})\partial_s \ln q \\
&\quad + 4\partial_s(h^{0s}/h^{00})]\,\eta^0; \\
(\hat{C}_1\eta)^1 &= -\eta^\lambda\partial_\lambda q^{23} + q^{2s}\partial_s\eta^3 + q^{3s}\partial_s\eta^2 - 2q^{23}\partial_s\eta^s \\
&\quad + \left\{-\frac{h^{0s}}{h^{00}}\partial_s q^{23} + q^{2s}\partial_s\left(\frac{h^{03}}{h^{00}}\right) + q^{3s}\partial_s\left(\frac{h^{02}}{h^{00}}\right) - 2q^{23}\partial_s\left(\frac{h^{0s}}{h^{00}}\right)\right\}\eta^0; \\
(\hat{C}_1\eta)^2 &= -\eta^\lambda\partial_\lambda q^{31} + q^{3s}\partial_s\eta^1 + q^{1s}\partial_s\eta^3 - 2q^{31}\partial_s\eta^s \\
&\quad + \left\{-\frac{h^{0s}}{h^{00}}\partial_s q^{31} + q^{3s}\partial_s\left(\frac{h^{01}}{h^{00}}\right) + q^{1s}\partial_s\left(\frac{h^{03}}{h^{00}}\right) - 2q^{31}\partial_s\left(\frac{h^{0s}}{h^{00}}\right)\right\}\eta^0; \\
(\hat{C}_1\eta)^3 &= -\eta^\lambda\partial_\lambda q^{12} + q^{1s}\partial_s\eta^2 + q^{2s}\partial_s\eta^1 - 2q^{12}\partial_s\eta^s \\
&\quad + \left\{-\frac{h^{0s}}{h^{00}}\partial_s q^{12} + q^{1s}\partial_s\left(\frac{h^{02}}{h^{00}}\right) + q^{2s}\partial_s\left(\frac{h^{01}}{h^{00}}\right) - 2q^{12}\partial_s\left(\frac{h^{0s}}{h^{00}}\right)\right\}\eta^0.
\end{aligned}\right\} \qquad (24.27)$$

The local factors in the products of differentials, together with the local factor arising from the integration with respect to π_{ik} and the differentials themselves, are collected in the expression

$$\prod_x (h^{00})^{-4} q^{-2} \prod_{\mu\leqslant\nu} dh^{\mu\nu}. \qquad (24.28)$$

The factors in front of the differentials can be reduced to the form

$$(h^{00})^{-1} h^{-5/2} q^{1/2}, \qquad (24.29)$$

where the last factor can be omitted as a consequence of the constraint condition $q = \exp\Phi$. As a result, our path integral takes the form

$$\int \exp(iS[h]) \det \hat{B}_1 \prod_x \left\{\left[\prod_a \delta(\chi_a)\right] h^{-5/2} \prod_{\mu\leqslant\nu} dh^{\mu\nu}\right\}, \qquad (24.30)$$

where the operator $\hat{B}_1$ differs from $\hat{C}_1$ by the local factor $(h^{00})^{-2}$.

We shall now show that the integral which we have written is an integral over classes of gravitational fields in the sense of §20, the classes being parametrized by the conditions (24.21) and the invariant measure having the form (23.13). For this purpose, it is sufficient to verify that $\det \hat{B}_1$ coincides with the factor $\Delta_\chi[h]$ determined by the condition

$$\Delta_\chi[h] \int \prod_x \left(\prod_a \delta(\chi_a^\Omega)\right) d\Omega(x) = 1. \qquad (24.31)$$

The integral in this expression can be calculated by the same technique as the integral in Eq. (23.17). This gives

$$\Delta_\chi [h] = \det \hat{B}, \tag{24.32}$$

where the operator $\hat{B}$ is defined as follows:

$$\left.\begin{aligned}
(\hat{B}\zeta)^0 &= -\zeta^\lambda \partial_\lambda \ln q - 4\partial_s \zeta^s + 4(h^{0s}/h^{00})\partial_s \zeta^0;\\
(\hat{B}\zeta)^1 &= -\zeta^\lambda \partial_\lambda q^{23} + q^{2s}\partial_s \zeta^3 + q^{3s}\partial_s \zeta^\lambda - 2q^{2s}\partial_s \zeta^s\\
&\quad -\{(h^{02}/h^{00})q^{3s} + (h^{03}/h^{00})q^{2s} - 2(h^{0s}/h^{00})q^{23}\}\partial_s \zeta^0;\\
(\hat{B}\zeta)^2 &= -\zeta^\lambda \partial_\lambda q^{31} + q^{3s}\partial_s \zeta^1 + q^{1s}\partial_s \zeta^3 - 2q^{31}\partial_s \zeta^s\\
&\quad -\{(h^{03}/h^{00})q^{1s} + (h^{01}/h^{00})q^{3s} - 2(h^{0s}/h^{00})q^{31}\}\partial_s \zeta^0;\\
(\hat{B}\zeta)^3 &= -\zeta^\lambda \partial_\lambda q^{12} + q^{1s}\partial_s \zeta^2 + q^{2s}\partial_s \zeta^1 - 2q^{12}\partial_s \zeta^s\\
&\quad -\{(h^{01}/h^{00})q^{2s} + (h^{02}/h^{00})q^{1s} - 2(h^{0s}/h^{00})q^{12}\}\partial_s \zeta^0.
\end{aligned}\right\} \tag{24.33}$$

It is easy to see that

$$\det \hat{B} = \det \hat{B}_1. \tag{24.34}$$

In fact, we can go over from one operator to the other by means of the triangular substitution

$$\xi^0 = \eta^0;\ \xi^i = \eta^i + (h^{0i}/h^{00})\,\eta^0,\quad i = 1, 2, 3. \tag{24.35}$$

Thus, after the formal changes of variables of integration, the explicitly unitary Hamiltonian form of the path integral is transformed to an integral over classes of equivalent fields with a certain concrete parametrization of the classes. The corresponding invariant measure has the form (23.13). This justifies the Lorentz-invariant expression (23.16) for the path integral, which is an expression of this same integral with another parametrization of the classes.

The next, more difficult, problem is the consistent application of the renormalization procedure based on invariant regularization. The difficulties are due to the cumbersome character of the theory and the fact that from a formal point of view it is not renormalizable.

§25. ATTEMPTS TO CONSTRUCT A GAUGE-INVARIANT THEORY OF THE ELECTROMAGNETIC AND WEAK INTERACTIONS

The examples of gauge-invariant theories considered above are constructed as part of a more general theoretical scheme incorporating all existing interactions. The construction of such a general model is the most complicated problem facing elementary-particle physics. A more specific idea is to unify the electromagnetic and weak interactions by means of a multiplet of gauge fields, and for a long time this has attracted the attention of theoreticians. The realization of this idea gave birth to certain models, of which the Weinberg—Salam model[39] has achieved the greatest fame. In this section, we shall briefly consider the Weinberg—Salam model, and also a gauge-invariant model of the electromagnetic and weak interactions of leptons proposed by Faddeev.[40] Like all theories of gauge fields, these models can be formulated most naturally in the language of path integrals.

The Weinberg—Salam model is based on the idea of spontaneous breakdown of an initially existing invariance with respect to gauge transformations of massless vector fields of Yang—Mills type. The gauge group of the model is the group U(2), which is isomorphic to the group of unitary 2 × 2 matrices and reduces to a product of the group U(1) of phase transformations and the group of unitary 2 × 2 matrices with unit determinant.

The connection generated by the group U(2) is formed by vector fields of two types — a multiplet A_μ^a $(a = 1, 2, 3)$ of Yang—Mills type and a field B_μ. In addition to these fields, the model contains lepton fields and fields of auxiliary scalar particles, which lead to spontaneous breakdown of the gauge U(2) invariance. Of the fields of lepton type, the model includes fields of electron type:

$$L = {}^1/_2 (1+\gamma_5) {}^{\nu_e}_{\psi_e}; \quad R = {}^1/_2 (1+\gamma_5)\, \psi_e, \tag{25.1}$$

where ψ_e is the electron field, and ν_e is the field of the electron neutrino. The scalar fields form the doublet

$$\varphi = \begin{pmatrix} \varphi_0 \\ \varphi_- \end{pmatrix}. \tag{25.2}$$

The Lagrangian of the model has the form

$$\begin{gathered} L = -{}^1/_4 (\partial_\mu A_\nu - \partial_\nu A_\mu + g[A_\mu, A_\nu])^2 - {}^1/_4 (\partial_\mu B_\nu - \partial_\nu B_\mu)^2 - \\ \bar{R}\gamma^\mu (\partial_\mu - ig' B_\mu) R - \bar{L}\gamma^\mu (\partial_\mu + igtA_\mu - (i/2) g' B_\mu) L - \\ {}^1/_2 (\partial_\mu \varphi - igtA_\mu + (i/2) g' B_\mu \varphi)^2 - G_e (\bar{L}\varphi R + \bar{R}\bar{\varphi} L) + \\ M_1^2 \varphi\varphi^+ - h(\varphi^+\varphi)^2, \end{gathered} \tag{25.3}$$

where g and g' are the coupling constants of the multiplet A_μ and the singlet B_μ, respectively.

The mechanism of spontaneous symmetry breaking and mass generation first proposed by Higgs[41] amounts to the appearance of an anomalous expectation value

$$\lambda = \langle \varphi^0 \rangle \tag{25.4}$$

for the zeroth component of the φ field. This mechanism is encountered in the theory of superfluidity. We go over from the initial fields to new "physical" fields by subtracting from the φ fields their anomalous expectation values. These fields can be taken to be the φ^- field and the fields

$$\varphi_1 = (\varphi^0 + \bar{\varphi}^0 - 2\lambda)/\sqrt{2}; \quad \varphi_2 = (\varphi^0 - \bar{\varphi}^0)/i\sqrt{2}. \qquad (25.5)$$

In first-order perturbation theory, the value of λ is determined by the condition of a maximum of the expression $-M_1^2\varphi^+\varphi + h(\varphi^+\varphi)^2$ with the substitutions $\varphi^0 = \lambda$ and $\varphi^- = 0$. This leads to the formula

$$\lambda^2 = M_1^2/2h. \qquad (25.6)$$

As a result, it turns out that the field φ_1 has mass M_1, while the fields φ_2 and φ^- remain massless. The appearance of massless excitations in models with spontaneous symmetry breaking was first clearly noted by Goldstone.[42] Here, however, these excitations do not have a direct physical meaning and can be eliminated by means of gauge transformations.

The mass of the φ_1 meson turns out to be very large (in comparison with the electron mass m_e), and for this reason the coupling of φ to the remaining fields can be neglected. As a result, the appearance of the anomalous expectation value (25.4) can, in the first approximation, be taken into account by the simple replacement of the field φ by its vacuum expectation value:

$$\langle\varphi\rangle = \lambda \begin{pmatrix} 1 \\ 0 \end{pmatrix}. \qquad (25.7)$$

With this substitution, the original Lagrangian (25.3) reduces to the expression

$$-{}^1/_4(\partial_\mu A_\nu - \partial_\nu A^\mu + g[A_\mu, A_\nu])^2 - {}^1/_4(\partial_\mu B_\nu - \partial_\nu B_\mu)^2 -$$
$$\bar{R}\gamma_\mu(\partial_\mu - ig' B_\mu)R - \bar{L}\gamma_\mu(\partial_\mu + igtA_\mu - {}^i/_2 g' B_\mu)L - {}^1/_8\lambda^2 g^2((A^1_\mu)^2 +$$
$$(A^2_\mu))^2 - {}^1/_8(gA^3_\mu + g' B_\mu)^2 - \lambda G_e \bar{\psi}_e \psi_e. \qquad (25.8)$$

The electron acquires a mass

$$m_e = \lambda G_e. \qquad (25.9)$$

The charged vector field

$$W_\mu = (A^1_\mu + i A^2_\mu)/\sqrt{2} \qquad (25.10)$$

describes an intermediate boson with mass

$$M_W = \lambda g/2. \qquad (25.11)$$

From the neutral fields A^3_μ and B_μ, we can form the combinations

$$\left.\begin{aligned} Z_\mu &= (g^2 + g'^2)^{-1/2}(gA^3_\mu + g' B_\mu); \\ A_\mu &= (g^2 + g'^2)^{-1/2}(-g' A^3_\mu + gB_\mu) \end{aligned}\right\} \qquad (25.12)$$

with masses

$$M_Z = {}^1/_2\, \lambda\, (g^2 + g'^2)^{1/2};\ \ M_A = 0. \qquad (25.13)$$

Thus, one of the components of the multiplet of vector fields A_μ has zero mass, and this component must be taken to be the photon field.

The term of interaction of the leptons with the vector fields can be written in the form

$$\frac{ig}{2\sqrt{2}} \overline{\psi}_e (1+\gamma_5)\, \nu W_\mu + \frac{igg'}{(g^2+g'^2)^{1/2}} \overline{\psi}_e \gamma_\mu \psi_e W_\mu + \frac{i\,(g^2+g'^2)^{1/2}}{4} \left[\frac{3\,(g'^2-g^2)}{g'^2+g^2} \overline{\psi}_e \gamma_\mu \psi_e - \overline{\psi}_e \gamma_\mu \gamma_5 \psi_e + \overline{\nu} \gamma_\mu (1 + \gamma_5)\, \nu \right] Z_\mu. \qquad (25.14)$$

The second term in (25.14) shows that the electron charge is

$$e = gg'\,(g^2 + g'^2)^{-1/2} \qquad (25.15)$$

and is thus smaller than each of the original charges g and g'. Assuming that W_μ is related in the usual way to the hadrons and the muon, we obtain the relation

$$G_W/\sqrt{2} = g^2/8\, M_W^2 = 1/2\lambda^2. \qquad (25.16)$$

It follows from (25.12) and (25.16) that the masses of the intermediate bosons are very large,

$$M_Z > 80 \text{ GeV},\ M_W > 40 \text{ GeV}, \qquad (25.17)$$

not only in comparison with the electron mass, but also in comparison with the hadron masses.

The Weinberg—Salam model leads to an interaction of neutral currents. To see this, we eliminate the field of the intermediate bosons by making the transformation $W_\mu \to W_\mu + W_\mu^0$, which eliminates the terms of interaction with the leptons linear in W_μ. This leads to a direct interaction of the lepton currents of the form

$$\sum_{a,b} \int d^4x\, d^4y\, j^a(x)\, j^b(y)\, D(x-y), \qquad (25.18)$$

where

$$D(x-y) = (2\pi)^{-4} \int (k^2 - M_W^2 - i0)^{-1} \exp\,[ik\,(x-y)]\, d^4k \qquad (25.19)$$

is the propagator of the W field with mass M_W. For large M_W, we can neglect k^2 in comparison with M_W^2 in the denominator of the integrand and replace the expression (25.19) by

$$-M_W^{-2}\, \delta\,(x-y). \qquad (25.20)$$

After making this substitution, the terms of lepton interaction are proportional to

$$\sum_{a,b} \int d^4 x j^a(x) j^b(x); \qquad (25.21)$$

the terms $\int d^4xj^a(x) j^b(x)$ correspond to an interaction of neutral currents. Such an interaction is characteristic of the Weinberg—Salam model and is absent in the $(V - A)$ variant of the weak-interaction model.

The existence of an interaction of neutral currents has been confirmed experimentally.[43,44] This is a strong argument in favor of the Weinberg—Salam model.

The methods of the theory of gauge fields[7-9,45-48] make it possible to justify the assumption that the model is renormalizable.[39] For simpler examples, this has been done by 't Hooft.[11] Here we shall not prove the renormalizability of the Weinberg—Salam model. We merely note that a correct perturbation theory must be constructed in accordance with the general scheme of quantization of gauge fields outlined in this chapter.

In spite of the qualitative agreement between the conclusions of the Weinberg—Salam model and experiment which has now been obtained, this model still cannot be regarded with complete confidence as the only acceptable variant of a unified gauge theory of the weak and electromagnetic interactions. This model has a number of esthetic defects, of which we mention two: 1) the use of a nonsimple group U(2) as the gauge group violates the idea of universality of the interaction; 2) the introduction of a linear multiplet of scalar fields with subsequent spontaneous symmetry breaking (the Higgs mechanism) does not have a natural interpretation in terms of the idea of gauge invariance.

A gauge-invariant model of the electromagnetic and weak interactions of leptons which is free of these defects has been proposed by Faddeev.[40] We shall consider this model, following mainly Ref. 40.

Faddeev's model is based on the simplest nontrivial gauge group O(3) (the three-dimensional rotation group) and contains only a triplet of vector fields. Instead of the scalar Higgs field, we have here a field of directions $n(x)$ in charge space, determining in it a neutral subspace. The set of physical particles in the model incorporates the known leptons, photons, and charged intermediate bosons.

We shall consider the geometrical ideas on which the model is based.

The classical fields are divided into two classes: 1) the sections of some fiber bundle over the space-time manifold, invariant with respect to the local action of the gauge group; 2) connections in this fiber bundle, giving parallel transport of the fields of the first class.

The fiber is usually a product of a linear space, in which there is realized a representation of the Lorentz group in accordance with the spins of the considered fields of the first class, and an internal (charge) space, in which there acts an internal symmetry group. If the effects of gravitation are neglected, a nontrivial connection is generated only by the latter group and is given by a set of Yang—Mills vector fields,[22] the number of which is equal to the dimension of the gauge group. In the model in question, the set of fields of the first class contains spinor lepton fields, and the gauge group is the group O(3). The internal space is taken to be a product of the linear space of its representation and the nonlinear manifold in which it acts. The linear component of the fiber is associated with a multiplet of spinor lepton fields. The simplest possibility for this is the space of the vector representation, R^3. The leptons, i.e., $\bar{\mu}$, e, and $\nu = \nu_e + \nu_\mu$ (the antimuon, electron, and neutrino), are combined into a three-dimensional vector, so that the space R^3 is also better from the physical point of view.

We can distinguish the neutral lepton from the charged lepton if we indicate which of the generators of the group O(3) is the charge. Thus, specification of the charge is equivalent to the introduction of a direction in R^3, which can be called the neutral direction. As a result, we have a manifold of directions S^2, which is assumed to be a direct factor in the internal space.

Thus, the model contains three fermions ψ_1, ψ_2, ψ_3 combined into an isovector $\psi \in R^3$, a set of scalar fields n_1, n_2, n_3 satisfying the condition

$$n_1^2 + n_2^2 + n_3^2 = 1 \qquad (25.22)$$

and forming a unit vector $n \in S^2$, and three vector fields $Z_\mu^1, Z_\mu^2, Z_\mu^3$ forming an isovector $Z_\mu \in R^3$.

It is convenient to introduce the three real antisymmetric matrices

$$V_1 = \begin{pmatrix} 0 & 0 & 0 \\ 0 & 0 & 1 \\ 0 & -1 & 0 \end{pmatrix}; \; V_2 = \begin{pmatrix} 0 & 0 & -1 \\ 0 & 0 & 0 \\ 1 & 0 & 0 \end{pmatrix}; \; V_3 = \begin{pmatrix} 0 & 1 & 0 \\ -1 & 0 & 0 \\ 0 & 0 & 0 \end{pmatrix}, \qquad (25.23)$$

which give a representation of the Lie algebra of the group O(3), and to consider the set V_1, V_2, V_3 as a three-dimensional vector V.

We write the action of the gauge group in infinitesimal

form:

$$\delta\psi = \psi \wedge \varepsilon;\ \delta n = n \wedge \varepsilon;\ \delta Z_\mu = \partial_\mu \varepsilon + Z_\mu \wedge \varepsilon, \qquad (25.24)$$

where $\varepsilon = \varepsilon(x)$ is the vector of an infinitesimal local rotation. We shall henceforth employ the notation (,) and $\wedge$ for the scalar and vector products in R^3, for example, $\psi \wedge \varepsilon = (V, \varepsilon)\psi$.

In terms of the fields Z_μ and n, we can construct two more triplets of vector fields

$$Y_\mu = \partial_\mu n + Z_\mu \wedge n;\ X_\mu = n \wedge Y_\mu, \qquad (25.25)$$

which transform as vectors under gauge transformations:

$$\delta Y_\mu = Y_\mu \wedge \varepsilon;\ \delta X_\mu = X_\mu \wedge \varepsilon. \qquad (25.26)$$

We can use these fields in addition to the field Z_μ in the definition of the covariant derivative of the field ψ, obtaining matrices by means of tensor multiplication of Z_μ and Y_μ by n.

Consider the specific combination

$$\nabla_\mu \psi = \partial_\mu \psi + (Z_\mu,\ V)\psi + (X_\mu \otimes n + n \otimes X_\mu)\gamma_5\psi, \qquad (25.27)$$

constructed by means of the matrix γ_5 ($\gamma_5^+ = -\gamma_5$, $\gamma_5^2 = -1$). It is distinguished by the additional condition of commutation of ∇_μ with the transformation

$$\left.\begin{aligned} &\delta\psi = \alpha n(n, \gamma_5\psi);\ \delta n = 0; \\ &\delta Z_\mu = \alpha Y_\mu;\ \delta X_\mu = \alpha Y_\mu, \end{aligned}\right\} \qquad (25.28)$$

where α is an infinitesimal constant. The expression (25.27) determines the infinitesimal connection associated with the group SU(3).

The lepton number L produced by the ordinary phase transformation

$$\delta\psi = i\beta\psi;\ \delta n = 0;\ \ \delta Z_\mu = 0, \qquad (25.29)$$

where β is a constant (independent of x), and the charge Q produced by the generator (V, n) determine the complete set of quantum numbers for the classification of the leptons. The transformation (25.28) then gives an additional classification of the neutral leptons, which gives an absolute distinction between their opposite helicities.

A Lagrangian which is invariant with respect to the transformations described above has the form

$$L = (1/2i)\{\bar\psi\gamma_\nu\nabla_\mu\psi - (\gamma_\mu\nabla_\mu\bar\psi)\psi\} + (1/e^2)L_{YM} + (m^2/2e^2)(Y_\mu, Y_\mu), \qquad (25.30)$$

where L_{YM} is the Yang—Mills Lagrangian for the field Z_μ:

$$L_{YM} = -{}^1/_4 (Z_{\mu\nu}, Z_{\mu\nu});\ Z_{\mu\nu} = \partial_\mu Z_\nu - \partial_\nu Z_\mu + Z_\mu \wedge Z_\nu. \qquad (25.31)$$

The constant e is dimensionless, and m has the dimensions of mass.

We write separately the terms of the Lagrangian (25.31) describing the interaction of the fermions with the vector field:

$$L_1 = i\,(Z_\mu, v_\mu) + i\,(X_\mu, a_\mu), \tag{25.32}$$

where

$$v_\mu = \bar{\psi}\gamma_\mu V\psi;\; a_\mu = (\bar{\psi}n)\,\gamma_\mu\gamma_5\psi + \bar{\psi}\gamma_\mu\gamma_5\,(n\psi). \tag{25.33}$$

The expression (25.33) is analogous to the standard $(V - A)$ structure in the ordinary weak-interaction theory. This analogy is confirmed by the behavior of L_1 with respect to the operators of spatial reflection P and charge reflection R_Q. Schwinger[50] defined the reflection R_Q geometrically as a transition to antiparticles, followed by a change of sign of the neutral direction in the charge space R^3. In explicit form, R_Q is defined as follows:

$$\left.\begin{aligned} &\psi^{\parallel} \to \psi^{*\parallel};\; \psi^{\perp} \to \psi^{*\perp};\; n \to -n;\\ &Z_\mu^{\parallel} \to -Z_\mu^{\parallel};\; Z_\mu^{\perp} \to Z_\mu^{\perp};\; Y_\mu \to Y_\mu;\; X_\mu \to X_\mu, \end{aligned}\right\} \tag{25.34}$$

where

$$Z_\mu^{\parallel} = (Z_\mu, n);\; Z_\mu^{\perp} = Z_\mu - (Z_\mu, n)\,n. \tag{25.35}$$

Note that $X_\mu^{\parallel} = Y_\mu^{\parallel} = 0$. The bilinear forms v_μ and a_μ transform as follows:

$$v_\mu^{\parallel} \to -v_\mu^{\parallel};\; v_\mu^{\perp} \to v_\mu^{\perp};\; a_\mu^{\parallel} \to a_\mu^{\parallel};\; a_\mu^{\perp} \to -a_\mu^{\perp}, \tag{25.36}$$

so that under charge reflection the second term in (25.32) changes sign. But it changes sign under spatial reflection, so that the interaction L_1 is invariant with respect to the transformation of combined parity $R_Q P$.

We now consider the possible structure of the mass terms for fermions, bearing in mind that they can be added to the initial Lagrangian (25.30) or calculated dynamically. We take the three matrices I, $i\,(V, n)$, and $n \otimes n$, a linear combination of which can be the matrix M of the mass term. The condition of invariance with respect to the transformation (25.29) gives a matrix M of the form

$$M = a\,(I - n \otimes n) + ib\,(Vn), \tag{25.37}$$

where the coefficients a and b have the dimensions of mass.

For the physical interpretation of the model, let us consider it in the particular gauge $n = n_0$, where n_0 is the constant vector (0, 0, 1). This condition means that the charge is associated with the matrix $-i\,(V_3)$ and the field ψ_3 is neutral. The fields $\psi_1 + i\psi_2$ and $\psi_1 - i\psi_2$ have masses $a + b$ and $a - b$ and charges 1 and -1, respectively. The neutral field ψ_3 has mass 0. The fields X_μ and Y_μ have the components

$$X_\mu^1 = Z_\mu^1;\ X_\mu^2 = Z_\mu^2;\ X_\mu^3 = 0;\ Y_\mu^1 = Z_\mu^2;\ Y_\mu^2 = -Z_\mu^1;\ Y_\mu^3 = 0, \quad (25.38)$$

so that the interaction L_1 takes the form

$$L_1 = Z_\mu^1 (v_\mu^1 + a_\mu^1) + Z_\mu^2 (v_\mu^2 + a_\mu^2) + Z_\mu^3 v_\mu^3. \quad (25.39)$$

The last term in the Lagrangian (25.30) gives the mass term for the vector mesons Z_μ^1 and Z_μ^2. As a result, the Lagrangian L_1 describes a weak $(V - A)$ interaction of electrons, muons, and neutrinos, mediated by intermediate massive vector bosons $W_\mu^\pm$, and an interaction of charged leptons with the electromagnetic field in the standard form

$$e\{A_\lambda(\overline{\psi}_e\gamma_\lambda\psi + \overline{\psi}_{(\mu)}\gamma_\lambda\psi_{(\mu)}) +$$
$$W_\lambda\,((\overline{\psi}_e\gamma_\lambda\,(1 + i\gamma_5)\,\psi_e + \overline{\psi}_{(\mu)}\,\gamma_\lambda\,(1 + i\gamma_5)\,\psi_{(\mu)}) + \text{c. c.}\}, \quad (25.40)$$

if we make the identification

$$\left.\begin{aligned} &Z_\mu^3 = eA_\mu;\ Z_\mu^1 \pm Z_\mu^2 = eW_\mu^\pm;\\ &\psi_1 + i\psi_2 = \psi_\mu^*;\ \psi_1 - i\psi_2 = \psi_e;\ \psi_3 = \psi_{\nu_e} + \psi_{\nu_\mu}^*. \end{aligned}\right\} \quad (25.41)$$

The constant e then has the meaning of electric charge, e/m is the weak-interaction constant, and m is the mass of the intermediate boson.

In quantizing this model, it is convenient to take the gauge condition to be the condition of transversality of the vector fields:

$$\partial_\mu Z_\mu = 0. \quad (25.42)$$

In accordance with the general methods of quantizing gauge fields, we must then add to the Lagrangian (25.30) a compensating term containing an interaction of the vector particles with fictitious scalar fermions. In the transverse gauge, the field $n(x)$ not only does not disappear from the Lagrangian, but it appears in it in an essentially non-linear way. The problem of correct quantization of the field $n(x)$ is of great importance. It is to be hoped that application of the methods of path integration will help us to obtain a correct solution.

§26. VORTEX-LIKE EXCITATIONS IN QUANTUM FIELD THEORY

The concept of excitations having the character of quantum vortices, which arose in the theory of superconductivity and superfluidity, has recently been carried over to relativistic quantum field theory. The basis for this is the hypothesis that the strongly interacting particles (if not all, then at least some of them) are vortex-like excitations. This hypothesis makes it possible to reduce the number of fundamental fields. The need for this has been felt particularly strongly in recent years, now that something like 100—200 strongly interacting particles and resonances have been discovered. In this situation, the standard scheme of field theory, which associates each particle with a fundamental field, becomes cumbersome, ineffective for practical application, and unattractive from the point of view of the elegance of the theory.

Vortex-like excitations exist both in exactly solvable models, where they have become known as *solitons*, and in other theories. We shall first consider one of the simplest models of relativistic field theory in which vortex-like excitations are possible — the Goldstone model[42] with one time and two space dimensions. We shall then consider the possible existence of vortex-like excitations in certain models of field theory in four-dimensional space-time and, in particular, in gauge theories with Yang—Mills fields.

The method of path integration seems most appropriate in this domain, where the excitations are essentially collective and are formed by many initial particles.

We turn to the treatment of the relativistic Goldstone model. It is convenient to write the action functional of this model in the Euclidean variables:

$$S = -\int (|\nabla\psi|^2 - \lambda|\psi|^2 + (g/4)|\psi|^4)\, dx. \qquad (26.1)$$

The functional (26.1) describes a complex scalar field having a self-interaction with coupling constant $g > 0$. The coefficient λ is positive. This corresponds to the fact that when the interaction is switched off $(g = 0)$ there exist particles with a negative value of the square of the mass (*tachyons*).

Goldstone drew attention to the fact that when the interaction is switched on the system undergoes a Bose condensation.[42] As a result, there appear particles with zero and finite mass (with positive square). As we shall see, the particles with finite mass are unstable and have a finite lifetime.

The form of the action functional (26.1) is reminiscent of the corresponding functional for a nonideal Bose gas. It is natural to conjecture that there may exist in the Goldstone model excitations similar to the quantum vortices that are characteristic of superfluid Bose systems.

Before giving a direct treatment of quantum vortices, we shall construct for the Goldstone model a perturbation theory which contains no infrared divergences and which is convenient, in particular, for the calculation of the lifetime of an unstable particle. For the time being, we shall not particularize the dimension n of the model.

The density of the condensate is determined in the first approximation by the condition of a maximum of the expression

$$\lambda |\psi|^2 - (g/4)|\psi|^4 \tag{26.2}$$

and is given by

$$\rho_0 = |\psi_0|^2 = 2\lambda/g. \tag{26.3}$$

In the action (26.1), we transform to the polar coordinates given by $\psi = \sqrt{\rho}\exp(i\varphi)$ and $\bar{\psi} = \sqrt{\rho}\exp(-i\varphi)$, and in place of ρ we introduce the variable $\pi = \rho - \rho_0$. In the variables φ and π, the action has the form

$$S = -\int [(\rho_0 + \pi)(\nabla\varphi)^2 + (\nabla\pi)^2/4(\rho_0 + \pi) + g\pi^2/4)]\,dx + (\lambda^2/g)\int dx. \tag{26.4}$$

Here the expression $(\lambda^2/g)\int dx$ is the contribution of the Bose condensate.

In the path integral with respect to the variables $\varphi(x)$ and $\pi(x)$, we make the transformation

$$\varphi \to \varphi/\sqrt{2\rho_0};\ \pi \to \sqrt{2\rho_0}\,\pi, \tag{26.5}$$

which converts the expression (26.4) to the form

$$S = -\int \Big[{}^1/_2(\nabla\varphi)^2 + {}^1/_2\sqrt{2/\rho_0}\,\pi(\nabla\varphi)^2 + \frac{1}{2}\Big(\frac{(\nabla\pi)^2}{1+\sqrt{2/\rho_0}\;\pi} + 2\lambda\pi^2\Big)\Big]\,dx + \frac{\lambda}{g}\int dx. \tag{26.6}$$

The constant $\sqrt{2}/\rho_0$ determines the strength of the interaction of the φ and π fields, as well as the self-interaction of the π field.

We note that Eq. (26.6) implies instability of the system of interacting tachyons with respect to an arbitrarily weak stabilizing perturbation. Indeed, if for fixed λ we consider in (26.6) the limit $g \to +0$, we find that the second term in (26.6) tends to infinity, while the first term becomes a quadratic form describing noninteracting massless particles and particles with a positive square of the mass, $m^2 = 2\lambda$. Allowance for the interaction makes the massive particle unstable.

We shall construct a perturbation theory in terms of the variables φ and π. The elements of the diagrams of the perturbation theory will be two lines corresponding to the fields φ and π, one vertex corresponding to the $\varphi\pi$ interaction, and an infinite number of vertices corresponding to the $\pi\pi$ interaction. We give the expressions for the lines, the $\varphi\pi$ interaction vertex, and the first of the $\pi\pi$ interaction vertices:

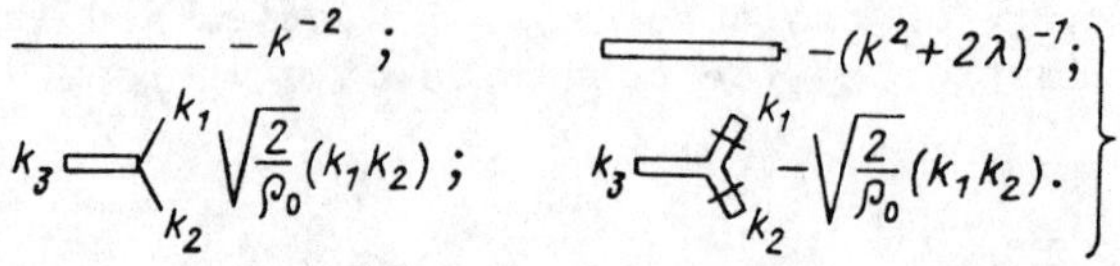

$$\begin{matrix} \underline{\qquad} -k^{-2}\,; & \quad\quad \overline{\rule{1cm}{0pt}} -(k^2+2\lambda)^{-1}; \\ k_3 \!\!\!\begin{array}{c}{\scriptstyle k_1}\\ {\scriptstyle k_2}\end{array}\!\!\! \sqrt{\tfrac{2}{\rho_0}}(k_1 k_2)\,; & \quad\quad k_3 \!\!\!\begin{array}{c}{\scriptstyle k_1}\\ {\scriptstyle k_2}\end{array}\!\!\! -\sqrt{\tfrac{2}{\rho_0}}(k_1 k_2). \end{matrix} \qquad (26.7)$$

Here the φ field is denoted by a single line, and the π field by a double line. In the $\pi\pi$ vertex, the bars indicate outgoing lines of the vertex carrying momenta k_1 and k_2. The scalar product of these momenta appears in the expression corresponding to this vertex.

The expression corresponding to a diagram of perturbation theory can be obtained by integrating the product of the expressions corresponding to the elements of the diagram with respect to the independent momenta and multiplying the result by the factor

$$(1/r)\,(-1/(2\pi)^n)^c, \qquad (26.8)$$

where r is the order of the symmetry group, and c is the number of independent loops of the diagram. Since the theory is constructed in the Euclidean variables, to obtain the physical results the expressions corresponding to the diagrams must be continued into the region of physical energies and momenta.

The resulting perturbation theory contains no infrared divergences, but it diverges at large momenta and is formally nonrenormalizable. Therefore it would be more consistent to integrate first over the rapidly varying components ψ_1 and $\bar{\psi}_1$ of the fields ψ and $\bar{\psi}$, defined as part of the Fourier expansions

$$\psi(x) = \int \exp(ikx)\,\psi(k)\,dk; \quad \bar{\psi}(x) = \int \exp(-ikx)\,\bar{\psi}(k)\,dk, \qquad (26.9)$$

namely, the integrals of $\psi(k)\exp(ikx)$ and $\bar{\psi}(k)\exp(-ikx)$ over the region $|k| > k_0$. In the integral over the slowly varying fields $(\psi_0(x) = \psi(x) - \psi_1(x)$ and $\bar{\psi}_0(x) = \bar{\psi}(x) - \bar{\psi}_1(x))$ we can now go over to polar coordinates. Integrating the functional $\exp S$ over the rapidly varying components of the fields ψ_1 and $\bar{\psi}_1$, we obtain a functional $\exp \tilde{S}$ containing only the slowly varying components of the fields ψ and $\bar{\psi}$. The expression for $\tilde{S}$ in the first approximation is the same as the expression for S, differing from it by corrections which cancel the divergences on integration over the slowly varying fields. We shall not consider these corrections here. The momentum k_0 which separates the large and small momenta can be estimated in order of magnitude.

The result of such an estimate can be formulated as the inequalities

$$\left.\begin{aligned}\sqrt{\lambda} \ll k_0 \ll \sqrt{\lambda/g} \text{ for } n=4;\\ \sqrt{\lambda} \ll k_0 \ll \lambda/g \text{ for } n=3,\end{aligned}\right\} \tag{26.10}$$

which can be satisfied if the coupling constant g is small.

As an example of the application of perturbation theory with the elements (26.7), we calculate the lifetime of a massive particle, which is determined by the second-order diagram

$$\Sigma_2 =$$ 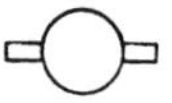 (26.11)

The diagrams resulting from the $\pi\pi$ interaction do not contribute to the imaginary part. Physically, this corresponds to the fact that a massive particle cannot decay into two other particles with the same mass. The expression corresponding to the diagram (26.11) has the form

$$-\frac{1}{2}\frac{2}{\rho_0}\frac{1}{(2\pi)^n}\int d^n k_1 \frac{(k_1 k_2)^2}{k_1^2 k_2^2} = -\frac{g}{2\lambda(2\pi)^n}\int d^n k_1 \frac{(k_1 k_2)^2}{k_1^2 k_2^2}. \tag{26.12}$$

Consider the imaginary part of this expression for $k^2 = -2\lambda$. As a consequence of the equality $k = k_1 + k_2$, we have

$$k_1 k_2 = {}^1/_2\,[(k_1+k_2)^2 - k_1^2 - k_2^2] = -\lambda - (k_1^2 + k_2^2)/2;$$

$$(k_1 k_2)^2 = \lambda^2 + \lambda\,(k_1^2 + k_2^2) + {}^1/_4\,(k_1^2 + k_2^2)^2.$$

Only the first term λ^2 on the right-hand side of the last equation contributes to the imaginary part. The remaining terms lead to real, but formally divergent, integrals of the form $\int d^n k_1/k_2^2$, $\int d^n k_1$, and $\int (k_2^2/k_1^2)\, d^n k_1$; thus,

$$\operatorname{Im}\Sigma = -\operatorname{Im}\,[\lambda g/2\,(2\pi)^n]\int d^n\, k_1/k_1^2 k_2^2. \tag{26.13}$$

The integral in this equation converges for $n = 3$ and diverges for $n = 4$. The imaginary part of the integral is finite for both $n = 3$ and $n = 4$. Making the analytic continuation $k^2 \to -2\lambda + i0$, we obtain

$$\operatorname{Im}\Sigma = g\sqrt{2\lambda}/2^5 \text{ for } n=3;\ \operatorname{Im}\Sigma = 2\lambda/2^5\pi \text{ for } n=4. \tag{26.14}$$

The corresponding equations for the lifetime of the π particle have the form

$$\tau = 2^6 g^{-1},\ n=3,\quad \tau = 2^7\pi g^{-1}(2\lambda)^{-1/2},\ n=4. \tag{26.15}$$

We now modify the path integral to allow for quantum vortices. Let us consider the case $n = 3$ (one-dimensional time and two-dimensional space). Quantum vortices correspond to lines in the three-dimensional space (x_0, x_1, x_2) on which we have zero values of the functions ψ and $\overline{\psi}$, with respect to which the integration is taken. The phase φ of the function ψ acquires an increment $2\pi n$ (where n is an integer) when we pass around the line. We consider here only vortices with a phase increment $\pm 2\pi$ ($|n| = 1$). States with $|n| > 1$ are unstable and decay into vortices with $|n| = 1$.

An individual vortex corresponds to a solution of the equation

$$-\Delta\psi - \lambda\psi + (g/2)\overline{\psi}\psi\psi = 0, \tag{26.16}$$

obtained by variation of the action S in (26.1) with respect to $\overline{\psi}$, which depends on the variables in the plane orthogonal to the world line and has the form $f(r) \exp(i\theta)$, where θ is the polar angle, and $f(r)$ is a real function of the distance r from the axis of the vortex. Equation (26.16) reduces to an ordinary differential equation for the function $f(r)$:

$$f'' + f'/r - f/r^2 + \lambda f - gf^3/2 = 0, \tag{26.17}$$

which is identical with the corresponding equation in the theory of a Bose gas considered by Pitaevskiĭ.[51] The solution of this equation vanishes for $r = 0$ and tends to $\sqrt{\rho_0}$ as $r \to \infty$. It is natural to refer to the characteristic length $\lambda^{-1/2}$ as the r a d i u s o f t h e v o r t e x t u b e.

To describe a situation with several vortices,[52] we enclose each world line in a tube of radius r_0 much larger than the radius $\lambda^{-1/2}$ of the vortex tube. In the first approximation, the sum of the integrals over the vortex tubes in the action functional S can be represented in the form

$$-\sum_i m_B(r_0) \int ds_i. \tag{26.18}$$

Here ds_i is the element of length of the vortex line i, and $m_B(r_0)$ is the mass (energy) of the vortex enclosed in the tube. The second quantity depends logarithmically on r_0 and is given by the formula

$$m_B(r_0) = 2\pi\rho_0 \ln(r_0/a), \tag{26.19}$$

where a is a parameter of the order of the radius of the vortex tube.

We separate the contribution to the action (26.4) from the vortex tubes and then make the change of variables (26.5). This gives the expression

$$-\int \left[\frac{1}{2}(\nabla\varphi)^2 + \frac{1}{2}\sqrt{\frac{2}{\rho_0}}\pi(\nabla\varphi)^2 + \frac{1}{2}\left(\frac{(\nabla\pi)^2}{1+\sqrt{2/\rho_0}\,\pi} + 2\lambda\pi^2 \right) \right] \times$$

$$dx - \sum_i m_B(r_0) \int ds_i + (\lambda^2/g) \int dx. \qquad (26.20)$$

The functional $\exp S$ must be integrated with respect to the fields φ and π and also over the trajectories of the centers of the vortices. The function $\varphi(x)$ in the action (26.20) is not unique and acquires an increment

$$\pm 2\pi \sqrt{2\rho_0} = \pm q. \qquad (26.21)$$

The quantity q, which is inversely proportional to the coupling constant of the $\varphi\pi$ and $\pi\pi$ interactions, has the meaning of electric charge. To prove this, we go over to a path integral with respect to a new variable having the meaning of a vector potential of the electromagnetic field. We make this transformation for a small constant $\sqrt{2/\rho_0}$ characterizing the $\varphi\pi$ and $\pi\pi$ interactions. Neglecting this constant, the contribution of the field φ to the functional (26.20) is given by the quadratic form

$$-{}^1/_2 \int (\nabla\varphi)^2\, dx. \qquad (26.22)$$

However, we must allow for the fact that the function φ is now nonunique and acquires an increment $\pm q$ when we pass around the vortex line. We restore uniqueness by making the displacement

$$\varphi(\mathrm{x}) \to \varphi(\mathrm{x}) + \varphi_0(\mathrm{x}) \qquad (26.23)$$

by a function $\varphi_0(\mathrm{x})$ — a solution of the three-dimensional Laplace equation, which "absorbs" the nonuniqueness. To find the function $\varphi_0(x)$, we note that its three-dimensional gradient $\nabla\varphi_0(\mathbf{x}) = \mathbf{h}(\mathbf{x})$ is a solution of the problem of magnetostatics in three-dimensional space given by the equations

$$\mathrm{curl}\,\mathbf{h} = q\mathbf{j};\ \mathrm{div}\,\mathbf{h} = 0. \qquad (26.24)$$

Here $\mathbf{j}$ is the sum of the unit linear currents flowing along the trajectories of the centers of the vortices. The function $\varphi_0(\mathrm{x})$ is the nonunique scalar potential of the magnetic field $\mathbf{h}$ produced by the system of linear currents. The square of the gradient $(\nabla\varphi)^2$ in the integral (26.22) reduces to the sum $(\nabla\varphi)^2 + (\nabla\varphi_0)^2$. The integral of the first term describes a noninteracting field and is of no interest. But the integral of $(\nabla\varphi_0)^2 = h^2$ is proportional to the energy of the magnetic field of the system of linear currents.

One usually solves the magnetostatic problem (26.24) by means of a vector potential $\mathbf{a}(\mathrm{x})$ ($\mathbf{h} = \mathrm{curl}\,\mathbf{a}$, $\mathrm{div}\,\mathbf{a} = 0$). The vector potential $\mathbf{a}$ for the system of linear currents is a sum of the contributions of the linear currents:

$$\mathbf{a}(\mathbf{x}) = (q/4\pi) \sum_i \int d\mathbf{l}(\mathbf{y})/|\mathbf{x}-\mathbf{y}|. \tag{26.25}$$

The expression obtained from (26.22) by means of the substitution $\varphi \to \varphi_0$ can be represented as a double sum of contributions from the various currents:

$$S_1 = -(q^2/8\pi) \sum_{i,k} \int\int d\mathbf{l}_i(\mathbf{x})\, d\mathbf{l}_k(\mathbf{y})/|\mathbf{x}-\mathbf{y}|. \tag{26.26}$$

The terms with $i = k$ in (26.26) diverge for $\mathbf{x}$ near $\mathbf{y}$. This divergence is a result of the approximation in which the vortices are regarded as point-like and their corresponding currents as linear. If we separate the centers of the vortices by circles of radius r_0 greater than the radius of the vortex tube but less than the mean distance between the vortices, we find that the double integral in (26.26) is actually taken over the region $|\mathbf{x}-\mathbf{y}| > r_0$ and the divergences disappear. The expression (26.26), which describes the interaction of vortices in nonlocal form, can be transformed by writing it in terms of a path integral according to the formula

$$\exp S_1 = \frac{\int \exp\{-dx\,[{}^1/_2(\mathrm{curl}\,\mathbf{A})^2 + iq\,(\mathbf{Aj})]\} \prod_x \delta\,(\mathrm{div}\,\mathbf{A}) \prod_i dA_i}{\int \exp\{-{}^1/_2 \int dx\,(\mathrm{curl}\,\mathbf{A})^2\} \prod_x \delta\,(\mathrm{div}\,\mathbf{A}) \prod_i dA_i} \tag{26.27}$$

Here $\mathbf{A}$ is the variable of path integration, having the meaning of a vector potential, and its expansion

$$\mathbf{A}(\mathbf{x}) = \int_{k<\tilde{k}_0} \exp(i\mathbf{kx})\, \mathbf{a}(\mathbf{k})\, d^3k \tag{26.28}$$

is limited above by momenta less than $\tilde{k}_0 \sim r_0^{-1}$. The gauge $\mathrm{div}\,\mathbf{A} = 0$ in the three-dimensional Euclidean theory is an analog of the Lorentz gauge $\partial_\mu A_\mu = 0$ in the four-dimensional pseudo-Euclidean case. Equation (26.27) is proved by making the displacement $\mathbf{A} \to \mathbf{A} + \mathbf{A}_0$, which eliminates the linear form in $\mathbf{A}$ in the exponential function of the integrand in the numerator.

Thus, the part of the action functional describing the interaction of vortices with the phase field φ can be transformed to the form

$$-m_B(r_0) \sum_i \int ds_i - iq \int (\mathbf{jA})\, d^3x - {}^1/_2 \int (\mathrm{curl}\,\mathbf{A})^2\, d^3x, \tag{26.29}$$

corresponding to the system of charged particles in the electromagnetic field. Note that the action (26.29) is gauge-invariant, unlike the original action (26.1).

Quantum vortices exist as particles in their own right. The obvious law of conservation of the difference between the numbers of vortices rotating in the positive and negative directions is an analog of the law of conservation of the difference between the numbers of particles and antiparticles.

Strictly speaking, the mass of a single vortex is infinite, owing to the energy of the φ field surrounding the vortex. One can speak only of a finite mass (energy) inside a finite volume. For example, the mass (energy) $m_B(r)$ inside a circle of radius r with center at the center of a vortex is given by Eq. (26.19) with the substitution $r_0 \to r$.

The Goldstone model with quantum vortices can be said to be the simplest model of the strong and electromagnetic interactions in (2 + 1)-dimensional space-time. Here the quantum vortices play the role of protons, the π particles play the role of pions, and the φ particles play the role of photons. This analogy is based on the properties of the particles and their corresponding fields. Indeed, the interaction between quantum vortices is mediated by the φ field at large distances and also the π field at small distances, just as the interaction between protons is mediated by photons at large distances and also pions at small distances. Moreover, the decay of a massive π particle into two massless φ particles can be regarded as an analog of the decay of a pion into two photons. Finally, the difference between the numbers of "particles" and "antiparticles" is conserved for quantum vortices. This difference has the meaning of electric charge, which in this model is equal to the baryon charge.

We now consider the possible existence of vortex-like excitations in certain models of field theory in four-dimensional space-time. A generalization of the Goldstone model is provided here by a model with three real scalar fields and an action functional of the form

$$S = -{}^1/_2 \int d^4 x \left[\sum_a (\nabla \varphi_a)^2 - \lambda \sum_a \varphi_a^2 + (g/2) \left(\sum_a \varphi_a^2 \right)^2 \right], \quad (26.30)$$

which is written here in the Euclidean variables. The condition $\delta S = 0$ is the equation

$$-\Delta_4 \varphi_a - \lambda \varphi_a + g \left(\sum_a \varphi_a^2 \right) \varphi_a = 0. \quad (26.31)$$

This equation has a constant solution $\varphi_a = \text{const}$ with the condition

$$\sum_a \varphi_a^2 = \lambda / g, \quad (26.32)$$

as well as a solution describing a vortex-like excitation independent of the "time" coordinate x_4 and having the form

$$\varphi_a = (x_a/r)\, f(r), \tag{26.33}$$

where $r = (x_1^2 + x_2^2 + x_3^2)^{1/2}$ is the distance from a distinguished origin in three-dimensional space. Equation (26.31) reduces to a second-order equation for the function f of the form

$$f'' + 2f'/r - 2f/r^2 + \lambda f - gf^3 = 0. \tag{26.34}$$

We are interested in the solution of this equation behaving in proportion to r as $r \to 0$ and tending to the constant $(\lambda/g)^{1/2}$ as $r \to \infty$. It can be shown that such a solution of this equation actually exists. However, the functional

$$^1/_2 \int_{r<r_0} \left[\sum_a (\nabla\varphi_a)^2 - \lambda \sum_a \varphi_a^2 + (g/2)\left(\sum_a \varphi_a^2\right)^2\right] dx, \tag{26.35}$$

which gives the energy of the excitation in the volume $r < r_0$, is proportional to r_0 in the limit $r_0 \to \infty$. Thus, a vortex-like excitation has infinite energy and cannot be interpreted as a new particle.

More complicated vortex-like solutions exist in models with Yang—Mills fields. For example, for a system with the action

$$-^1/_2 \int \left[\sum_{a,\mu} \left(\partial_\mu \varphi_a + \varepsilon\varepsilon_{abc}\, b_\mu^b\, \varphi_c\right)^2 - \lambda \sum_a \varphi_a^2 + (g/2)\left(\sum_a \varphi_a^2\right)^2\right] d^4x -$$
$$^1/_2 \int \sum_{a,\mu} \left(\partial_\mu b_\nu^a - \partial_\nu b_\mu^a + \varepsilon\varepsilon_{abc}\, b_\mu^b\, b_\nu^c\right)^2 d^4x \tag{26.36}$$

one can seek a solution of the form

$$\varphi_a(x) = x_a u(r) r^{-1};\quad b_i^a(x) = \varepsilon_{iab} x_b \left[a(r) - (\varepsilon r^2)^{-1}\right];$$
$$b_0^a(x) = 0. \tag{26.37}$$

Such a solution was proposed and studied independently by 't Hooft[53] and Polyakov.[54] It was shown that this solution has a finite energy functional and can thus be associated with a new particle. The energy density falls off like $\sim r^{-4}$ when $r \to \infty$, like the energy density of a point charge. However, in this case the energy is an analog of the energy of a magnetic field, and the solution itself is therefore called a magnetic monopole. Note also that in accordance with (26.37) the isotopic vector $\varphi_a(x)$ tends to different limits as $r \to \infty$, depending on the direction of the vector $\mathbf{r}$ (it has a "hedgehog" structure).

The search for other, more realistic, models of field

theory with vortex-like solutions is a very topical problem at the present time. We cannot exclude the possibility that this approach provides the key to the construction of a consistent theory of the strong interactions.[55-64]

REFERENCES

[1]N. N. Bogolyubov and D. V. Shirkov, Vvedenie v teoriyu kvantovannykh poleĭ, Gostekhizdat, Moscow (1957) [English translation: Introduction to the Theory of Quantized Fields, Wiley, New York (1959)].
[2]A. I. Akhiezer and V. B. Berestetskiĭ, Kvantovaya élektrodinamika, Nauka, Moscow (1969) [English translation of 2nd ed.: Quantum Electrodynamics, Interscience, New York (1965)].
[3]S. S. Schweber, An Introduction to Relativistic Quantum Field Theory, Row, Peterson and Co., Evanston, Ill. (1961).
[4]R. P. Feynman, Acta Phys. Pol. 24, 697 (1963).
[5]B. S. DeWitt, Phys. Rev. 160, 1113 (1967); 162, 1195, 1239 (1967).
[6]L. D. Faddeev and V. N. Popov, Phys. Lett. 25B, 29 (1967).
[7]V. N. Popov and L. D. Faddeev, "Perturbation theory for gauge fields" [in Russian], Preprint, Institute of Theoretical Physics, USSR Academy of Sciences, Kiev (1967).
[8]L. D. Faddeev, "Hamiltonian form of the theory of gravitation" [in Russian], in: Tezisy 5-ĭ Mezhdunarodnoĭ konferentsii po gravitatsii i teorii otnositel'nosti (Abstracts of the 5th Intern. Conf. on Gravitation and the Theory of Relativity), Tbilisi (1968).
[9]L. D. Faddeev, Teor. Mat. Fiz. 1, 3 (1969).
[10]G. 't Hooft, Nucl. Phys. B33, 173 (1971).
[11]G. 't Hooft, Nucl. Phys. B35, 167 (1971).
[12]A. A. Slavnov and L. D. Faddeev, Vvedenie v kvantovuyu teoriyu kalibrovochnykh poleĭ, Nauka, Moscow (1978) [English translation: Gauge Fields: Introduction to Quantum Theory, Benjamin/Cummings (1980)].
[13]G. W. Mackey, Mathematical Foundations of Quantum Mechanics, Benjamin, New York (1963).
[14]E. Cartan, Lecons sur les Invariants Intégraux, Hermann, Paris (1922).
[15]R. P. Feynman, Rev. Mod. Phys. 20, 367 (1948).
[16]R. P. Feynman, Phys. Rev. 84, 108 (1951).
[17]M. A. Evgrafov, Dokl. Akad. Nauk SSSR 191, 979 (1970).
[18]S. M. Bilen'kii, Vvedenie v diagrammnuyu tekhniku Feĭnmana, Atomizdat, Moscow (1971) [English translation: Introduction to Feynman Diagrams, Pergamon, Oxford (1974)].
[19]F. A. Berezin, Metod vtorichnogo kvantovaniya, Nauka,

Moscow (1965) [English translation: The Method of Second Quantization, Academic Press, New York (1966)].
[20]I. Bialynicki-Birula, J. Math. Phys. 3, 1094 (1962).
[21]J. S. Ward, Phys. Rev. 77, 293 (1949); Phys. Rev. 78, 182 (1950).
[22]C. N. Yang and R. L. Mills, Phys. Rev. 96, 191 (1954).
[23]J. Schwinger, Phys. Rev. 125, 1043 (1962); 127, 324 (1962).
[24]D. J. Gross and F. Wilczek, Phys. Rev. Lett. 30, 1343 (1973).
[25]H. D. Politzer, Phys. Rev. Lett. 30, 1346 (1973).
[26]H. Weyl, Phys. Rev. 77, 699 (1950).
[27]V. A. Fock, Teoriya prostranstva, vremeni i tyagoteniya, Nauka, Moscow (1965) [English translation: The Theory of Space, Time and Gravitation, Pergamon, London (1964)].
[28]E. S. Fradkin and G. A. Vilkovisky, Phys. Rev. D 8, 4241 (1973).
[29]M. Kaku, Nucl. Phys. B91, 99 (1975).
[30]C. Aragone and J. Chela-Flores, Nuovo Cimento 25B, 225 (1975).
[31]Y. Takahashi and H. Umezawa, Prog. Theor. Phys. 9, 14, 501 (1953).
[32]T. D. Lee and C. N. Yang, Phys. Rev. 128, 885 (1962).
[33]P. A. M. Dirac, Proc. R. Soc. London A246, 333, 326 (1958).
[34]R. Arnowitt, S. Deser, and C. W. Misner, 117, 1595 (1960).
[35]J. Schwinger, Phys. Rev. 130, 1253 (1963); 132, 1317 (1963).
[36]P. G. Bergmann, Rev. Mod. Phys. 33, 510 (1961).
[37]J. L. Anderson, Rev. Mod. Phys. 36, 929 (1964).
[38]J. Favard, Cours de Géométrie Différentielle Locale, Gauthier-Villars, Paris (1957).
[39]S. Weinberg, Phys. Rev. Lett. 19, 1264 (1967).
[40]L. D. Faddeev, "Gauge-invariant model of the electromagnetic and weak interaction of leptons" [in Russian], Preprint, Leningrad Division, Mathematics Institute, USSR Academy of Sciences (1972).
[41]P. W. Higgs, Phys. Rev. 145, 1156 (1966).
[42]J. Goldstone, Nuovo Cimento 19, 154 (1961).
[43]F. J. Hasert et al., Phys. Lett. 46B, 121 (1973).
[44]F. J. Hasert et al., Phys. Lett. 46B, 138 (1973).
[45]A. A. Slavnov and L. D. Faddeev, Teor. Mat. Fiz. 3, 18 (1970).
[46]S. Mandelstam, Phys. Rev. 175, 1580, 1604 (1968).
[47]E. S. Fradkin, "Functional method in quantum statistics and in many-body theory" [in Russian], in: Problemy teoreticheskoĭ fiziki (Problems of Theoretical Physics), Nauka, Moscow (1969), p. 386.

[48]E. S. Fradkin and I. V. Tyutin, Phys. Lett. 30B, 562 (1969); Phys. Rev. D 2, 2841 (1970).
[49]M. Veltman, Nucl. Phys. B7, 637 (1968).
[50]J. Schwinger, Proc. Nat. Acad. Sci. USA 44, 956 (1958).
[51]L. P. Pitaevskii, Zh. Eksp. Teor. Fiz. 40, 646 (1961) [Sov. Phys. JETP 13, 451 (1961)].
[52]V. N. Popov, "Quantum vortices in the relativistic Goldstone model," in: Proc. of the 12th Winter School of Theoretical Physics in Karpacz, p. 397.
[53]G. 't Hooft, Nucl. Phys. B79, 276 (1974).
[54]A. M. Polyakov, Pis'ma Zh. Eksp. Teor. Fiz. 20, 430 (1974) [JETP Lett. 20, 194 (1974)].
[55]L. D. Faddeev, Pis'ma Zh. Eksp. Teor. Fiz. 21, 141 (1975) [JETP Lett. 21, 64 (1975)].
[56]I. Ya. Aref'eva and V. E. Korepin, Pis'ma Zh. Eksp. Teor. Fiz. 20, 680 (1974) [JETP Lett. 20, 312 (1974)].
[57]V. E. Korepin, P. P. Kulish, and L. D. Faddeev, Pis'ma Zh. Eksp. Teor. Fiz. 21, 302 (1975) [JETP Lett. 21, 138 (1975)].
[58]V. E. Korepin and L. D. Faddeev, Teor. Mat. Fiz. 25, 147 (1975).
[59]R. F. Dashen, B. Hasslacher, and A. Neveu, Phys. Rev. D 10, 4125 (1974).
[60]N. H. Christ and T. D. Lee, Phys. Rev. D 12, 1607 (1975).
[61]E. Tomboulis, Phys. Rev. D 12, 1678 (1975).
[62]R. Jackiw and G. Woo, Phys. Rev. D 12, 1705 (1975).
[63]L. D. Faddeev, "Vortex-line solutions of a unified model of electromagnetic and weak interactions of leptons," Preprint MPI-PAE/Pth 16, Munich (1974).
[64]L. D. Faddeev and V. E. Korepin, Phys. Rep. 42C, 1 (1978).

APPENDIX. ON THE STRUCTURE OF PHYSICAL THEORIES*

1. Hilbert's Sixth Problem

Among the problems formulated by Hilbert at the turn of the century, there is a problem which has hitherto not been solved. This is the sixth problem: the mathematical formulation of the axioms of physics. Hilbert wrote: "To construct the physical axioms according to the model of the axioms of geometry, one must first try to encompass the largest possible class of physical phenomena by means of a small number of axioms and then, by adding each subsequent axiom, to arrive at more special theories, after which there may arise a classification principle which can make use of the deep theory of infinite Lie groups of transformations. Moreover, as is done in geometry, the mathematician must bear in mind not only the facts of actual reality, but also all the logically possible theories, and must be particularly careful to obtain the most complete survey of the totality of consequences which follow from the adopted systematization." It is necessary to take note of a distinction between purely mathematical reasoning and the usually employed physical reasoning. Mathematical reasoning is analytic, i.e., it is done in accordance with definite logical rules on the basis of the adopted definitions and axioms. No additional information which is not contained in the initial definitions and axioms is admitted in the process of logical deduction. Otherwise, it would be possible to obtain arbitrary consequences. Mathematical assertions are valid for the abstract objects introduced by means of the definitions, these being logical atoms of the theory.

The scheme of deduction of mathematical (analytic) reasoning is as follows:

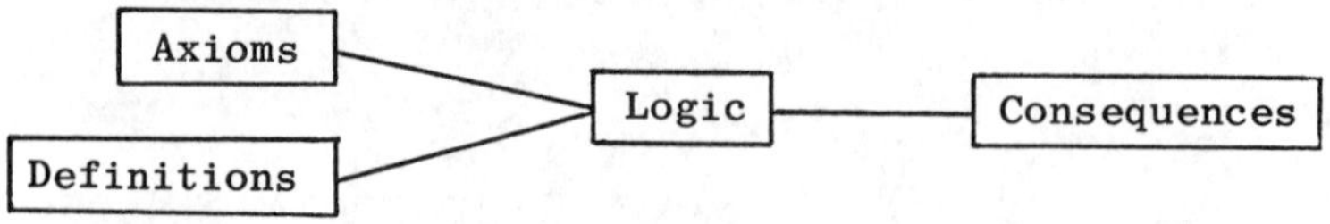

As a rule, a physical theory is based on concepts which are poorly defined from the point of view of mathematics. These bear the marks of the various methods of experimental study of physical objects, as well as the sense and even emotional perception of them by the experimenter. Therefore, for the

*Translation of a paper by N. P. Konopleva, in: Teoretiko-gruppovye metody v fizike. Trudy mezhdunarodnogo seminara (Group-Theoretical Methods in Physics. Proc. of the Intern. Seminar), Zvenigorod, 28—30 November 1979, Vol. I, Nauka, Moscow, 1980, p. 337.

axiomatization of physics it is necessary, first and foremost, to go over from concrete ideas to general concepts. The general concepts usually reflect a small part of the properties of real objects, but then the distinguished properties are inherent in many real objects, so that arguments based on the general concepts have a certain degree of generality, which is necessary for scientific inferences. Thus, the logical atoms of a physical theory are abstract objects which possess properties that are common to some class of real physical objects. Consequently, under different conditions one and the same physical object can serve as a model of the logical atoms of physical theories which differ from the standpoint of their mathematical apparatus. Conversely, one and the same mathematical apparatus can be used to describe phenomena which are completely different in their physical nature (for example, D'Alembert's equations and all possible periodic processes).

If a correspondence has been established between mathematical concepts and real objects such that under definite conditions a mathematical theory provides a correct description of the behavior of these real objects, we shall say that we have found a physical realization of the mathematical theory (or the mathematical concepts). A mathematical theory becomes physical if a physical realization of its basic concepts has been found.

2. General Requirements on Physical Theories

If a physical theory, like any other symbolic system, is to convey information about the external world, it must possess a number of properties which turn it into a language. For example, it must contain as characteristic objects: 1) constants, i.e., semantic elements; 2) variable elements, depending on the situation; 3) ontological elements, which make it possible to answer questions of the type: "does there exist an object with enumerated properties?"

The structure of any physical theory reflects the process of obtaining information about the external world known as experimental investigation. A distinctive feature of experimental investigation is the requirement of reproducibility of the results. This means that it is implicitly assumed that there exist a class of mutually identical objects of investigation, a class of identical situations in which these objects can occur, and a class of identical frames of reference or instruments by means of which the measuring procedure is implemented. Regardless of how the identity of the studied objects or frames of reference is established in practice, the identity relation has the structure of a group. Therefore, any physical theory must contain among its axioms some symmetry principle specified by a group of transformations.

The symmetry group of a theory reflects the properties of the measuring instruments and procedures used in experiment. As a rule, the measuring procedure consists in comparison of a studied object and a standard. The symmetry group can be used to convert the results of measurements performed in one frame of reference to any other frame of reference (in the class of equivalent frames with respect to the transformations of the given group). Thus, the symmetry group specifies a principle of relativity and, consequently, the degree of generality of the theory, and this determines the type of results which do not depend on the choice of the frame of reference in the class of equivalent frames. Results that are independent of the choice of the frame of reference are formulated in terms of the invariants of the symmetry group of the theory.

The constant (semantic) characteristics of physical objects are determined by the set of invariants of the symmetry group of the theory. These can be divided into two classes: space-time invariants, and invariants of so-called internal symmetries. Space-time invariants enable us to "recognize" a given object and to identify objects at different space-time points. Invariants of internal symmetries are usually various types of charges that characterize the interaction between physical objects.

Variables depending on the situation, the characteristics of the objects, are formulated in terms of quantities which are noninvariant but which transform according to a definite law under the symmetry group of the theory. These reflect, in particular, the choice of the frame of reference and coordinate effects. Quantities which do not have a definite transformation law in a given theory cannot be reproduced unambiguously in the class of frames of reference specified by its symmetry group and must be regarded as theoretical images of accidental (within the framework of the given theory) phenomena. The criterion for existence can be formulated differently in different theories, but it always fixes a condition which is invariant with respect to the choice of the frame of reference and which determines whether an object is unambiguously observable experimentally (for example, a nonzero field tensor or curvature tensor).

Thus, the fundamental general principles on which any consistent physical theory is based are principles of invariance and symmetry. They determine the structure of the concepts used to convey information about the external world. At the same time, they constitute a theoretical image of the instruments used in experiment. An analogous situation also exists in geometry. Indeed, two-dimensional Euclidean geometry can be regarded as a theory of the invariants of the group of motions of the plane. At the same time, two-dimensional rotations and displacements constitute the group of motions of the implements used to construct geometrical figures, to prove the congruence (i.e., equality) of some of them, and to prove theorems. These implements are the compass and ruler without divisions. The use of other implements (for example, a ruler with divisions) would take us beyond the scope of Euclidean geometry (the conformal group would become the group of motions of the implements and the symmetry group of the theory). Thus, every physical theory contains in the structure of its axioms, in embryonic form, the properties which the instruments used to test it must possess. Conversely, the choice of the instruments and scheme of an experiment predetermines the possible type of symmetry of the theory describing the given experiment. Although the physical experiment is far more complex than planimetric constructions and it is difficult to determine directly an experimentally adequate type of symmetry, the logical connection between experimental and theoretical methods of investigating the world is the same in physics as in geometry. As Heisenberg said, "We must remember that what we observe is not Nature itself, but Nature which appears in the form in which it reveals itself as a result of our manner of asking the questions."

3. Structure of a Consistent Physical Theory

In order to discuss the means of solving Hilbert's sixth problem, we classify the various types of theoretical assertions according to their degree of generality and present the result in the form of the following scheme:

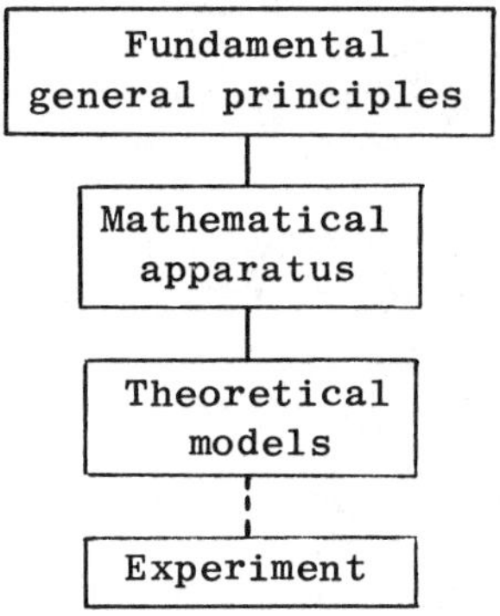

This scheme is reminiscent of Russell's hierarchy of types of propositions, which makes it possible to avoid logical paradoxes. Going from top to bottom, we obtain assertions which are more concrete, i.e., less general, than at the preceding level. The assertion of each level is valid for the classes of different concrete assertions of the level below it and is general in relation to them. Thus, using exactly the same fundamental general principles, it is possible to construct different forms of mathematical apparatus. At the present time, the following are known in physics: 1) the Lagrangian formalism; 2) the Hamiltonian formalism; 3) the axiomatic approach in quantum field theory; 4) the geometrical formulation of field theory. With each mathematical apparatus, the final objective is to obtain equations of motion, field equations, and conservation laws. However, by themselves neither the equations of motion, nor the field equations, nor the conservation laws describe any concrete physics. At this level, the degree of generality of the theory is too great. The behavior of concrete physical systems is described by particular solutions of the field equations and conservation laws. To find these solutions, it is necessary to construct a theoretical model of the studied physical system; for example, one must specify initial data, determine boundary conditions, eliminate superfluous variables, particularize the frame of reference, and so forth. Thus, Maxwell's equations become solvable when a concrete model of the current is specified: 1) the Lorentz current $j = \rho v$ of the discrete classical charges and the supplementary equations for the new variable v; 2) the Ohmic current $j = \sigma E$; 3) the Hall current $j = \kappa H$; 4) the superconducting London current $j = \lambda A$. In the general theory of relativity, it is necessary to fix the frame of reference in order to solve the field equations. In the theory of gauge fields, the superfluous field components are eliminated by means of gauge conditions when going over to models, and when the fields are quantized even the vacuum is redefined: instead of the usual vacuum as the ground state of the system (the "origin"), a spontaneously broken vacuum is chosen.

Every solution of the field equations (or of other general relations) which is found actually describes not a single real physical system, but a whole class of similar physical systems with definite relationships. Therefore it is possible in principle to model some physical processes by others if the corresponding equations are the same for both. Every theoretical model is capable of describing a more or less broad class of experiments.

Thus, the breadth of the class of physical phenomena encompassed by a given theory is determined by the set of admissible means of measurement. The largest possible breadth is determined by the postulates about the symmetry group of the theory. This symmetry group is the symmetry group of the space of states of the instruments and, at the same time, provides us with a fundamental principle for constructing the theory. By adding the definitions of the field variables and axioms such as the principle of least action or the extremality of the Hamiltonian (or of any other functional), the mathematical apparatus of the theory can be constructed according to definite rules. By means of particularizing assumptions and axioms, one then constructs models which yield relations that can be tested experimentally. The problem of "recognizing the images" arises, in essence, in the last stage of identifying the experimentally observed phenomena and the theoretical (abstract) concepts. This is the problem of finding the physical realization of a theory. For theories constructed in a deductive manner (following the foregoing scheme from top to bottom), it can be extremely complex. For example, in general relativity the problem of the physical realization of the theoretical concepts was approached 50 years after the creation of the theory.

Until the beginning of the twentieth century, the traditional method of constructing the physical theory was the inductive method, which proceeds from experiment, i.e., from the particular to the general (following the scheme from bottom to top). Individual fields of physics (Newton's mechanics, Maxwell's electrodynamics) were axiomatized only after having been sufficiently well studied experimentally. An understanding of the final form to be taken by a physical theory and of the rules for constructing any theory makes it possible to construct a physical theory axiomatically, going from the general to the particular, as Hilbert wanted to do. As is well known, Hilbert attempted to construct a unified theory of gravitation and electromagnetism on the basis of general principles. From a variational principle he obtained the equations which are now known as the Einstein equations (Einstein obtained them at the same time as Hilbert, but in a different, equivalent form). However, the unified theory was not further developed by Hilbert. Therefore general relativity can be regarded as the first physical theory constructed without recourse to experiment. At the present time, there exists a physical theory which was constructed axiomatically prior to experiment and subsequently found its physical realization. This is the theory of gauge fields. As Hilbert predicted, it makes use of the deep theory of infinite Lie groups to classify interactions. Moreover, it admits a purely geometrical formulation, in which the analogy between the axiomatic structures of geometry and physics becomes clear.

4. Local Symmetries and Classification of Interactions

The principles of invariance and symmetry are among the fundamental general principles for constructing a physical theory. A symmetry can be local, valid in the neighborhood of a point, or global, valid over all space-time. The first possibility means that the frame of reference used in an experiment has finite dimensions (of the order of the region in which the symmetry holds). The second possibility presupposes the existence of an absolute frame of reference extended over all space-time. This assumption entails the assumption that there exists a long-range interaction which relates the individual elements of the real frame of reference. Therefore the postulate of locality of the symmetries seems more physical. It reflects more adequately the real experimental situation. But within the experimental accuracy the postulate that a given symmetry is global may not lead to contradictions with the experiment if the scale of the studied phenomena is much smaller than the characteristic dimensions of the frame of reference.

The postulate of locality of the symmetries makes it necessary to introduce the idea of fields which relate local frames of reference referred to different space-time points. Allowance for these fields makes it possible to convert results obtained in one local frame of reference to another local frame of reference and to compare them with one another. The fields associated with local symmetries are called gauge fields. These include the gravitational and electromagnetic fields, as well as the fields which mediate the strong and weak interactions. A gauge field is a mathematical concept signifying an object with definite transformation properties. Among the real physical systems, it can correspond to both single objects (the photon) and multiplets of particles (the vector mesons). There can also be more complex realizations in the form of collective excitations, and so forth.

Whereas global symmetries permit the classification of elementary particles according to their properties, i.e., according to the values of the invariants of the symmetry group, local symmetries make it possible to classify interactions according to their associated local gauge groups and to construct a hierarchy of interactions. Local symmetries are a particular case of infinite Lie groups. Thus, the theory of gauge fields actually leads to a classification principle which makes use of the theory of infinite Lie groups.

5. Types of Geometries and Types of Physical Theories

There exist three methods of constructing geometry: 1) Klein's approach, which assumes that space is homogeneous; all properties of geometrical objects in Klein's geometry are described by sets of invariants of the symmetry group of space; 2) Riemann's approach, which does not assume any symmetry of the space; in this case, the characteristics of geometrical objects are constructed step by step from local differential expressions; connection coefficients are required for the construction of the space as a whole; 3) Cartan's approach, in which the space as a whole constitutes a set of local homogeneous Klein spaces associated with each point of a Riemannian space and interrelated by generalized connection coefficients.

The geometrical approaches in physics can be classified in a natural way in accordance with the foregoing conceptions of geometry. Klein's point of view, which originated in the axiomatics of Euclidean geometry, is used, for example, in classical and relativistic mechanics. The images of Riemannian geometry — the metric, connection coefficients, and curvature — were used in general relativity. Cartan's approach, which was developed in the modern geometry of fiber spaces, made it possible to geometrize the theory of gauge fields.

The connection between physics and geometry is determined by Poincaré's symbolic formula $G = G_0 + F$, where G represents the dynamical geometry, G_0 the geometry of the "background," and F the forces of interaction. The meaning of this formula is that physics and geometry do not occur separately in experiment; only the combination of geometry and physical laws is subject to experimental verification. The decomposition of the sum G into a purely geometrical background G_0 and an interaction F depends on us, or, more precisely, on the choice of the means of measurement.

As long as physical phenomena are described as occurring at some place and time, space-time ideas cannot be excluded from the theoretical description of experiment. But the idea of forces which produce an interaction is not essential. A force-free description of interactions renders the theory purely geometrical. The actually observed bending of trajectories of particles is described by means of the concept of connection coefficients of a nonholonomic space, which replaces the concept of force. If one and the same phenomenon is described in two different ways, there must exist a "principle of equivalence" which permits the transition from one description to the other. Poincaré's formula can be understood as the assertion that every description by means of forces can be associated with a purely geometrical description. But in view of the relation between the form of the theory and the choice of the means of measurement, we must remember that the scheme of an experiment to test the geometrical theory must be different from one to test the ordinary theory of interactions in terms of forces.

Any geometrical theory is a theory of the motion of test bodies. A test body is defined as a body subject to the action of an external field, but itself exerting no inverse action. Real bodies can satisfy the condition only approximately. Among extended bodies, the best test bodies are those for which the ratio of the surface forces to the volume forces is minimal. The volume forces are usually geometrized.

The geometry of space-time is determined by the axiom about inertial (force-free) motion. The inertial trajectories are related by the transformations of the symmetry group, which specifies the principle of relativity of the theory. If the symmetry group is global, space-time is a homogeneous Klein space with a rigid (i.e., not related to dynamics) geometry. Local symmetries lead to a dynamical geometry of Riemann or Cartan type.

The choice of the class of inertial motions is not the result of any experiments, but is always one of the postulates of a physical theory. This postulate reflects the choice of the type of measuring instruments which must show a zero reading when in inertial motion from the point of view of the given theory. For example, the classes of inertial motion defined by Newton and by Einstein are not identical. Therefore the instruments which determine experimentally the position of a body must be arranged differently, depending on which of these theories we use.

Thus, a geometrical description equivalent to a description in terms of forces always exists, but for an experimental verification of the geometrical form of the theory the test bodies and instruments must be correctly chosen.

INDEX